Buildings and fire

BUILDINGS AND FIRE

T. J. Shields
G. W. H. Silcock

Copublished in the United States with
John Wiley & Sons, Inc., New York

Longman Scientific & Technical
Longman Group UK Limited
Longman House, Burnt Mill, Harlow
Essex CM20 2JE, England
Associated companies throughout the world
Copublished in the United States with
John Wiley & Sons, Inc., 605 Third Avenue, New York, NY 10158

First published 1987

British Library Cataloguing in Publication Data
Shields, T.J.
Buildings and fire.
1. Fire prevention
I. Title II. Silcock, G.W.H.
628.9′22 TH9145

ISBN 0-582-30524-1

Library of Congress Cataloging in Publication Data
Shields, T.J., 1943–
Buildings and fire.
Includes bibliographies and index.
1. Building, Fireproof. 2. Fire prevention.
I. Silcock, G.W.H., 1943– II. Title.
TH1065.S48 1987 693.8′2 86-21344
ISBN 0-470-20750-7

Set in 10/11 pt Monophoto 2000 Times
Produced by Longman Singapore Publishers (Pte) Ltd.
Printed in Singapore.

Contents

Preface

In the past the topic of fire safety in the design and construction of buildings has been included in some curricula under various subject headings. These have included design technology, building technology, environmental science and materials.

The investigations that follow each fire and subsequent research point conclusively to the need for studies in fire safety to be coherent and founded on a sound understanding of elementary fire dynamics. Growing awareness of the importance of fire safety considerations in environmental design has led to the emergence and recognition of fire safety engineering as a discipline in its own right.

This book introduces the reader in a logical and integrated fashion to the subject of fire safety engineering. Expanding technology has moved far in advance of the development of comprehensive performance standards for whole buildings. However, if the overall performance of a building can be taken as a function of the building's components performance, a building's fire safety performance can be similarly treated.

Traditionally an acceptable level of fire safety has been achieved by the use, sometimes indiscriminate use, of fire protection methods. Technology has developed such that even with passive fire protection methods, low-level and high-level technologies are being utilised to provide solutions to essentially similar problems, i.e. methods and techniques are available which, although they may be similar in principle, are quite different in character. For example, the general problem formulated may be the retarding of the passage of fire from one volume to other volumes within the same building. In an existing building a low-level technology solution has been employed, in part, in that the panels only in existing doors have been successfully treated with an intumescent paint system so as to upgrade the doors and provide the required level of performance. In a new building, for example an atrium building, a high level technology solution might employ automatically-operated descending, fire-resisting screens to prevent the ingress of fire into the atrium space.

As previously stated, reliance in the past has been almost entirely based on passive protection methods. In recent years, however, there has been a growth in intelligent systems, i.e. a means other than human, whereby the presence of fire can be detected, attacked, controlled (if not extinguished) and the occupants alerted to the threat of fire.

At present, current prescriptive legislation, largely by omission, precludes the use of such intelligent systems as a component of fire safety provision, thus failing to recognise the concept of trade-off between the various components in the formulation of the overall fire safety package. Nevertheless, any cost effective solution to a fire safety problem may include the use of such systems in preference to, or to complement, passive protection and must necessarily include an analysis of the occupancy, including the people, and the nature and utilisation of the premises.

Thus the process of fire safety engineering can be best represented by a series of matrices from which the various components contributing to a cost effective solution to any problem can be obtained.

This book emphasises the complex and integrative nature of fire safety engineering. The treatment of the subject matter is at various levels so as to be of use to diploma and undergraduate students as well as the practitioners. In particular, students pursuing courses in Architectural Technology, Building and Civil Engineering, Building Services Engineering, Building and Quantity Surveying, Environmental Health and Environmental Engineering will find this book relevant to the academic and vocational aims of the course on which they are engaged.

Acknowledgements

We are grateful to the following for permission to reproduce copyright material: Academic Press Inc. (Fig 4.5); American Society of Testing and Materials (Fig 13.8); The Blackie Publishing Group (Tables 6.7 and 6.8); British Standards Institution, from *BS 476: Part 10: 1983* (Figs 6.2, 6.3 and 6.13), from *CP 112 Part 2: 1971 B.S. Code of Practice for the Structural Use of Timber* (Tables 8.9 and 8.10), which has been superseded by *BS 5268: Part 2*. Extracts from British Standards are reproduced by permission of the British Standards Institution. Complete copies can be obtained from BSI at Lindford Wood, Milton Keynes, MK14 6LE; Elsevier Applied Science Publishers Ltd (Figs 2.10, 2.11, 2,12, 8,25 and 8.26) from *Fire and Buildings* by T. I. Lie; Elsevier Sequoia S.A., Switzerland (Figs 13.3 and 13.4); Fire Prevention Association (Figs 1.2, 1.3 and 5.20, Tables 1.4, 1.5 and 1.6), Fire Safety Engineering Unit (Fig 5.1); Reproduced with permission of the Controller of Her Majesty's Stationery Office (Figs 5.9, 5.26, 5.27, 5.28, 5.29, 8.18, 8.19, 8.22, 8.23, 8.24, 10.8, 10.10, 10.12, 10.14, 10.18, 10.19 and 12.32, Tables 9.4 and 12.8); A. J. Hinks (Table 8.11); Institution of Structural Engineers (Tables 8.6, 8.7 and 8.8); Dr Bernard Lewis (Fig 4.5); London Transport Museum (Table 12.7); National Fire Protection Association, U.S.A. (Figs 12.33 and 12.34); National Research Council of Canada (Fig 8.20 and Table 8.11); New Civil Engineering (Institute of Civil Engineering) (Fig 9.37); Risk and Insurance Group Ltd (Table 1.3); Swedish Institute of Steel Construction (Fig 8.12 and Tables 8.2(a), 8.2(b) and 8.4); UNISAF (Figs 12.35, 12.36, 12.37 and 12.38); The U.S. Bureau of Mines (Fig 4.8); John Wiley and Sons Limited, from *Fire and Human Behaviour* by I. Appleton, Ed. D. Canter, 1980 (Tables 1.2 and 12.9); Professor J. Witteveen (Figs 8.15, 8.16 and 8.17).

CHAPTER 1

Evolution of fire safety in buildings

1.1 HISTORICAL BACKGROUND

Fires in buildings are nearly always man-made, i.e. resulting from error or negligence.

Primitive man used heat for cooking, warming and lighting his dwelling with the inherent risk that misuse or accident in his control of fuel might precipitate disaster. Today, as in primitive society, that risk has not been eliminated despite the apparent sophistication of modern living. With the development of habitations, attitudes to fire protection/fire precautions also developed, sometimes subtly, but mostly from bitter experience.

The principal aims of fire precautions are simply to safeguard life and property and are achieved by:

1. Reducing fire incidence
2. Controlling fire propagation and spread
3. Providing adequate means of escape for occupants of buildings.

In medieval times, dwellings were of mostly timber-framed construction with thatched roofs, and within the walled townships overcrowding, narrow lane ways, overhanging eaves and indiscriminate use of combustibles provided all the necessary ingredients for the conflagrations which followed.

In 1136 London, Bath and York suffered severe fire damage, and another disastrous fire in London in 1212 led to the introduction of ordinances regarding certain building uses and materials to be used for new or restored roofs.

The Great Fire of London in 1666 destroyed four-fifths of the city before being brought under control and as a result the first real positive steps were taken, which in effect provided for the forerunners of building control as we know it. Sir Christopher Wren, Hugh May and Roger Platt were appointed as Royal Commissioners. City Surveyors were also appointed and drafted various regulations which were embodied in the Rebuilding Act of 1667.

The Rebuilding Act of 1667[1] was further consolidated in the Act of 1774 which required that:

1. Buildings be divided into seven classes
2. Minimum thicknesses of external and party walls specified to be complied with
3. Party walls to be carried up through the roof for minimum height specified
4. Parapet walls to be treated in a similar manner to party walls
5. No recesses to be provided in party walls other than for chimneys and flues
6. Party walls to be free from openings except to connect two warehouses or stables and any such opening to be closed by an iron door
7. Timbers in party walls to be separated by 8 in. (212 mm) of solid brickwork from one another, or from any chimney or flue
8. Chimney backs to be 13 in. (325 mm) thick in cellars and 8½ in. thick above
9. Chimney breasts to be 8½ in. (212 mm) thick in cellars and at least 4 in. (100 mm) thick elsewhere
10. Hearths to be of brick and stone at least 18 in. (450 mm) wide and 12 in. (300 mm) longer than the fireplace opening in the breast
11. Materials suitable for use were specified
    (a) *for walls:* brick, stone, lead, tin, slate or tile
    (b) *for roofs:* glass, copper, lead, tin, slate, tile or stone
    (c) *dangerous trades*, e.g. turpentine distilling – only permitted in premises 50 ft (15 m) away from any other building.

1.2 THE DEVELOPMENT OF CURRENT FIRE SAFETY REGULATIONS

In effect the data obtained through experience and embodied in the Act of 1774 form the basis of current regulations. Concepts such as:

1. Purpose grouping
2. Space separation between buildings
3. Ignition prevention
4. Compartmentation
5. Isolation
6. Segregation
7. Constructional component integrity,

although not expressly stated in the Act of 1774 are embodied in current fire safety legislation, and by the application of science and scientific methods, the broad principles contained in the Act of 1774 have been validated, modified and adapted to meet the needs and complexities of modern society.

Towards the end of the nineteenth century it was possible to construct large multi-storey buildings, the structural elements of which were of non-combustible materials that should have been capable of prolonged fire resistance. However, even in the most advanced buildings of the period, fires still occurred. Fire-load density was not yet considered as a constraining factor and the storage of combustibles in large quantities was a risk that had to be taken. This fire loading was perhaps unnecessarily added to by using timber flooring, panelling, partitions and staircases.

Many 'fire-proof' buildings were constructed only to be destroyed by fire, simply because architects had not yet realised that providing structural elements with a degree of fire resistance was not enough. Often the designers' intention, in providing fire-resisting floors for example, was negated because continuous ducts, staircases, lift shafts, etc., were allowed to penetrate the floors without the provision of fire stops. Eventually after many buildings were gutted the lesson was learned.

At the beginning of the twentieth century the use of reinforced concrete and steel-frame constructions assumed importance, but like cast iron, steel, while being non-combustible, provided negligible fire resistance without adequate protection. The regulations in force did not require minimum periods of fire resistance, indeed, the very term fire-resisting was not adequately defined. Little work had been done in attempting to identify the factors which contributed to fire severity and consequently influenced the nature and quality of protection afforded to buildings.

1.3 THE INTRODUCTION OF FIRE TESTING

The British Fire Protection Committee was legally incorporated in 1899 and immediately set about commissioning tests on current forms of construction and materials.

The first test carried out was on a floor construction and the time–temperature graph is reproduced here also showing the development through to BS 476: Part 1 and BS 476: Part 8: 1972 Fig. 1.1.

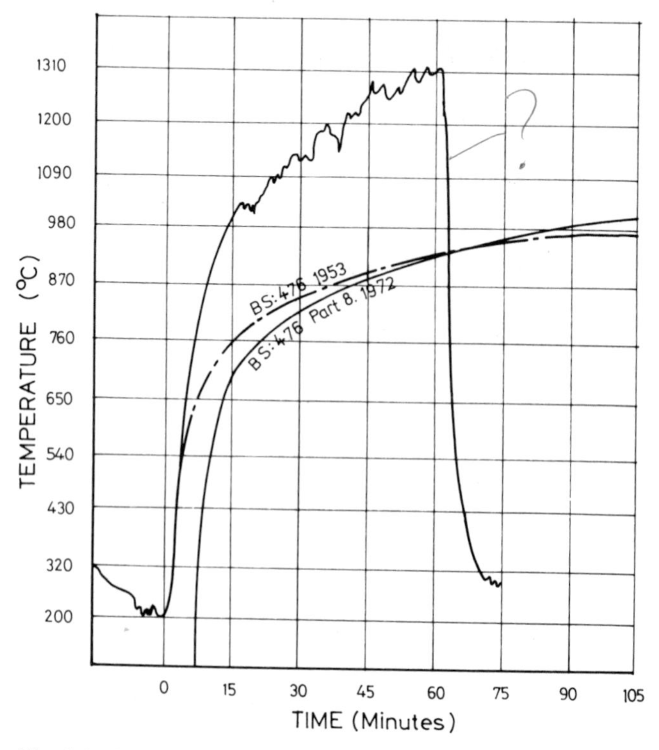

Fig. 1.1 Development of standard fire curve

1.4 THE FIRST INTERNATIONAL FIRE PREVENTION CONGRESS

By this time fire protection was not only of national concern but also of international interest and in 1903 the First International Fire Prevention Congress was convened.

This Congress:

1. Recommended the use of the term 'fire-resisting'
2. Confirmed proposed International Standards
3. Advocated the establishment of National Testing Stations using uniform testing methods
4. Supported the view that every fire should be investigated and reports submitted for record and publication.

By the time the British Fire Prevention Committee had issued its First Quinquennial Report in 1904, it had conducted 79 tests, issued 57 publications containing test results and 22 other publications in case studies of fire prevention.

1.5 THE DEVELOPMENT OF FIRE SAFETY POLICIES AND STATISTICS

It was now becoming increasingly clear that the notion of 'fire protection' which had evolved as a necessity from past experience was becoming increasingly complex in an evolving society involved in technological advancement. In 1921 a Royal Commission was appointed 'to enquire' into the existing provision for:

1.(a) the avoidance of loss from buildings, including the regulations dealing with the construction of buildings
  (b) dangerous processes
  (c) advisory role with regard to fire prevention, and
2.(a) the extinction of outbreaks of fire, including the control, maintenance organisation, equipment and training of Fire Brigades in Great Britain, and
  (b) to report whether any, and if so what, changes were necessary whether by statutory provision or otherwise, in order to secure the best possible protection of life and property against fire risks, due regard being paid to considerations of economy as well as efficiency.

The Commission reported in 1923 and an analysis of 19,000 fires recorded in 1919 was presented.

Table 1.1 shows the Commission's findings.

Obviously this kind of breakdown is unsatisfactory as too much is left undecided. For example:

1. How many fires were not reported?
2. The term 'business use' is too broad
3. Other causes require definition.

The figure of fire incidence against flues provoked consideration with regard to the effect of installing new heating appliances in buildings with defective flues, e.g. central-heating plant.

Table 1.1 Incidence of fire

Cause	Number
Carelessness, particularly with matches	3,300
Structural defects, mainly flues	2,200
Lighting, heating, business use	5,600
Incendiarism, proved or suspected	1,600
Other causes known or unknown	6,300

1.6 THE ORIGINS OF FIRE TERMINOLOGY

In 1932 the first BS 476 entitled *British Standard Definitions for Fire Resistance, Incombustibility and Non-flammability of Building Materials and Structures (including Methods of Test)* was issued.

BS 476: 1932 was primarily concerned with definitions and remained unchanged until 1953 when a new standard was issued, BS 476: 1953 – Fire Tests on Building Materials subsequently revised in 1972. It is because of the importance of BS 476 and its influence on the effective use of materials, components and elements of structure that a whole chapter of this book is devoted to a study of the development and analysis of testing, and perhaps more significantly to an analysis of the data obtained from these tests.

1.7 FIRE GRADING

Not until 1946 when the first report on the Fire Grading of Buildings was published in the *Post-War Building Studies*[2] series, was the subject of fire grading, relative to the fire load of a building, i.e. the number of units of heat which would be liberated per unit floor area by the complete combustion of the structure and contents, seriously studied in this country. It is surprising that data compiled in 1946 are still being used today as a basis for fire grading in buildings.

1.8 FIRE SAFETY FEEDBACK

A study of the background of structural fire precautions will show a progressive trend, perhaps as a result of some disaster, but it would be a folly to fall into the trap of thinking that the design, construction and management of buildings is a static science and art. It is a dynamic activity producing buildings of increasing complexity for multiple usage and occupancy in an atmosphere of managerial and technological change.

The principal lessons to be learned from a study of the development of structural fire protection are:

1. That the protection of life and property from a fire hazard is a complex problem to which there can be no simple solution

2. New materials and new methods of new construction demand new solutions and approaches to an old problem
3. That the introduction of structural fire precautions has contributed significantly to reducing fire hazards and that further developments with regard to testing of materials and components is a natural progression
4. That building codes contain only minimum requirements
5. That building regulations and codes of practice which embody minimum standards because of the legislative process lag behind research developments and findings
6. That the best intentions of designers may be entirely frustrated by lack of attention to detail
7. That good housekeeping is essential.

1.9 COST OF FIRE

One very good reason that the lessons referred to earlier be reinforced and consolidated is manifested in the cost of fire in the United Kingdom.

Appleton[3] gave the total cost of fire in the United Kingdom in round figures as approximately £1,000 million per annum. The various components of fire cost are given in Table 1.2.

The Bland Report[4] considered the productivity losses in the United States to be somewhat higher than the direct losses. This may simply reflect the state of the economy in a developing society. Much useful data in fire losses are compiled by insurance organisations and published regularly.

Table 1.3 gives a comparison of fire losses over a twelve-month

Table 1.2 Breakdown of average annual cost of fire in United Kingdom

	£m
Direct losses	250
Consequential losses	75
Fire Brigades	300
Fire protection	200
Insurance	200
Enforcement costs	50
Total	£1,075

Table 1.3 Estimated fire damage in Great Britain (England, Scotland and Wales)

	1982 (£m)	1981 (£m)		1983 (£m)	1982 (£m)
October	31.0	25.8	January	37.3	34.7
November	30.4	42.8	February	33.3	37.3
December	22.9	30.5	March	34.9	41.5
			April	22.7	25.1
	84.3	99.1	May	26.6	22.7
			June	194.8†	36.2
			July	46.1	39.7
			August	42.7	40.6
			September	23.5	28.6
				451.9	306.4
				536.2	405.5

† Includes £165 million for Army Ordnance Depot.

period. This information was first published in *Foresight*[5] and is given in Table 1.3 in its original form.

The data given include insured and uninsured damage, but does not take into account consequential loss, lost production, lost orders and exports.

A close scrutiny of available fire statistics will indicate that the cost of fire damage has not decreased (Fig. 1.2). For the period 1970–79 (Fig. 1.3) the cost of fire damage has remained at a level of approximately 0.25 per cent of the gross national product.[6] The method of costing employed takes into account inflation and industrial performance over the fiscal year.

It is possible also to compare the annual fire losses of different countries on this basis, as shown in Table 1.4.

Analysis of the cost of fire damage[7,15] has established that major fires which correspond to less than 0.5 per cent of the total number of fires reported represent some 65 per cent of the total cost of fire damage. This information would tend to suggest a strong correlation between the cost of fire damage, fire spread and occupancy of the fire-affected premises. Table 1.5 gives a breakdown of major industrial fires in 1978 in terms of occupancy and cost.

Some sections of industry can be identified as being more susceptible to the occurrence of fire than others and Table 1.6 gives an indication of the frequency of the incidence of fire in certain sectors of industry for 1978.

In addition to the total cost of fire, e.g. in the United Kingdom of approximately £1,000 million per annum, there is the loss of life and injury to persons to consider. It is perhaps a little-known fact that

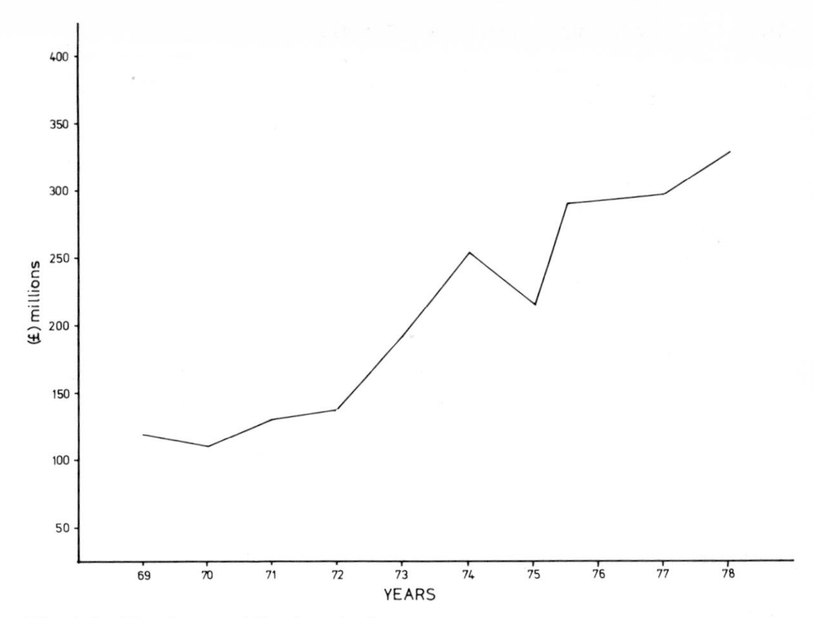

Fig. 1.2 Total annual fire loss in £m

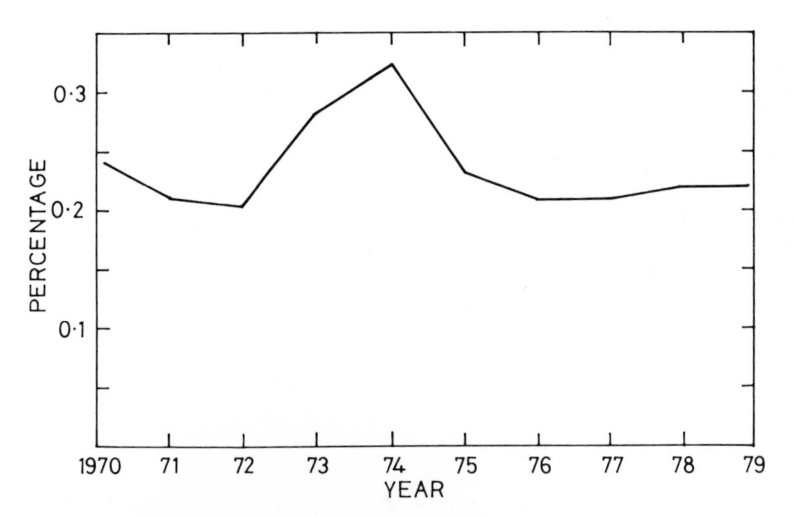

Fig. 1.3 Fire damage expressed as a percentage of GNP

there are approximately 1,000 deaths and 6,000 non-fatal casualties per annum in the UK as a consequence of fire. An examination of the UK Fire Statistics[8] will show that the majority – over 50 per cent – of fatal and non-fatal casualties from fires fall within the age groups 1–19 years and 60 years plus. Consequently in terms of

Table 1.4 Fire losses as a percentage of GNP

	1974	1975	1976	1977	1978	Average
Australia	0.35	0.48	0.43	0.44	–	0.43
Finland	–	0.17	0.21	0.17	0.15	0.18
France	0.21	0.23	0.28	0.28	0.27	0.25
W. Germany	0.16	0.18	0.20	0.19	–	0.18
Great Britain	0.34	0.24	0.25	0.24	0.23	0.26
Netherlands	0.16	0.18	0.18	0.19	–	0.18
Norway	0.32	0.32	0.47	0.41	0.43	0.39
South Africa	–	0.20	0.35	0.23	0.21	0.25
Sweden	0.22	0.25	0.27	0.20	0.22	0.23
USA	0.34	0.27	0.20	0.32	0.19	0.26

earnings and earnings potential, a person within the age bands quoted may not in monetary terms be worth very much. In some cases the cost attached to the loss of a life in a fire may be considerably less than injury which would require hospitalisation and constant care, perhaps for the remainder of the injured person's life.

It could be argued that in national terms the costs associated with the loss of life and injury from fire are not very significant. However, the loss of life or injury to an individual may have a severe financial attachment or social consequence of a high order locally.

In France the cost of a life lost in a road accident[8] has been estimated at FF1,400,000 per death and FF100,000 per injury. These estimates include costs associated with death, medical services, legal services, police services and costs suffered by society through loss of production.

So far it has not been possible to include loss of life and injury to persons in the cost studies of fire damage and a major part of the problem is the determination of an appropriate cost to attach to the loss of life in a fire.

However the costs related to road deaths and casualties as indicated for France above could be used to give a first approximation of the total costs associated with fire casualties.

Moulens[8] reported the seven criteria established by the World Fire Statistics Centre for evaluating the overall cost of fire. These were

direct costs
indirect costs
cost of insurance
cost of emergency services
cost of prevention
cost of research
cost of information

Table 1.5 Breakdown of fires by occupancy for 1978

Occupancy	In buildings	
	No. of fires	Cost £'000
1. Agricultural/Forestry/Fishing	62	2,786
2. Mining and Quarrying	–	–
3. Food/Drink/Tobacco	47	7,734
(a) Food	35	4,455
(b) Drink	12	3,279
(c) Tobacco	–	–
4. Coal and Petroleum Products	2	195
5. Chemical and Allied Industries	28	5,764
(a) Synthetic resins and plastics materials and synthetic rubber	8	1,732
(b) Other	20	4,032
6. Metal Manufacture	17	3,207
(a) Wrought and cast iron manufacture	4	191
(b) Steel and steel tube manufacture	4	1,230
(c) Other	9	1,786
7. Mechanical Engineering	12	2,307
8. Instrument Engineering	5	352
9. Electrical Engineering	25	8,010
10. Engineering/Metal Goods Unspecified	12	9,163
11. Shipbuilding/Marine Engineering	1	76
12. Vehicles	14	4,563
(a) Motor vehicles manufacture	8	923
(b) Other	6	3,640
13. Metal goods not elsewhere specified	23	3,765
14. Textiles	45	8,377
(a) Production of man-made fibres	6	565
(b) Cotton/linen/flax	7	918
(c) Woollen and worsted	4	364
(d) Hosiery	8	807
(e) Carpets	2	130
(f) Other	18	5,593
15. Leather/Leather Goods/Fur	6	1,061
16. Clothing and Footwear	19	2,508
(a) Clothing	18	2,388
(b) Footwear	1	120
17. Bricks/Pottery/Glass/Cement	8	793
(a) Bricks/fire clay/refractory goods	–	–
(b) Pottery	5	641
(c) Glass	2	82
(d) Other	1	70
18. Timber and Furniture	34	2,629
(a) Timber	14	859
(b) Furniture/upholstery/bedding	16	1,483
(c) Other	4	287

(*cont.*)

Table 1.5 (*cont.*)

Occupancy	In buildings	
	No. of fires	**Cost £'000**
19. Paper/Printing/Publishing	31	5,241
(a) Paper and board packaging materials	19	3,346
(b) Printing	12	1,895
20. Other Manufacturing Industries	28	4,920
(a) Rubber	11	3,203
(b) Plastics	11	1,112
(c) Other	6	605
21. Construction	11	1,134
22. Gas/Electricity/Water	1	3,500
(a) Gas	–	–
(b) Electricity	1	3,500
(c) Water	–	–
	131	28,456

Table 1.6 Incidence of major fires in industry

1978 Occupancy	Frequency of major fires
Clothing and Footwear	1 in 11
Electrical Engineering	1 in 12
Instrument Engineering	1 in 12
Leather/Leather Goods/Fur	1 in 13
Food/Drink/Tobacco	1 in 13
Timber/Furniture	1 in 18
Paper/Printing/Publishing	1 in 18
Other Manufacturing Industries	1 in 18
Engineering/Metal Goods Unspecified	1 in 21
Chemical and Allied Industries	1 in 22
Textiles	1 in 22

From these seven criteria the components of losses and expenses can be distinguished

Losses: direct costs
indirect costs
Human losses

Expenses: prevention
cost of insurance
emergency services.

1.10 FIRE STATISTICS

The fire statistics for the United Kingdom[8] are published annually and Fig. 1.4 shows the fire incidence trend for occupied buildings for 1971–1981.

Figure 1.5 shows the trend for fires in dwellings with the peak occurring in 1979. This increase was partly due to the cold winter of 1979, but it is interesting to note that the return to pre-1979 levels is proving particularly slow.

Figure 1.6 shows the trends for fires in hospitals and boarding houses. From the graph it can be seen that apparently following the introduction of the Fire Precautions Act 1971,[9] the incidence of fires

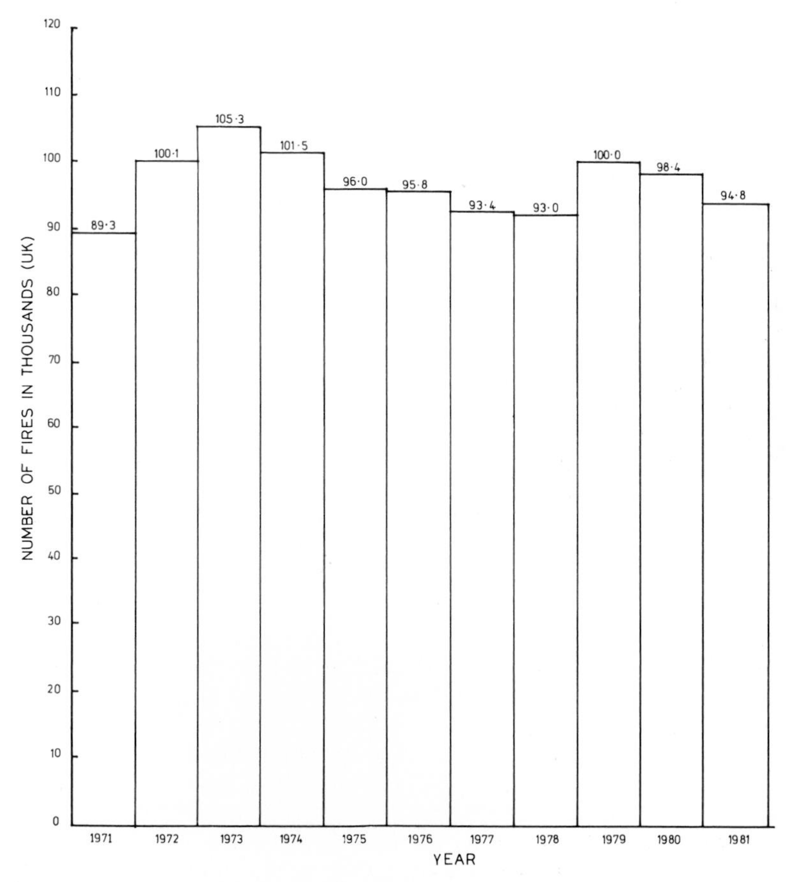

Fig. 1.4 Fires in occupied buildings 1971–81

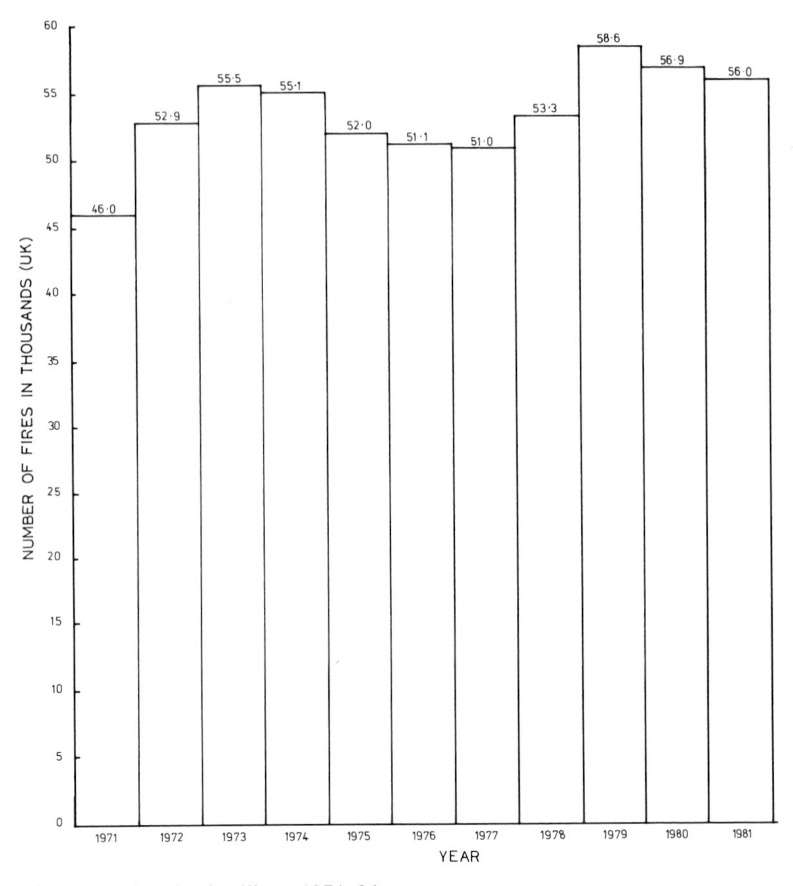

Fig. 1.5 Fires in dwellings 1971–81

in hotels and boarding houses increased. It would be a gross misinterpretation to suggest that the increase in fire incidence was related to the introduction of the Fire Precautions Act. The introduction of the Fire Precautions Act in itself may have caused previously unreported fires to be reported to the Fire Authorities and thus apparently distort the statistics. Reference to Fig. 1.7 will show that from 1971 onwards the number of fatalities in occupied buildings other than dwellings has fallen from 152 in 1971 to 80 fatalities in 1981.

By comparison Fig. 1.8 shows that the number of fatalities in dwellings as a consequence of fire has steadily increased from 574 in 1971 to 780 fatalities in 1979.

Consider Fig. 1.9 which shows that the number of people killed in fires as a consequence of being overcome by fire gases or smoke

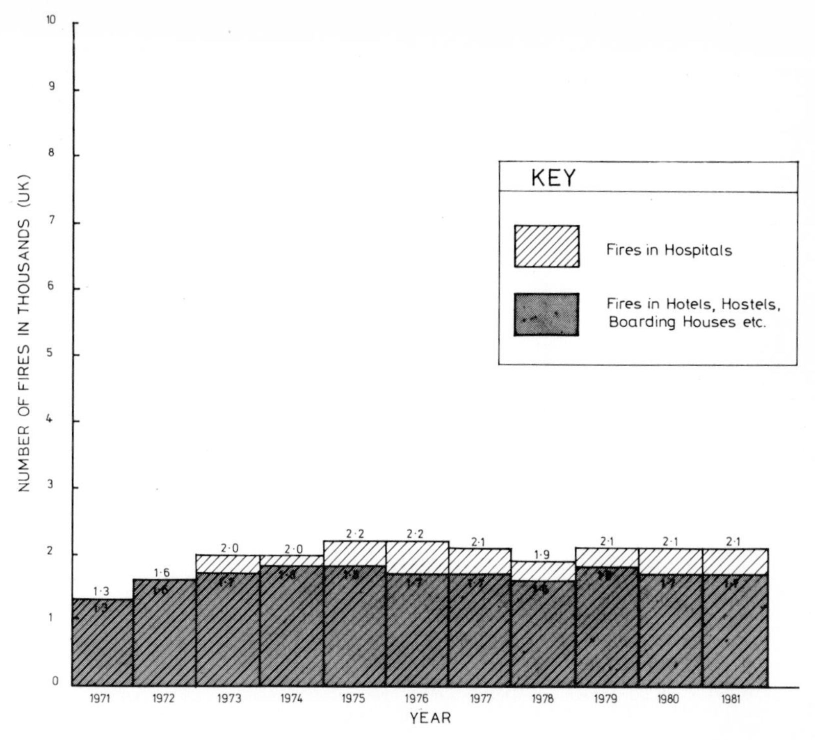

Fig. 1.6 Fires in hospitals against fires in hotels, hostels, boarding houses, etc.

increased from 381 in 1971 to 572 in 1981. Also the number of non-fatal casualties overcome by fire gases and smoke rose from 659 to 2,331 over the same period. Many of these casualties occurred in areas remote from the fire.

In the USA the National Fire Prevention Association operate the Fire Incident Data Organization (FIDO) System. This is a computerized data base which contains detailed information on individual fires of high technical interest. From this data base statistical reports are published annually.[11]

It is interesting to note that while the number of fire fatalities in dwellings in the UK rose over the period 1971–1981 the number of fire fatalities in the home in the USA over a similar period significantly decreased.[12]

O'Brien and Redding[11] in their analysis of the USA fire statistics for the year 1983, show that one in every 7,500 reported fires resulted in a large loss, i.e. direct property damage of $1 million or more. For 310 large loss fires selected over nine categories of

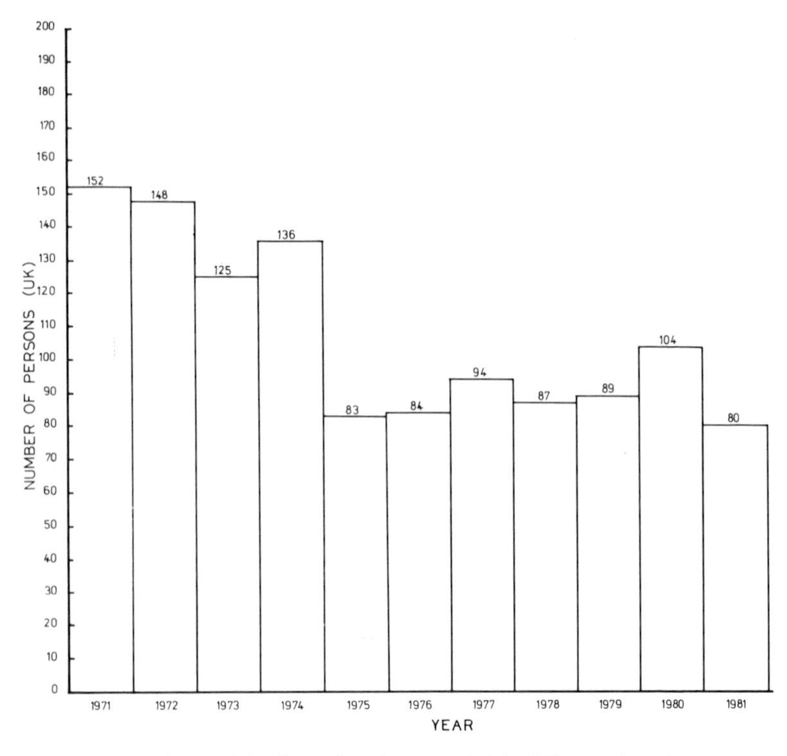

Fig. 1.7 Fatal casualties from fires in occupied buildings other than dwellings 1971–81

property use ranging from public assembly to special properties to total property loss was calculated as $854,855,000. These 310 fires represent a very small proportion of the total 2,326,500 reported fire incidents in 1983, i.e. 0.01 per cent, but nevertheless accounted for almost 13 per cent of the $6.6 billion in fire loss estimated for the same year.

It should be borne in mind that figures quoted in the fire statistics are generally accepted as an underestimate, e.g.

Fire incidence: fire statistics are compiled from fire authority reports of fires attended. Many fires go unreported.

Casualties: fatalities as a consequence of inhalation of toxic fumes and gases may occur some considerable time after exposure to the fire environment. Many non-fatal casualties go unrecorded.

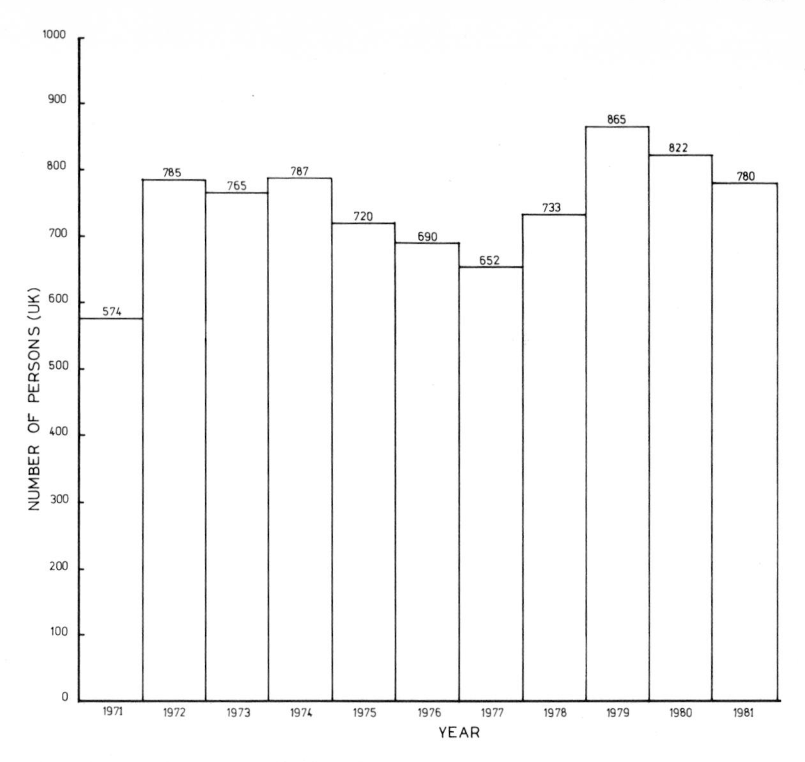

Fig. 1.8 Fatal casualties in dwellings 1971–81

1.11 FUTURE DEVELOPMENTS

1.11.1 Legislation

Building Regulations[13] are passive by nature and prescriptive by design. Consequently the application of the content of building regulations to a complex building can be fraught with difficulties, none more so than attempting to determine the adequacy of the level of fire protection achieved. Since 1980 the system of building control within the United Kingdom has been under review and it is anticipated that the form of future building regulations will be considerably altered, i.e. there will be a movement away from the immensely detailed and fairly inflexible system of statutory building controls of the 1980s to a set of shorter, broader functional regulations supported by precise statutory performance standards, coupled with a non-statutory series of approved documents which would contain the bulk of the detail. The fundamental propositions are:

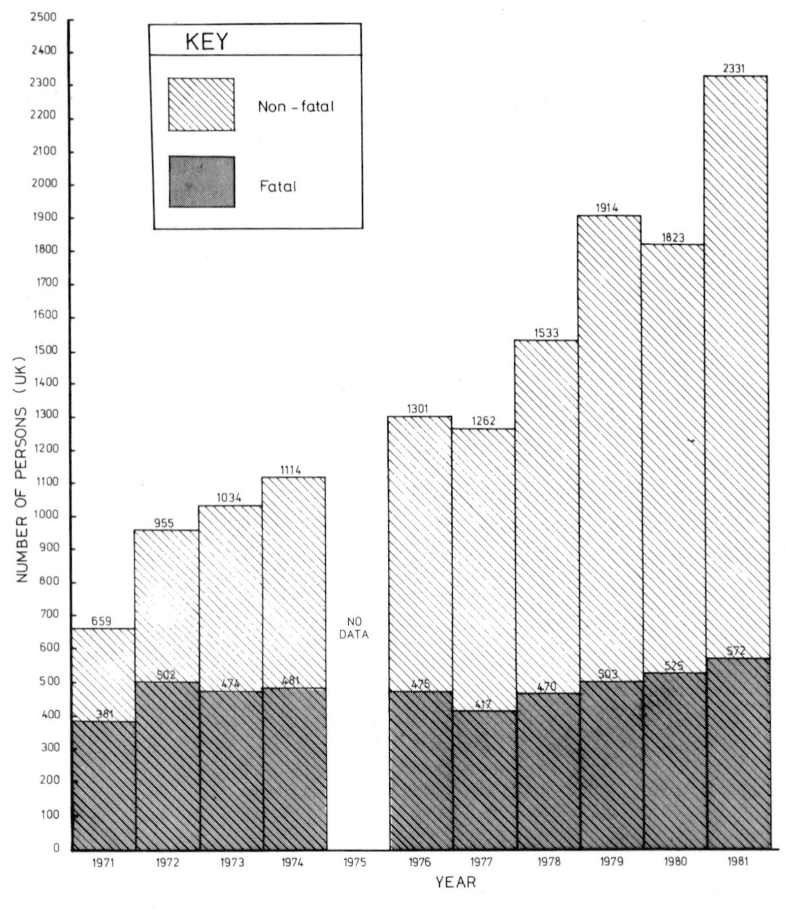

Fig. 1.9 Fatal and non-fatal casualties overcome by smoke from fires 1971–81

1. Regulations themselves should be reduced to broad *functional regulations*, clearly expressive of their purposes
2. Functional requirements should be supported wherever appropriate by precise *performance standards* (also within a formal Statutory Instrument) which would stipulate the performance to be achieved, and which would then form the basis for enforcement
3. Detailed technical material currently in the regulations should be removed from the Statutory Instrument and thus be more readily presented in a style acceptable to its users, and easier to keep up-to-date
4. *Approved Documents* should be given status by the Secretary of State's approval, and compliance with such approved documents

would be a defence against a charge of contravening the regulations
5. The link between the Statutory Regulations/performance standards and the non-statutory Approved Documents should be made by a *Manual to the Building Regulations*.

1.11.2 The performance concept

In order to appreciate the potential impact of the introduction of performance specifications and standards on fire safety engineering it will be useful here to restate what is meant by these terms.

Performance specifications state in precise terms the characteristics desired by the users of a product's or system's performance without regard to the specific means to be employed in achieving the results. They do not describe dimensions, materials, finishes or methods of manufacture or assembly; rather they describe the performance required by the user. Simply stated any material, composite, component or building must be fit for its intended purpose. To establish fitness for intended purpose, consideration must be given to the user's needs (present and future) and the conditions under which the component, etc., will be used. Thus an idea of how the component must perform is formulated which in turn leads to the precise details to be looked for when selecting from available components, etc., or when designing new.

The development of performance specifications and standards for each material, composite, component and building would necessitate the following:

1. A definition of user requirements
2. The establishment of conditions of use
3. A definition of key criteria for assessment
4. The development of methods of assessment
5. The setting of levels of performance
6. The relating of levels of performance set to user requirements and conditions of use.

As an exercise the reader should consider the problems associated with the development of performance standards for the provision of adequate means of escape, in the event of fire, from any familiar building. Further consideration of this particular problem will be given in Chapter 12.

1.11.3 Systemic approach to fire safety

The occurrence of an unwanted fire and subsequent injury and damage represents a failure of the system which ultimately encompasses all of society. Fire safety technology is by definition integrative in that a systems approach is required, which includes all of the components, e.g. economics, user requirements and fire dynamics, in order to determine optimum solutions to problems. Clearly it is essential to have an approach to fire safety engineering which considers fire-related components as an integrated whole, in the widest sense. Such an approach may be called 'systemic'.

However, a systemic approach is not the same as a systematic approach.[14] To see things in a systemic way is to perceive system within the situation; to be fully aware of the dynamism and interconnectedness within the entirety, whereas the systematic approach may be thought of as being methodical or tidy. A systemic approach should be systematic, but a systematic approach may not be systemic. The systemic approach requires:

1. A statement of the objectives of the fire safety programme
2. An assessment of the current level of fire safety provision and associated costs
3. A definition of the level of fire safety to be achieved.

Thus the systemic approach requires ways of measuring fire safety. Methods have been developed such as points schemes and mathematical models which may be utilised in order to evaluate the level of fire safety provision and Chapter 13 (Fire safety evaluation) considers these and other developments more fully.

1.11.4 Fire safety technology

The descriptive model

This model (Fig. 1.10) uses current industrial practices and solutions to problems as the basis for future design, i.e. yesterday's solution is imposed upon tomorrow's problems without a great deal of variation in applications. There is an inviolate core of fire safety components, usually prescribed in fire safety legislation, which must be utilised in any design solution to a particular problem. This core material embraces in reality the principles contained in the Act of 1774 and grows very slowly indeed. Surrounding this core of fire safety components are additional fire safety considerations which can be called 'The State of the Art'; what is currently practised – how it is practised; against what historical setting it has been developed (usually in response to a disastrous fire); and what practical problem(s) it solves. This model in itself is an explanation of the status quo, albeit that there is an expectation of change as a consequence of practical experience. Given the historical development of fire safety technology this model has some virtues, but slavish users of the model, particularly those charged with enforcing fire safety legislation, may have their prejudices derived from practical experience reinforced.

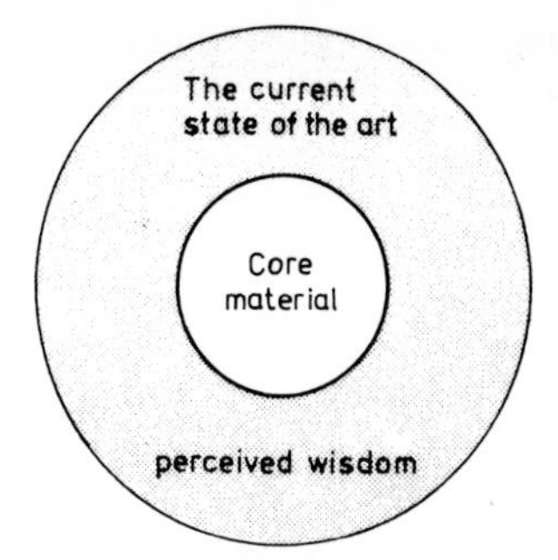

Fig. 1.10 Descriptive fire safety model

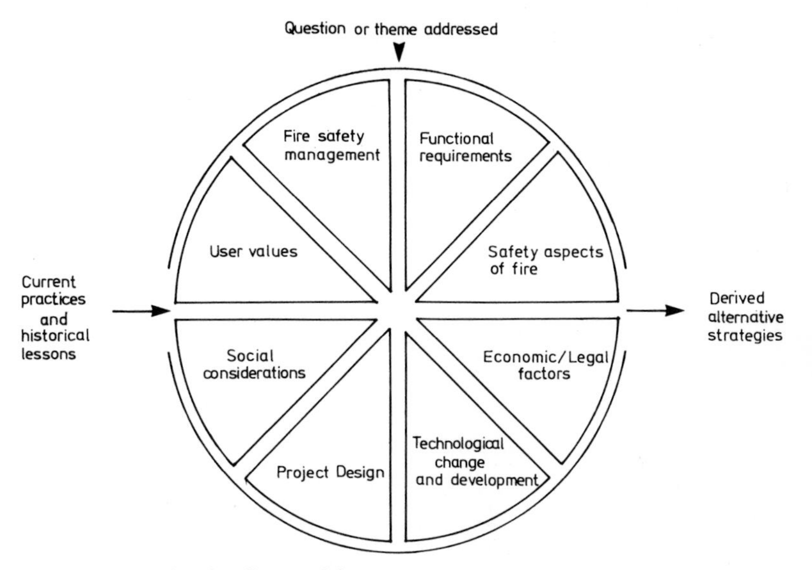

Fig. 1.11 Derivative fire model

The derivative model

This model (Fig. 1.11) is in fact an explanation of the integrative nature of fire safety technology and is consequently wider in scope and intellectually more demanding.

A major input into the system is an explanation of current practice, against the historical backcloths which formulated them, together with the development of an attitude of non-dependence towards the status quo. This non-dependence is reinforced by use of a contextual framework against which all current practices are criticised, analysed and evaluated, namely:

Functional requirements: functional efficiency determined via fire safety performance standards.

Aspects of fire safety: mandatory requirements, moral values, trade-offs, cost–benefit considerations.

Economic and legal considerations: influence of building legislation and associated costs with alternative design strategies.

Fire safety management: management of buildings in use, contents and people considered with the client for a particular design strategy.

Project design: the apportionment of risk between the client and design team with regard to innovative aspects of the design and the establishment of a continuing responsibility for the fire safety performance of the building and its component parts.

Technological change and development: current research and developments in all aspects of fire safety engineering.

Social considerations: the interaction between present and future needs of society and the technological means of satisfying them.

User values: cost–benefit analysis – value for money.

From this model a number of alternative strategies can be developed. The model could also be represented in matrix form, Fig. 1.12, where

(1) = objectives to policy vector
(2) = tactics to objectives matrix
(3) = tactics to policy vector
(4) = components to tactics matrix
(5) = components to policy matrix
(6) = interaction matrix
(7) = components ranked in terms of contribution to the attainment of policy

In conjunction with the establishment of the integrative nature of fire safety engineering, future developments will include the following:

1. *Prevention of fire occurrence:* ignition prevention, materials and composites
2. *Limitation of fire growth:* control of fire properties of materials, rate of heat release
3. *Use of intelligent systems:* communication systems, fire control

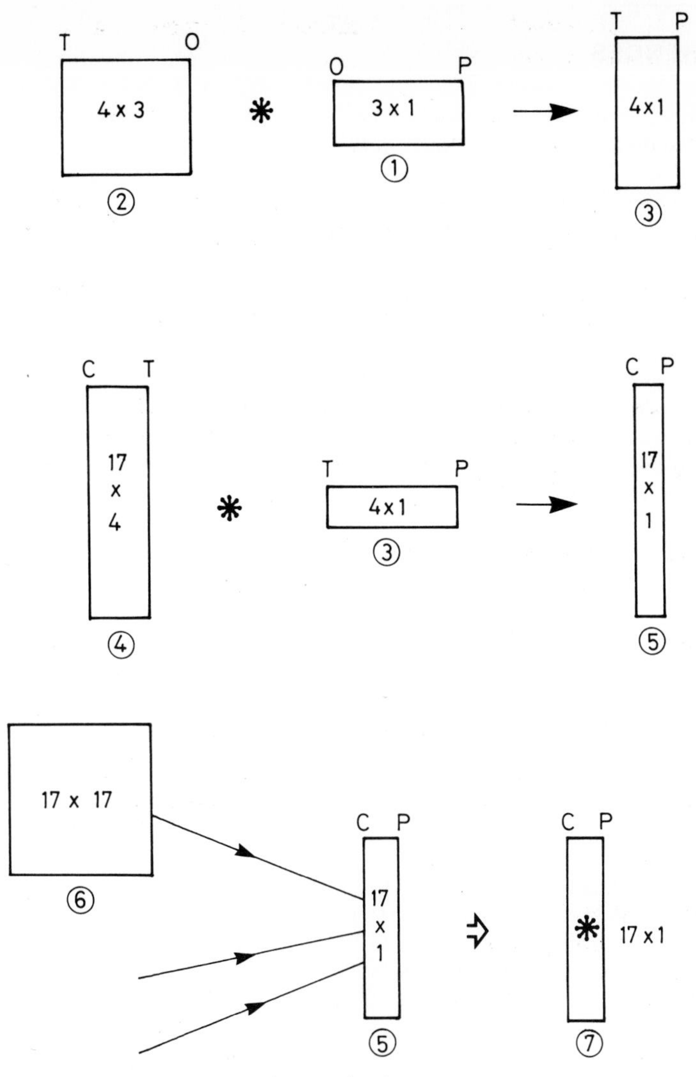

Fig. 1.12 Matrix interaction mechanisms

systems, smoke control systems, evacuation systems, refuge systems.

4. *Fire testing:* development of a suite of fire tests which measure some aspect of a materials/components exposure to a fire situation.

The future developments mentioned here will be discussed more fully in the text later, but by inclusion here serve to highlight the current trends in fire safety engineering.

REFERENCES

1. Hamilton S B, '*A Short History of the Structural Fire Protection of Buildings*', National Building Studies Special Report No. 27. HMSO London, 1958.
2. Ministry of Works, *Post-War Building Studies No. 20*. HMSO London, 1946.
3. Appleton I, 'The requirements of research into the behaviour of people in fires' in *Fires and Human Behaviour*, D Canter (ed.). Wiley 1980.
4. *American Burning Report of National Commission in Fire Prevention and Control 1973*. Washington DC. Superintendent of Documents US Gov. Printing Office 1973.
5. 'Reduction in fire losses', *Foresight*, **6**, Dec. 1983, p. 37. Foresight Publication Limited.
6. 'Management of fire risks, MR.3', *Fire Facts and Figures*. Fire Prevention Association, London 1980.
7. *Facts and Figures about Large Fires*, Fire Prevention No. 136, Fire Prevention Association, London 1984.
8. Moulens, Beaufort P, *The Cost of Fire*, *Proceedings of a European Symposium, Luxembourg*, September 1984. R Moevareau and M Thomas (eds), pp. 51–8. Elsevier Applied Science 1984.
9. *Fire Statistics United Kingdom 1981*, Home Office S3 Division, 50 Queen Anne's Gate, London.
10. Fire Precautions Act 1971, IIR 1971. CAO. HMSO London.
11. O'Brien A and Redding D. *Large Loss Fires in The United States During 1983*. Fire Journal Nov. 1984 Vol. 78 Part 6 pp. 17–58.
12. Hom S and Karter M J. *Fire Casualties, Fire Protection Handbook, Fifteenth Edition*, National Fire Protection Association 1981.
13. Building Regulations England and Wales 1976. HMSO. Building Regulations Northern Ireland 1977. HMSO. Building Standards (Scotland) Regulations. Scottish Development Department.
14. Beard A, 'A systemic look at fire safety', *Fire*, 1980.
15. Ewart K, Shields T J and Silcock G W, *An Analysis of the Cost of Compliance with the Fire Services Order*, (N.I.) 1985 Building Management and Economics December, 1986.

CHAPTER 2

Thermal properties of materials

2.1 INTRODUCTION

This chapter considers the thermal properties of materials which influence their behaviour during a fire. These properties are: (1) thermal conductivity; (2) specific heat capacity; (3) thermal diffusivity; (4) thermal inertia; (5) thermal expansion.

2.2 THERMAL CONDUCTIVITY

This quantity is a measure of a material's ability to pass heat energy through its bulk when subjected to a transient or steady-state heat flux. In building materials there are up to three distinct modes of energy transfer possible, these being:

1. Thermal conduction by atomic and molecular vibration
2. Thermal conduction by radiation
3. Thermal conduction due to mass transfer (gaseous conduction).

2.2.1 Thermal conduction (molecular and atomic processes)

The ease at which thermal energy can be transferred down a rod or other structure depends on the type of bonding between the molecules. The materials that have a rigidly-bound structure pass more energy than those with a weakly-bound structure. This is due to the fact that the frequency of vibration for rigidly-bound molecules is high, thus the rate at which the energy is transferred is also high. The energy is transmitted down the rod by very-high-frequency elastic waves. These waves can be compared to sound waves, but have much higher wave frequencies. The quantum of energy associated with such wave motion is called the phonon, which is passed from molecule to molecule in turn all the way down the length of the rod.

In an insulation-type material, the conduction process is dominated by these waves and their phonons. In comparison, the conductivity of a good thermal conductor, such as a metal, is due largely to electron drift where electrons move from regions of high temperature to regions of lower temperature. Thus the actual thermal resistance of an insulation-type material may be considered to be due to phonon–phonon collisions or phonon–lattice framework collision processes.

So if the interaction between phonons increases, as it must do with an increase in temperature, this will cause the thermal resistance to increase, hence lowering the thermal conductivity. Most masonry materials are polycrystalline, that is they consist of more than one crystal type, and since the phonons are scattered by crystal boundaries and by porosity it is not surprising that such materials have a lower phonon-conductivity than the single-crystal types. The situation of an amorphous type material, such as glass, is one in which the phonon-scattering process dominates thus making the conductivity of an amorphous material independent of temperature for practical purposes.

2.2.2 Thermal conduction due to radiation

Surfaces having a temperature above absolute zero are capable of emitting and absorbing radiation. The rate $\dot{q}$ at which this can occur is given by Stefan's Law

$$\dot{q} = A\sigma\Phi(T_B^4 - T_s^4)$$

where Φ is the configuration factor which is dependent on the geometry and ε where:

ε = emissivity of the surface
A = surface area of emitter or absorber
σ = Stefan's constant
$T_{B_{absol}}$ = temperature of the body
$T_{s_{absol}}$ = temperature of the surroundings.

Radiation passes through solids and undergoes scattering at structure imperfections, crystal boundaries and at pores. In a highly porous material this type of conductivity can be shown to be proportional to T^3 and becomes quite significant at temperatures above 500 °C. However, in opaque-type materials this type of conductivity only starts to become important at 1,000 °C or above.

2.2.3 Thermal conductivity due to fluid mass transfer

In this case a gas, such as air, which fills the pores within the material can add to the ability of the material to pass heat energy through the material under a thermal gradient or by a buoyancy effect.

For a porous material or fibrous-type insulation, the gaseous conduction component of the thermal conductivity has been shown to have a moderate contribution to make to the overall conductivity compared to the radiant component. This contribution is shown clearly in Fig. 2.1.

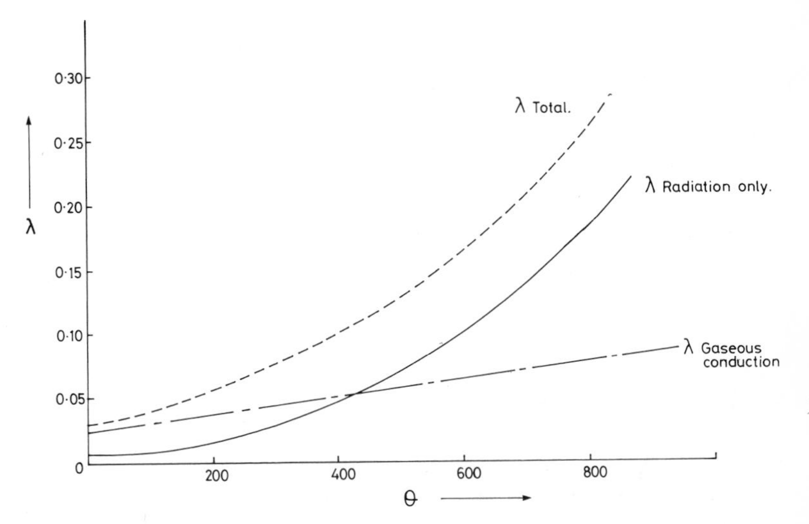

Fig. 2.1 Variation of the thermal conductivity with temperature for a medium-density fibrous insulant

It would now be useful to consider the variation of thermal conductivity of building materials with: (i) temperature; (ii) density; (iii) moisture content.

(i) Temperature

Experimental determination[1] of the thermal conductivity of concrete containing crystalline and amorphous-type aggregate is shown in Fig. 2.2, which supports the theoretical predictions expressed in the previous paragraphs.

The role of radiation is quite well illustrated in the variation of medium-density fibrous insulation where it shows the thermal

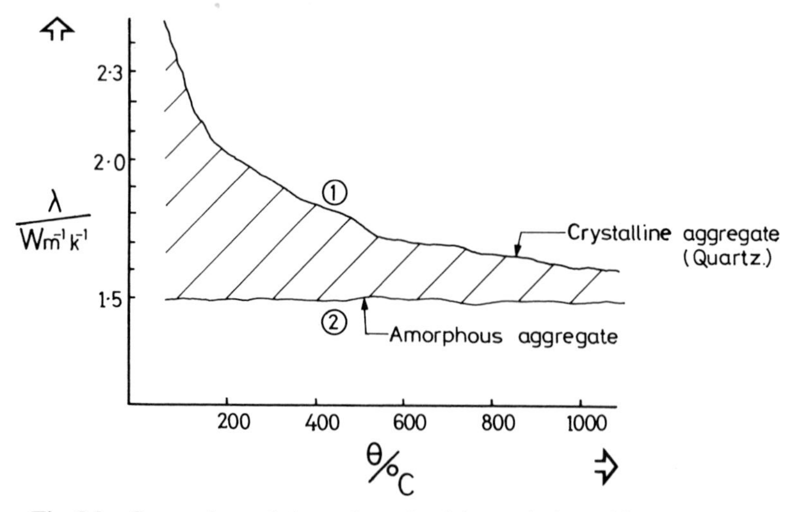

Fig. 2.2 Comparison of thermal conductivity variation with temperature for amorphous and crystalline materials

conductivity (λ) to be more or less proportional to T^3 at fire temperatures (Fig. 2.1).

(ii) Density variation (ρ)

This variation is quite well illustrated in Fig. 2.3 where the variation of thermal conductivity with density is shown.

(iii) Moisture content

Here, as the water displaces the air in the pore spaces, the ability of porous masonry-type material to conduct is enhanced since water is a good thermal conductor compared to stationary air. Thus as more of the air voids become filled with water, the greater the overall thermal conductivity becomes according to the following relationship:[2]

$$\lambda_x = \frac{f_x}{f_{1\%}} (\lambda_{1\%})$$

where $f_{(x)}$ = moisture factor which can be read from Table 2.1.

Thus moisture-laden masonry materials can have an overall thermal conductivity which will change as they warm up and dry out.

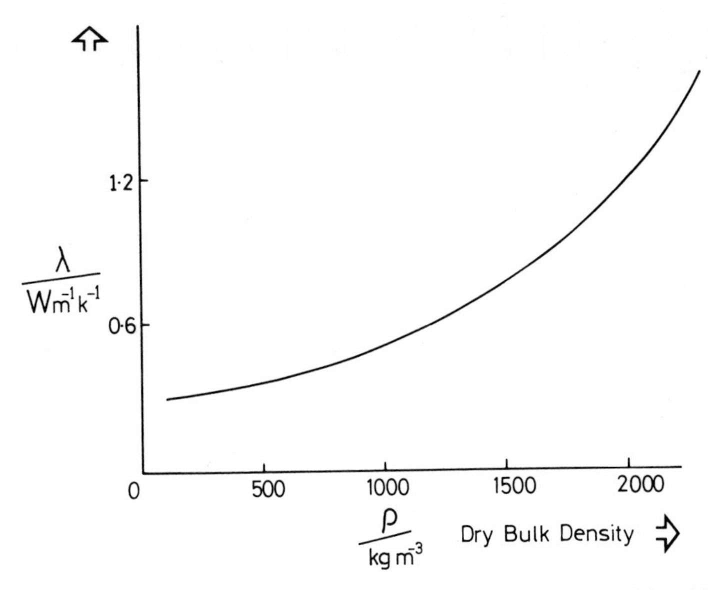

Fig. 2.3 Variation of thermal conductivity for porous materials with density

Table 2.1

Water content (%)	1	2.5	5	10	15	20	25
Moisture factor $f_{(x)}$	1.3	1.55	1.75	2.10	2.35	2.55	2.75

2.3 SPECIFIC HEAT CAPACITY (C_P)

This is usually defined as the amount of thermal energy needed to raise the temperature of 1 kg of a material by 1 degree kelvin.

Building materials will, in general, contain water or water vapour – this is especially so during the initial stages before the drying-out process has occurred under normal climatic conditions. Thus if the temperature of such a material is raised to temperatures in excess of the ambient temperature, the remaining free water and water vapour will start to be removed. This situation can arise during the early stages of a fire. At 100 °C (boiling point), water and vapour in the pores will be removed thus making the energy needed to raise 1 kg of material by 1 K at this temperature much greater than would

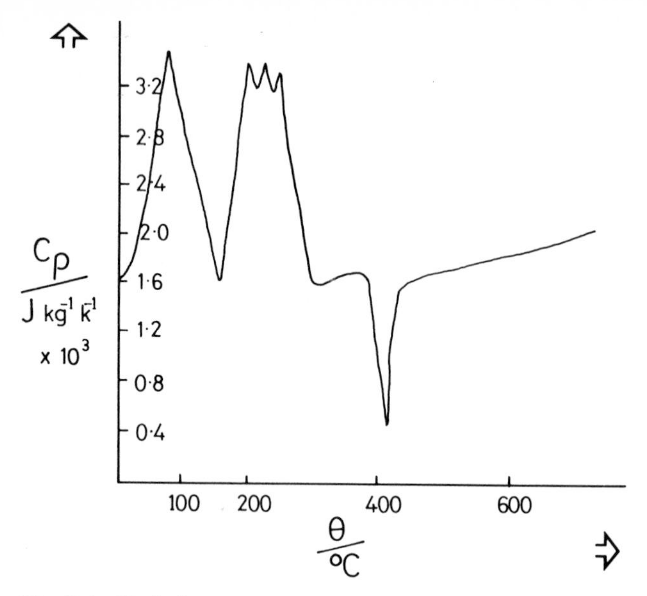

Fig. 2.4 Variation of specific heat capacity of gypsum with temperature

normally be the case at temperatures greater than or less than 100 °C. This sort of behaviour is shown in Fig. 2.4 which shows how the specific heat capacity (C_P) for gypsum varies with temperature. At much higher temperatures, chemical reactions tend to control the overall variation of C_P. This behaviour is more complex than that of the more chemically stable materials such as bricks, concrete and mineral wool (Fig. 2.5).

2.4 DERIVATION OF HEAT FLOW EQUATION

Consider a slab of insulating material as shown in Fig. 2.6. Here the heat crossing the plane AA′ may be considered to:

(a) heat up the slab of material between plane AA′ and plane BB′
(b) flow out across BB′ into the next section of the slab.

Expressing this heat flow process in mathematical terms

$$\lambda A \frac{d\theta}{dx} = \lambda A \frac{d}{dx}(\theta - \delta\theta) + A\,\delta x\,\rho C_p \frac{d\theta}{dt}$$

Rate at which energy is conducting across AA′	Rate at which energy is conducting across BB′	Rate at which slab of thickness δx is heating up

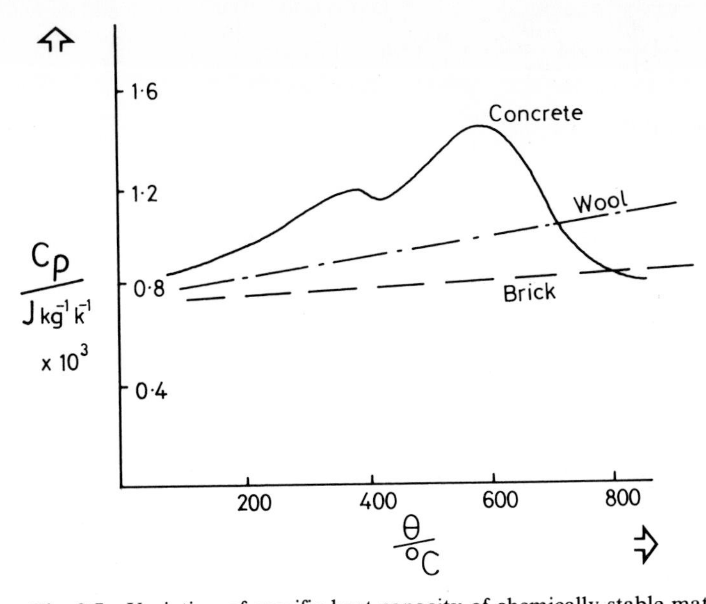

Fig. 2.5 Variation of specific heat capacity of chemically stable materials

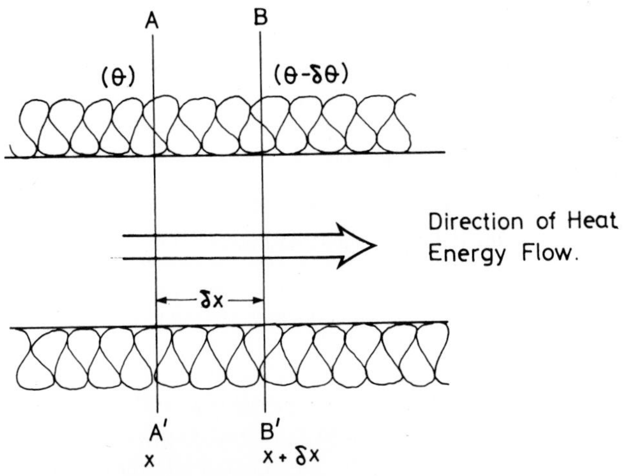

A = Cross section Area

λ = Thermal Conductivity of Material.

ρ = Density of Material.

Cp = Specific Heat Capacity of Material.

Fig. 2.6

$$\ldots \lambda A \frac{d\theta}{dx} = \lambda A \frac{d\theta}{dx} - \lambda A \frac{d}{dx}(\delta\theta) + A\,\delta x\,\rho C_p \frac{d\theta}{dt}$$

$$\ldots \lambda A \frac{d}{dx}(\delta\theta) = A\,\delta x\,\rho C_p \frac{d\theta}{dt}$$

Using the fact that $\delta\theta = \frac{d\theta}{dx} \cdot (\delta x)$

Then $\lambda A \frac{d^2\theta}{dx^2} \cdot \delta x = A\,\delta x\,\rho C_p \frac{d\theta}{dt}$.

$$\therefore \qquad \frac{d^2\theta}{dx^2} = \frac{\rho C_p}{\lambda} \frac{d\theta}{dt}$$

or $$\nabla^2\theta = \frac{1}{\alpha} \cdot \frac{d\theta}{dt}$$

where $$\alpha = \frac{\lambda}{\rho C_p}$$

and $$\nabla^2\theta = \frac{d^2\theta}{dx^2} + \frac{d^2\theta}{dy^2} + \frac{d^2\theta}{dz^2} = \frac{1}{\alpha} \cdot \frac{d\theta}{dt}$$

for a three-dimensional situation.

2.5 THERMAL DIFFUSIVITY (α)

Thermal diffusivity may be defined as:

$$\frac{\lambda}{\rho C_p} = \frac{\text{thermal conductivity of material}}{\text{density of material} \times \text{specific heat of material}}$$

The thermal diffusivity is also a measure of a material's ability to conduct heat energy in relation to its thermal storage capacity. Also the thermal diffusivity of a material gives an indication of the rate at which thermal energy can travel through a material, which in turn controls the rate of temperature rise within the material.

Materials having large diffusivity values respond quickly, whereas low diffusivity materials respond more slowly to a thermal heat flux. The temperature variation of α for steel is shown in Fig. 2.7(a) while Fig. 2.7(b) shows how α varies for masonry materials. It is essential that this temperature variation of the diffusivity of a material is known since situations may arise where the solution of the equation

$$\nabla^2\theta = \frac{1}{\alpha} \cdot \frac{\partial\theta}{\partial t}$$

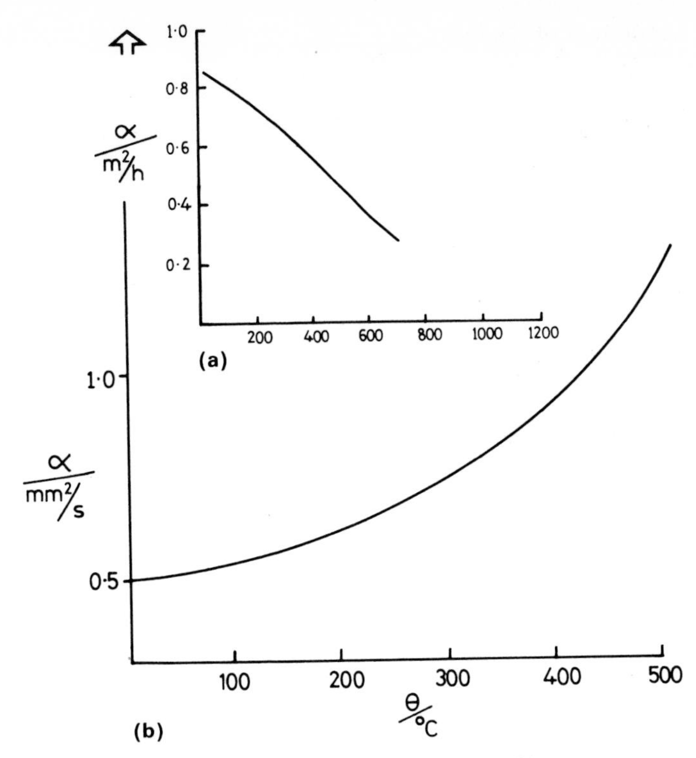

Fig. 2.7(a) Variation of the diffusivity of steel with temperature

Fig. 2.7(b) Variation of the diffusivity of masonry materials with temperature

which is derived in previous paragraph is required in the fire temperature range (see example * in Ch. 3). This may well be the case when a theoretical estimation of the fire resistance of a structure is needed (see Ch. 8).

2.6 SOLUTION OF TRANSIENT HEAT FLOW EQUATION

Solution of: $$\frac{d\theta}{dt} = \alpha \frac{d^2\theta}{dx^2} \qquad \ldots [1]$$

when a slab is subject to a sinusoidal temperature at the surface,

i.e. $\theta = \theta_0 \sin(2\pi ft)$ at $x = 0$ and $t = t$

A choice of solution of the form

$$\theta = C\,e^{-px} \sin(2\pi ft - qx)$$

where C, p and q are constants to be determined.

Rewriting eqn [1],

$$\frac{d\theta}{dt} = 2\pi f C\,e^{-px} \cos(2\pi ft - qx)$$

$$\frac{d\theta}{dx} = -pC\,e^{-px} \sin(2\pi ft - qx) - qC\,e^{-px} \cos(2\pi ft - qx) \qquad \ldots[2]$$

$$\frac{d^2\theta}{dx^2} = p^2C\,e^{-px} \sin(2\pi ft - qx) + pqC\,e^{-px} \cos(2\pi ft - qx)$$
$$+ pqC\,e^{-px} \cos(2\pi ft - qx) - q^2C\,e^{-px} \sin(2\pi ft - qx)$$

Substituting these expressions into eqn [2]

$$2\pi f \cos(2\pi ft - qx) = \alpha(p^2 \sin(2\pi ft - qx) + 2pq \cos(2\pi ft - qx)$$
$$- q^2 \sin(2\pi ft - qx))$$

$$2\pi f = 2pq\alpha \qquad \text{and} \qquad \alpha(p^2 - q^2) = 0$$

Since no sine terms on lhs, $(p^2 - q^2) = 0$

$$\therefore \quad 2\pi f = 2pq\alpha \qquad \text{i.e.} \quad pq = \frac{\pi f}{a}$$

$$\text{also} \quad p^2 = q^2 \qquad \text{i.e.} \quad p = q$$

$$\text{so} \qquad p = q = \pm\sqrt{\frac{\pi f}{\alpha}}$$

$\therefore$ solution can be written as

$$\theta = C \exp\left\{-x\sqrt{\frac{\pi f}{\alpha}}\right\} \sin\left(2\pi ft - x\sqrt{\frac{\pi f}{\alpha}}\right)$$

Inserting boundary condition

$$\theta = \theta_0 \sin(2\pi ft) \qquad \text{at } x = 0 \quad \text{and} \quad t = t$$

$$\therefore \qquad C = \theta_0$$

$$\therefore \qquad \theta = \theta_0 \exp\left\{-x\sqrt{\frac{\pi f}{\alpha}}\right\} \sin\left(2\pi ft - x\sqrt{\frac{\pi f}{\alpha}}\right)$$

$$\therefore \quad (\theta_0)_x = \theta_0 \exp\left\{-x\sqrt{\frac{\pi f}{\alpha}}\right\} \qquad \ldots[3]$$

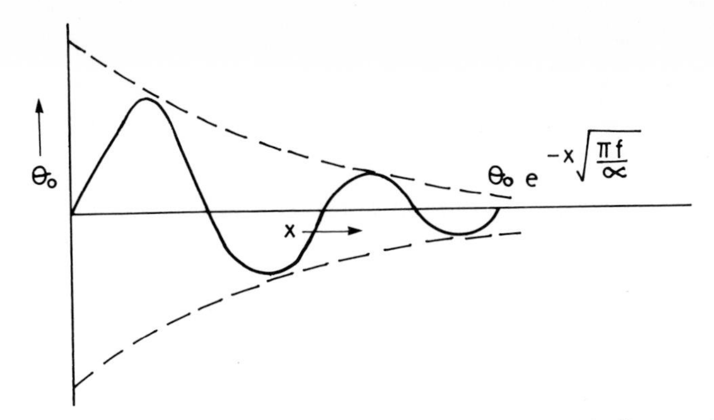

Fig. 2.8 Variation of temperature for heat wave with distance through a slab of material

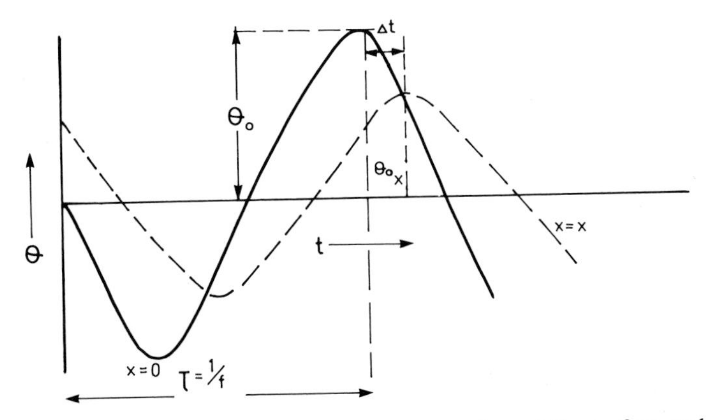

Fig. 2.9 Time lag between temperature maxima at the surface and within the slab of material

This means the amplitude of the heat energy wave as it moves through the material suffers damping according to eqn [3] (Fig. 2.8).

Also as this temperature wave propagates through the solid there is a lag in time between the *peaks* at depth x compared to those at the surface. This phase difference can be directly found from the $\sin(2\pi ft - x\sqrt{\pi f/a})$ term, i.e. $x\sqrt{\pi f/\alpha}$ is Φ, the phase lag.

The time lag Δt between the occurrence of a peak at $x=0$ compared to one at $x=x$ can be easily shown to be
$\Delta t = (x/2)\sqrt{1/\pi f\alpha}$.

From the above sketch the phase lag is equal to $2\pi f \cdot \Delta t$ (Fig. 2.9), i.e. $2\pi f\,\Delta t = x\sqrt{\dfrac{\pi f}{\alpha}}$

$\therefore \qquad \Delta t = \dfrac{x}{2}\sqrt{\dfrac{1}{\pi f \alpha}}$

The period τ for such a temperature wave can now be calculated, $\tau = 1/f$

$\therefore \quad 2\pi\dfrac{1}{\tau}\cdot\tau = \lambda\sqrt{\dfrac{\pi f}{\alpha}}$ i.e. when $\Delta t = \tau$

$\therefore \qquad \lambda = 2\sqrt{\dfrac{\pi\alpha}{f}}$

Using $V = f\cdot\lambda$ the temperature wave velocity can be determined,

i.e. $V = f\cdot 2\sqrt{\dfrac{\pi\alpha}{f}} = 2\sqrt{\pi f \alpha}$

i.e. $V = 2\sqrt{\pi f \alpha}$ ... [4]

WORKED EXAMPLE

Which of the following materials A, B or C (the physical and thermal properties of which are given in the Table 2.2 below) would make the most suitable insulating screen to withstand the penetration of a high temperature wave for as long as possible.

Solution

From the theory in the previous paragraph it was shown that the velocity of a temperature wave is given by eqn [4],

i.e. $V = 2\sqrt{\pi f \alpha}$

This implies that the material possessing the smallest value of thermal diffusivity will provide the best material to insulate the effects of a high temperature wave. Thus evaluating the diffusivity for each material according to $\alpha = \lambda/\rho C_p$

$A = 4.3 \times 10^{-6}\,m^2\,s^{-1}$

$B = 5.2 \times 10^{-7}\,m^2\,s^{-1}$

$C = 4.8 \times 10^{-7}\,m^2\,s^{-1}$.

Therefore material C is most suitable material of the *three* available.

Table 2.2

Material	λ (W m^{-1} K^{-1})	ρ (kg m^{-3})	C_p (kJ kg^{-1} K^{-1})
A	6.0×10^{-1}	11,500	0.84
B	1.0	1,200	1.60
C	3.5×10^{-1}	750	1.00

2.7 THERMAL INERTIA

Thermal inertia may be defined as the product of $(\lambda \rho C_p)$.

It is well known that the transient response of the surface lining of a structure depends on its thermal properties, i.e. the inertia of the surface lining nearest to the heat flux input. In Table 2.3 the thermal inertia of some building materials is given at ambient temperature. It is clear that the inertia variation can be easily determined provided the thermal conductivity and specific heat capacity variation with temperature are known.

It has been shown that[3] the time to flashover or for the fire growth period t_G depends on the thermal inertia $(\lambda \rho C_p)$ of the surface covering of a compartment wall,

i.e. $t_G = \phi(\lambda \rho C_p)$

or $t_G = \text{const}\,(\lambda \rho C_p)^n$

where $0 \leqslant n \leqslant 0.5$.

Assuming a value for $n = 0.5$

$$t_G = \text{const}\sqrt{\lambda \rho C_p}$$

Table 2.3

Material	Thermal conductivity λ (W m^{-1} K^{-1})	Density ρ (kg m^{-3})	Specific heat capacity C_p (J kg^{-1} K^{-1})	Thermal inertia $\lambda \rho C_p$ (W^2 m^{-4} K^{-2})
Brick	0.80	2,600	800	1.66×10^6
Plasterboard	0.16	950	840	1.276×10^5
Plaster finish	0.50	1,300	1,000	6.5×10^5
Cellular type insulation	0.03	20	1,500	900
Fibrous insulation	0.05	240	1,250	1.5×10^4

WORKED EXAMPLE

Compare the predicted fire growth periods for two identically-constructed enclosures having the same shape and dimensions but differing internal wall linings. Enclosure A has a plasterboard finish while enclosure B has a fibrous insulation lining.

Solution

Using the information provided in Table 2.3, i.e.

$(\lambda \rho C_p)_A = 130 \times 10^3$

$(\lambda \rho C_p)_B = 15 \times 10^3$

and the fact that $t_G = \text{const}\sqrt{\lambda \rho C_p}$

$$\text{then} \quad \frac{t_{GA}}{t_{GB}} = \frac{(\lambda \rho C_p)_A^{1/2}}{(\lambda \rho C_p)_B^{1/2}} = \sqrt{\frac{130}{15}} \simeq 3$$

$$\therefore \quad t_{GB} = \tfrac{1}{3} t_{GA}$$

This suggests that if a fire in enclosure A takes 20 minutes to reach flashover, then the same fire condition in enclosure B would grow to flashover in 7 or 8 minutes.

2.8 THERMAL EXPANSION

It is well known from practical experience that metallic materials will invariably expand when heated. This fact has been shown to be the case using simple expansion experiments at school level.

The tendency of a material to expand with an increase in temperature is associated with the increase in the internal kinetic energy of the atoms and molecules within the material. The co-efficient of expansion gives an indication of the expansion that is produced with change in temperature. This expansion co-efficient may be defined as:

> 'The expansion per unit length of a material when its temperature is raised by 1 °C or 1 K.'

In elementary physics courses, α is assumed to be constant over a large temperature range especially where metals are being considered. Thus a simple relationship can be derived relating the expansion $\Delta \ell$ to the original length ℓ and the temperature rise $\Delta \theta$

via the co-efficient of linear expansion α,

i.e. $\Delta\ell = \alpha\ell\,\Delta\theta$

Table 2.4 gives a list of the expansion co-efficients for common materials which assume α to be independent of temperature:

Table 2.4

Material	α/°C
Copper	1.67×10^{-5}
Steel	1.10×10^{-5}
Glass	9.00×10^{-6}
Aluminium	2.55×10^{-5}

It must be noted that at elevated temperatures well in excess of 100 °C, the assumption the α is temperature independent is no longer valid.

It has been shown for materials such as steel that the co-efficient of expansion increases quite slowly with temperature Fig. 2.10.

It must be noticed that at temperatures greater than 700 °C, steel

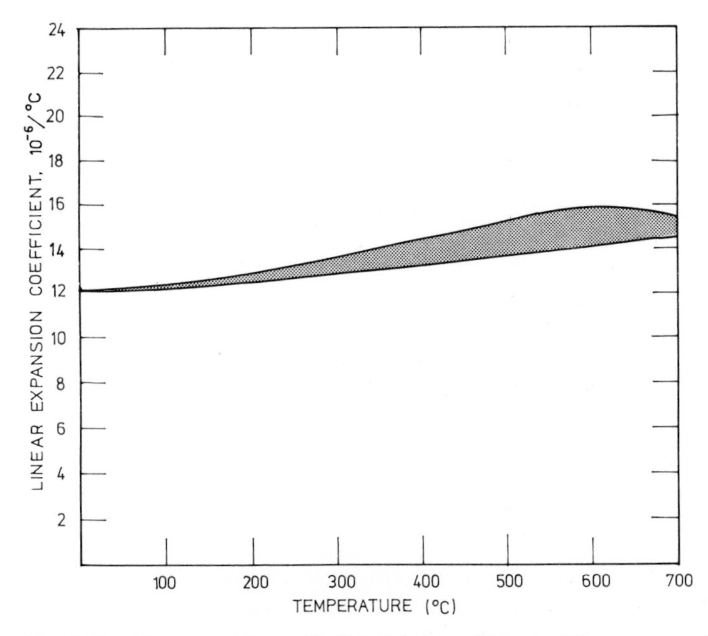

Fig. 2.10 Upper and lower limits of the co-efficient of linear expansion for steel

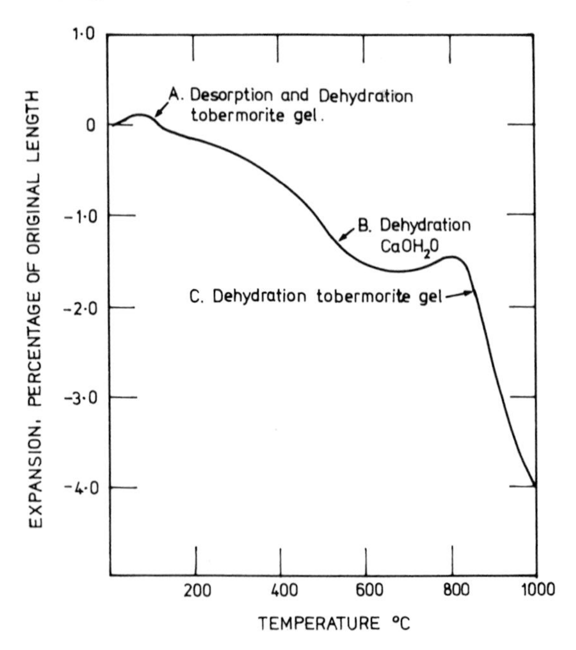

Fig. 2.11 Linear shrinkage of cement paste with increasing temperature. (Heating rate 5 °C per minute)

starts to contract; this is of little importance since at 700 °C steel has lost most of its useful strength.

For non-ferrous materials the variation in α is, however, irregular over the lower temperature ranges. The masonry-type materials, such as cement paste, show this tendency which is due to water removal which causes shrinkage and chemical transformations which can cause both expansion and shrinkage in the materials (Fig. 2.11).

For composite materials such as concrete the resulting expansion, if any, depends on the type of aggregate used. Soft aggregates such as perlite have a relatively small influence on the expansion of the concrete where the cement paste matrix controls the overall process. When hard aggregates such as granite are used, the influence exerted by the aggregate is much stronger and the concrete expands as shown (Fig. 2.12).

Table 2.5 gives a few of the co-efficients of linear expansion of concrete which use different aggregate types over a temperature range up to 100 °C.

Thus a knowledge of how the co-efficients of linear expansion of building materials vary with temperature is essential if the reader is to take full advantage and understand the content of the following chapters which deal with issues such as fire resistance where thermal stresses induced by fire temperatures are to be evaluated.

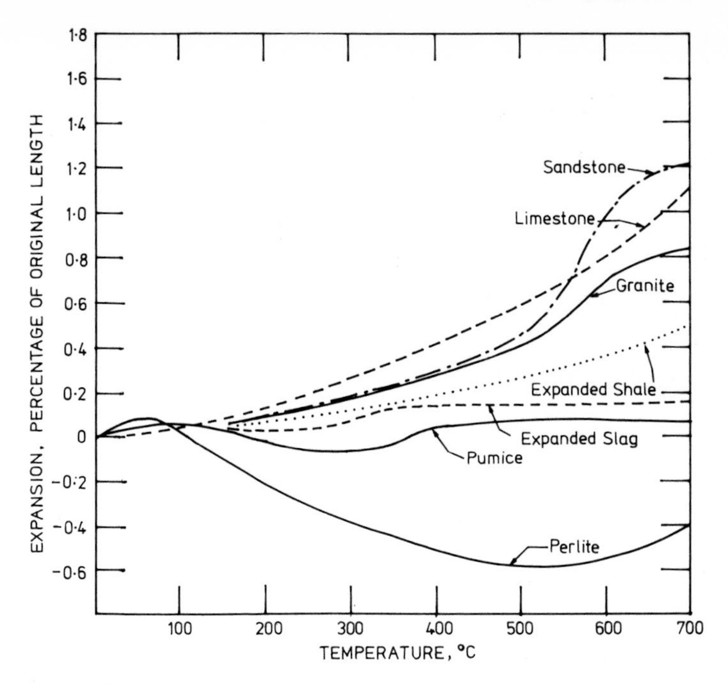

Fig. 2.12 Linear expansion with temperature of concretes made with different aggregates

Table 2.5

Aggregate type	**α/°C**
Granite	9.50×10^{-6}
Limestone	6.80×10^{-6}
Quartz	1.20×10^{-5}
Sandstone	1.12×10^{-5}

WORKED EXAMPLE

A zinc rod exactly 1.0 m long at 10 °C is 2.33 mm shorter than a copper rod at the same temperature. On heating both rods to 260 °C it is found that they are now of the same length. Given that the co-efficient of linear expansion of the copper rod is 1.67×10^{-5}/°C, calculate the co-efficient of linear expansion for the zinc rod.

Solution

$$\underset{10\,^\circ\mathrm{C}}{\ell(\mathrm{Zn})} = 1.0\,\mathrm{m} \qquad \underset{10\,^\circ\mathrm{C}}{\ell(\mathrm{Cu})} = 1.0023\,\mathrm{m}$$

$$\Delta\ell(\mathrm{Zn}) = \alpha_{\mathrm{Zn}} \times 1.00 \times 250$$

$$\Delta\ell(\mathrm{Cu}) = 1.67 \times 10^{-5} \times 1.0023 \times 250$$

$$\therefore \quad \underset{260^\circ\mathrm{C}}{\ell(\mathrm{Zn})} = 1.00 + 1.00 \times \alpha_{\mathrm{Zn}} \times 250$$

$$\underset{260^\circ\mathrm{C}}{\ell(\mathrm{Cu})} = 1.0023 + 1.0023 \times 1.67 \times 10^{-5} \times 250$$

Equating length and solving for α_{Zn}

$\alpha_{\mathrm{Zn}} = \underline{2.594 \times 10^{-5}}$ (*Ans.*)

2.9 CONCLUSIONS

It should be clear from the behaviour of materials in a fire situation that a detailed knowledge of the properties described in the foregoing paragraphs is essential, in particular the thermal diffusivity, inertia and expansion of the common building materials.

REFERENCES

1. Lee T T, *Fire and Buildings*. Applied Science Publishers, 1972.
2. Arnold P J, *Thermal Conductivity of Masonry Materials*. BRE Current Paper 1/70, 1970.
3. Thomas P H and Bullen M L, 'On the role of $K\rho C$ of room lining materials on the growth of room fires', *Fire and Materials* 1979, **3**, pp. 68–73.

CHAPTER 3

Physics of fire

3.1 INTRODUCTION

There are three clearly-defined physical processes whereby heat energy can travel from one region to another, namely conduction, convection and radiation.

3.2 CONDUCTION

This is a process in which thermal energy can transfer through a solid or liquid that is under the influence of a thermal stress or gradient. In most building materials it is a molecular process by which the thermal energy can travel like a sound wave along a solid. However, in the metallic materials this energy transfer is caused by electron movement within the solid material. A more detailed description of the molecular process is developed in Chapter 2 dealing with the thermal properties of materials. In a fire situation, a good conductor can carry thermal energy from the fire zone to heat up a combustible material in contact with it (Fig. 3.1).

3.3 STEADY HEAT FLOW

Steady-state heat flow occurs when the temperature gradient within the material has settled down to a constant value. The period before steady state is referred to as the transient state where the temperature within a material is changing with respect to time as shown in Fig. 3.2 which shows that lightweight materials A reach the steady-state condition much earlier than the heavyweight-type materials B, i.e. $\tau_A < \tau_B$ where τ is the time period before the material reaches the steady-state condition. The rate of heat energy flowing

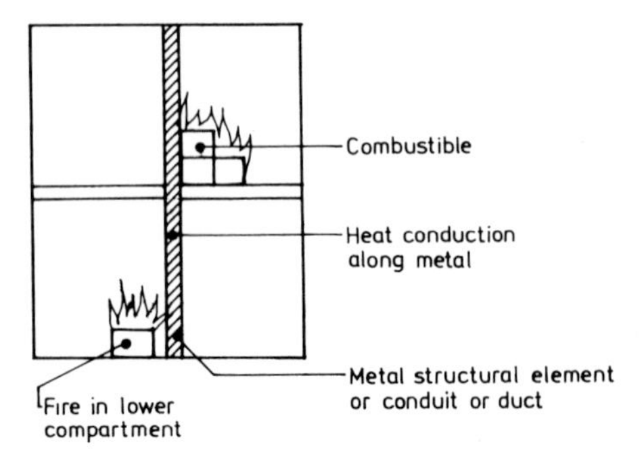

Fig. 3.1 Thermal conduction routes within a building structure

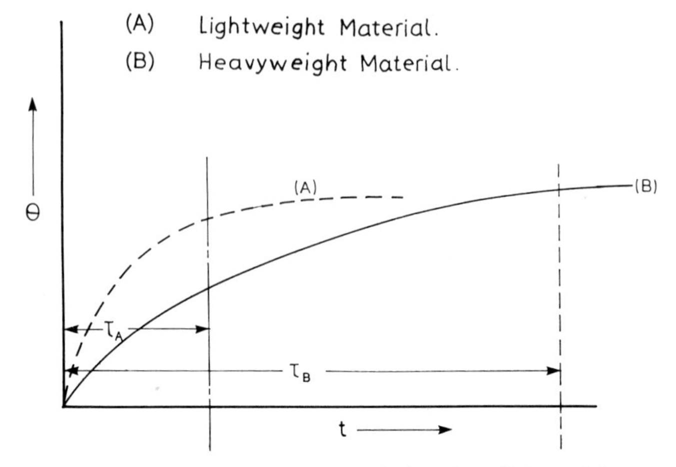

Fig. 3.2 Transient temperature variation for a light and heavy weight component

through a material under steady-state condition may be expressed as follows:

$$\dot{q} = \frac{\lambda \cdot A \cdot \Delta\theta}{\ell} \qquad \ldots[1]$$

where $\dot{q}$ = heat flow rate (W)
λ = thermal conductivity of material (W m^{-1} K^{-1})
A = cross-sectional area of sample (m^2)
ℓ = thickness of the sample (m)
$\Delta\theta$ = the temperature difference across the faces (see Fig. 3.3).

N.B. Heavyweight and lightweight in this context relate to the Thermal Capacity of the materials (MC_p) where:

M = the mass of material
C_p = the specific heat capacity of the material

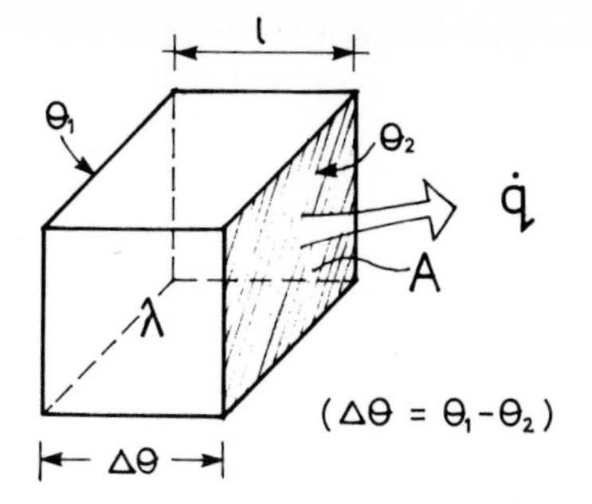

Fig. 3.3 Thermal conduction within a slab of material

Most structures are of a composite nature, i.e. they comprise several layers of materials all of which have a different thermal conductivity and thickness. Consequently eqn [1] must be modified to take these factors into account.

The equation:

$$\dot{q} = \frac{A \cdot \Delta\theta}{R_{\mathrm{TOT}}} \qquad \ldots [2]$$

describes a steady-state transfer process Fig. 3.4 where R_{TOT} is referred to as the total thermal resistance of the composite structure and is defined as follows:

$$R_{\mathrm{TOT}} = \frac{\ell_1}{\lambda_1} + \frac{\ell_2}{\lambda_2} + \frac{\ell_3}{\lambda_3}$$

If the heat conduction is from air to air then the expression used to describe such a process is given as:

$$\dot{q} = u \cdot A \cdot \mathrm{Air}^{(\Delta\theta)}\mathrm{Air} \qquad \ldots [3]$$

where $\mathrm{Air}^{(\Delta\theta)}\mathrm{Air} = (\theta_0 - \theta_5)$

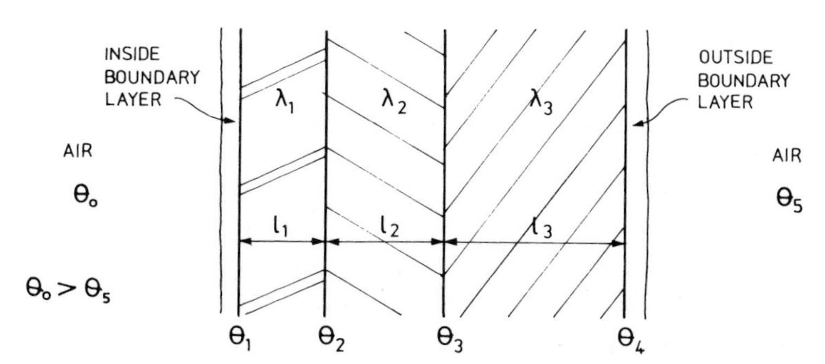

Fig. 3.4 Composite structure consisting of layers of different material

is called the air-to-air temperature difference and μ is referred to as the total thermal transmittance of the composite structure and is defined as:

$$u = \frac{1}{r_{IN} + \frac{\ell_1}{\lambda_1} + \frac{\ell_2}{\lambda_2} + \frac{\ell_3}{\lambda_3} + r_{OUT}}$$

which takes into account the role played by the stationary surface layers of air:

i.e. r_{IN} = inside surface resistance

r_{OUT} = outside surface resistance.

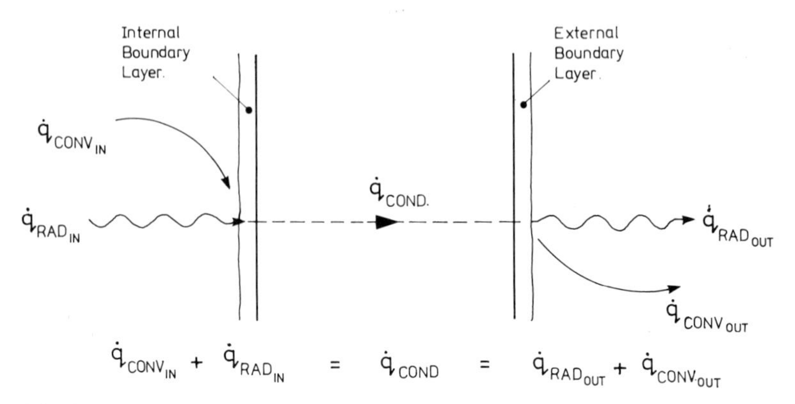

Fig. 3.5 Heat transfer mechanisms within and at the surface of a slab of material

These surface resistances are controlled by the convection and radiation exchanges that are responsible for dissipating conducted energy on the outside surface and transferring energy to the inside surface, see Fig. 3.5.

Normal values of the surface resistances are:

$r_{IN} = 0.12\,\text{m}^2\,\text{K W}^{-1}$

$r_{OUT} = 0.05\,\text{m}^2\,\text{K W}^{-1}$

The value for the external surface resistance will vary depending on the degree of exposure of surface and the climatic effects present at any period of time.

In the fire situation, where temperatures are much higher than usual, it can be shown that the surface resistance in these circumstances are very small and can be ignored.

Figure 3.6 shows the electrical analogue of the heat flow process

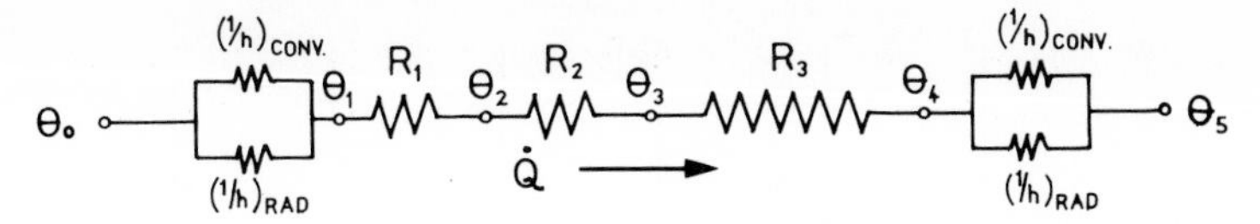

Fig. 3.6 Electrical analogue of heat transfer process through a composite slab

under steady-state conditions through the composite structure, including the surface resistance effects. Where h_{conv} and h_{rad} are the heat transfer coefficients of surfaces for convection and radiation respectively.

3.4 TRANSIENT HEAT FLOW

Since the fire process is invariably a transient process yielding a varying heat energy output, any conduction occurring as a direct result of its influence must also be transient.

It has been shown that the temperature variation of a point or plane within a material in which a transient heat flow process is occurring must be described by the equation:

$$\frac{\partial^2\theta}{\partial x^2}=\frac{1}{\alpha}\frac{\partial\theta}{\partial t} \qquad \ldots[4]$$

which has been derived in the previous chapter, where α is referred to as the thermal diffusivity and defined as:

$$\alpha=\frac{\lambda}{\rho C_p}$$

where ρ = density of material ($kg\,m^{-3}$)
C_p = specific heat capacity of the material ($J\,kg^{-1}\,K^{-1}$)

Equation [4] can be solved for various boundary conditions using analytical and numerical techniques. The following paragraph develops a simple numerical method which can be used to solve the transient heat from eqn [4].

3.5 NUMERICAL SOLUTION FOR TRANSIENT HEAT FLOW PROBLEMS

The solution of $\dfrac{\partial^2\theta}{\partial x^2}=\dfrac{1}{\alpha}\dfrac{\partial\theta}{\partial t}$

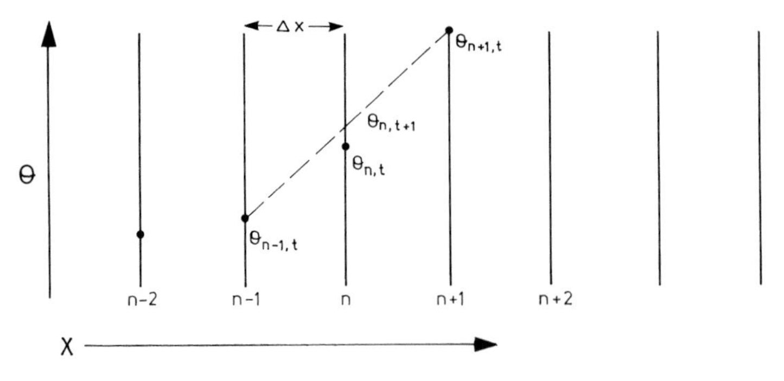

Fig. 3.7 Explicit method for the determination of temperature of a plane within a homogeneous material

can be achieved by finite differences. Consider a homogeneous slab of material as shown in Fig. 3.7 which can be considered to consist of many thin slabs of thickness x; allowing the temperature difference across each slab to be given by the approximation:

$$\frac{\Delta\theta}{\Delta x} = \frac{\theta_{n+1,t} - \theta_{n,t}}{\Delta x} \qquad \ldots[5]$$

This equation expresses the thermal gradient at any given time t for point or plane n.

The change of temperature with respect to time at this plane n is given by the following expression:

$$\Delta\theta\Delta t = \frac{\theta_{n,(t+\Delta t)} - \theta_{n,t}}{\Delta t} \qquad \ldots[6]$$

By substituting eqn [5] and eqn [6] into eqn [4], the following expression results:

$$\frac{\Delta}{\Delta x}\left(\frac{\Delta\theta}{\Delta x}\right) = \left(\frac{\theta_{n+1,t} - \theta_{n,t}}{\Delta x} - \frac{(\theta_{n,t} - \theta_{n-1,t}}{\Delta x}\right)$$

$$= \frac{1}{\alpha}\left(\frac{\theta_{n,t+\Delta t} - \theta_{n,t}}{\Delta t}\right)$$

On simplifying, the following results:

$$(\theta_{n,t+\Delta t}) - \theta_{n,t}) = \frac{\alpha\,\Delta t}{(\Delta x)^2}(\theta_{n+1,t} - 2\theta_{n,t} + \theta_{n-1,t})$$

Thus if a suitable time interval t is chosen so that $\theta_{n+1,t}$, $\theta_{n,(t+\Delta t)}$ and $\theta_{n-1,t}$ can be joined by a straight line (Fig. 3.7), then:

$$\theta_{n,(t+\Delta t)} = \tfrac{1}{2}(\theta_{n+1,t} + \theta_{n-1,t}) \qquad \ldots[7]$$

Thus $\frac{\alpha \Delta t}{\Delta x^2} = \frac{1}{2} =$ **Fourier number** ... [8]

This then allows the temperature at point n, after a time interval t, i.e.: $\theta_{n,(t+\Delta t)}$ to be determined very easily be drawing a straight line between the adjacent points as shown in Fig. 3.7. This method in a modified form can now be used to solve a simple transient heat flow problem.

WORKED EXAMPLE

Estimate the thickness of insulation sandwiched between two thin steel sheets, see Fig. 3.8, so that the temperature of the unexposed face of the composite slab should not exceed 150 °C after a period of one hour if the exposed face is subjected to a constant temperature of 815 °C. It may be

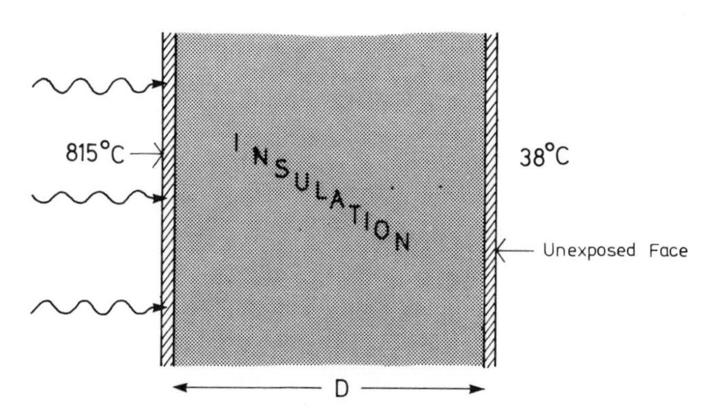

Fig. 3.8 Sketch of wall construction for worked example (3.5)

assumed that the thermal conductivity of the material remains constant over this period and that the insulation remains rigid. Since the steel plates are thin, their thermal conductivity and capacity may be ignored for this calculation.

Data

Thermal conductivity of insulation $\lambda = 0.15\ \mathrm{W\,m^{-1}\,K^{-1}}$
C_p = specific heat capacity of insulation = $1{,}045\ \mathrm{J\,kg^{-1}\,K^{-1}}$
ρ = density of insulation = $5{,}777\ \mathrm{kg\,m^{-3}}$
$\theta = 38$ °C when time $t = 0$.

It may be assumed that these thermal values are the average over the complete temperature range.

Solution

Thermal diffusivity of insulation $= \dfrac{\lambda}{\rho C_p}$

$$\therefore \qquad \alpha = 2.49 \times 10^{-7}\,\mathrm{m^2\,s^{-1}}$$

Making an assumption that $\Delta t = 10$ minutes $= 600$ secs is a reasonable time interval.
Substituting into the 'Fourier Number' condition, i.e.

$$F = \frac{1}{2} = \frac{\alpha\,\Delta t}{(\Delta x)^2}$$

a value for the slab, width Δx, can be made:

$$\Delta x^2 = 2 \times 2.49 \times 10^{-7} \times 600$$

$$\therefore \quad \Delta x = 1.4 \times 10^{-2}\,\mathrm{m}$$

A tabular solution is now prepared as shown in Table 3.1 and a manipulation based on eqn [7] made whereby the temperature of a layer at $t + \Delta t$ can be estimated using the average of the temperature at time t for the immediately adjacent layers.

Table 3.1

Time	**Layer number**						
	1	**2**	**3**	**4**	**5**	**6**	**7**
0	815	38	38	38	38	38	38
10	815	426	38	38	38	38	38
20	815	426	232	38	38	38	38
30	815	524	232	135	38	38	38
40	815	524	330	135	87	38	38
50	815	572	330	209	87	62	38
60	815	572	391	209	135*	62	50
70	815	603	391	263	135	93	50

Note: the 135 °C at 60 mins determined by

$$\frac{209 + 62}{2} = 135\,^{\circ}\mathrm{C}$$

Consulting Table 3.1 it would appear that 4 layers of insulation, each of thickness* $x = 1.4 \times 10^{-2}$ m, would give the necessary thickness of insulation for a fire endurance of one hour; i.e. $4 \times 1.4 \times 10^{-2} = 56$ mm thick.

However, the conduction mechanism of heat flow plays only a minor role in the propagation of fire; nevertheless, its effects are very important when considering fire-resistance characteristics of steel and steel-reinforced structures. This is developed later in Chapter 8 which is devoted solely to the calculation of fire resistance for structural elements.

3.6 RADIATION

This physical process unlike conduction or convection does not require a mechanical medium to enable energy to be transferred from a hot radiant surface to a combustible material.

Radiation, which may be in the ultra-violet, visible or heat radiation (infra-red) range, is an electromagnetic wave. The energy spectrum for a perfect radiator (black body) is as shown in Fig. 3.9. A fire may be considered to be a black-body radiator for almost all cases.

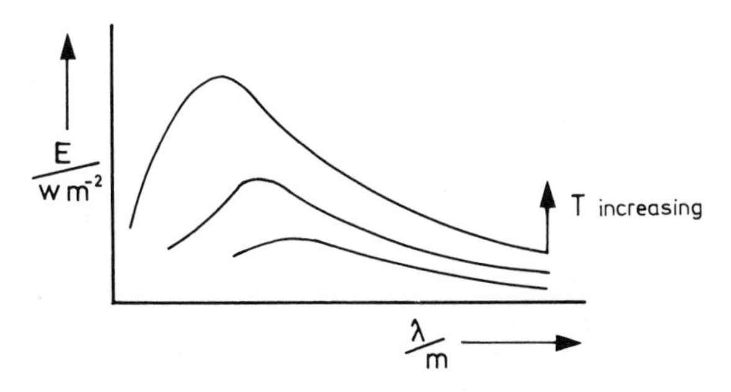

Fig. 3.9 Energy spectrum for black body radiation at various temperatures

It has been shown by Stefan[1] that the total rate of emission of radiant energy for a black body $\dot{Q}_{RAD}$ at a given absolute temperature T_B surrounded by a black-body enclosure at temperature T_S is given by the equation:

$$\dot{Q}_{RAD_{(BB)}} = A\sigma(T_B^4 - T_S^4) \qquad \ldots [9]$$

$$= A(E_B - E_S)$$

where A = total area of radiant surface (m^2)
σ = Stefan's constant = 5.67×10^{-8} W m^{-2} K^{-4}
E_B = Emissive power of body at temperature $T_B = \sigma T_B^4$
E_S = Emissive power of surface at temperature $T_S = \sigma T_S^4$

For a real situation or non-perfect black-body emission, a grey body is assumed to emit radiation, but less efficiently compared to a black body at the same temperature. The level of efficiency of a grey body is expressed by the emissivity of a surface

$$\varepsilon = \frac{\dot{q}_{GB}}{\dot{q}_{BB}} = \frac{\text{rate of radiation emission of a grey body}}{\text{rate of radiation emission for black body at same temperature}}$$

$$= \frac{E_{GB}}{E_{BB}}$$

then: $$\dot{Q}_{RAD} = \varepsilon A\sigma(T_B^4 - T_S^4) \qquad \ldots[10]$$

$$= \varepsilon A(E_B - E_S)$$

This equation gives the rate of emission of radiation for a grey body in black body surroundings (Fig. 3.10).

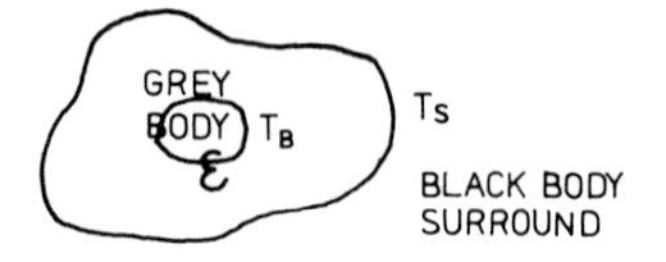

Fig. 3.10 A grey body within a black body enclosure

In eqn [10] the absorption rate is controlled by the absorption coefficient a, where a is defined as

$$a = \frac{\text{Rate of absorption of energy of surface at } T_S}{\text{Rate of absorption of energy of black-body surface at same temperature}}$$

It has been shown that $a = \varepsilon$ for most surfaces.

In most calculations in fire engineering, a fire is assumed to have an emissivity of 0.9; however, a value of $\varepsilon = 1$ is a good approximation for severe fires.

It is usually necessary to determine the radiation intensity level at some distance from a fire (Fig. 3.11).

It has been shown[2] that for a hot radiator surface 1 (Fig. 3.12) emitting radiation, the intensity of radiation flux incident at a point

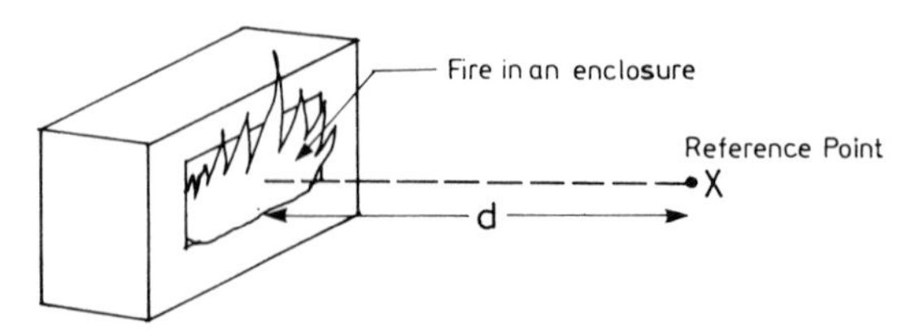

Fig. 3.11 A fire considered as a black body radiator at a given temperature

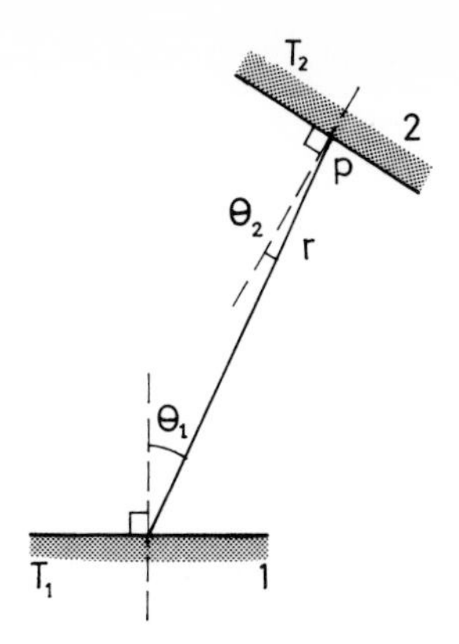

Fig. 3.12 View angles for surfaces 1 and 2

P on the surface 2 is expressed as:

$$\dot{q}_{1,2} = \varepsilon_1 \sigma T_1^4 \phi_{1,2} \qquad \ldots [11]$$

Here $\phi_{1,2}$ is the **Configuration Factor** and is defined as:

$$\phi_{1,2} = \frac{\text{Intensity of energy incident at point P on surface 2 from 1}}{\text{Intensity of energy emission by surface 1}} \qquad \ldots [12]$$

It has been shown[3] that:

$$\phi_{1,2} = \int_0^{A_1} \frac{\cos\theta_1 \cdot \cos\theta_2}{\pi r^2} \cdot \mathrm{d}A_1 \qquad \ldots [13]$$

Thus knowing the configuration factor $\phi_{1,2}$ it is possible to calculate the total radiant flux incident at P at some distance r from surface 1. The configuration factors for certain well-defined shapes and geometries can be found in many texts.[3]

Since the configuration factors are additive algebraically (Fig. 3.13), ϕ for more complicated shapes and geometries can be deduced as shown in Fig. 3.14 from the sum of the component parts. The configuration factors for a rectangular vertical surface Fig. 3.15 can be calculated solving:

$$\phi(\alpha, S) = \frac{1}{2\pi}\left(\sqrt{\frac{\alpha S}{1+\alpha S}} \times \tan^{-1}\left(\frac{\alpha/S}{1+\alpha S}\right) + \sqrt{\frac{\alpha/S}{1+\alpha S}} \times \tan^{-1}\sqrt{\frac{\alpha S}{1+\alpha/S}}\right) \qquad \ldots [14]$$

where $S = \dfrac{\text{height of radiator}}{\text{width of radiator}}$

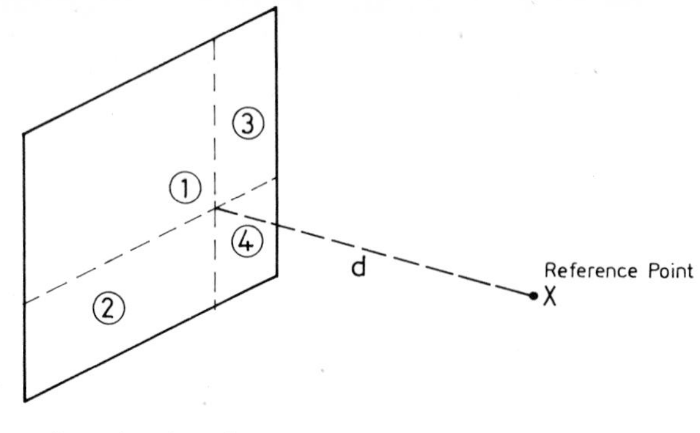

$$\phi_x = \phi_1 + \phi_2 + \phi_3 + \phi_4$$

Fig. 3.13 Algebraic addition of configuration factors

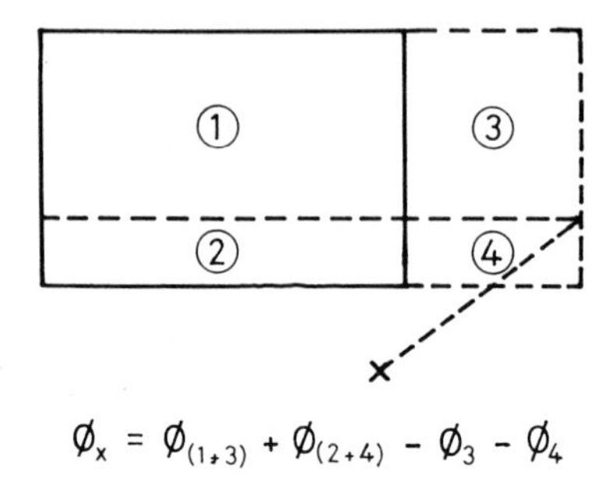

Fig. 3.14 Algebraic addition law used to determine configuration factor for a more complex situation

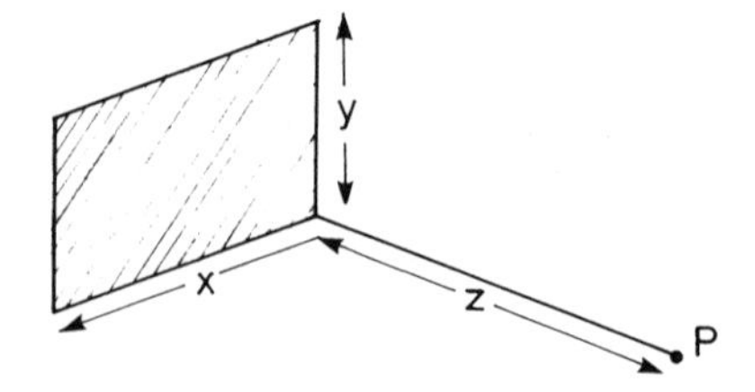

Fig. 3.15 Reference rectangular surface for use with configuration factor Table 3.2

and $$\alpha = \frac{\text{area of radiator}}{(\text{separation})^2} = \frac{xy}{Z^2}$$

A solution of the above is given in tabular form in Table 3.2.

An example which employs a vertical rectangular radiating surface

Table 3.2 Values of $\Phi(\alpha, S)$ for various values of α and S

Φ / α	$S=1$	$S=0.9$	$S=0.8$	$S=0.7$	$S=0.6$	$S=0.5$	$S=0.4$	$S=0.3$	$S=0.2$	$S=0.1$
2.0	0.178	0.178	0.177	0.175	0.172	0.167	0.161	0.149	0.132	0.102
1.0	0.139	0.138	0.137	0.136	0.133	0.129	0.129	0.113	0.099	0.075
0.9	0.132	0.132	0.131	0.130	0.127	0.123	0.117	0.108	0.094	0.071
0.8	0.125	0.125	0.124	0.122	0.120	0.116	0.111	0.102	0.089	0.067
0.7	0.117	0.116	0.116	0.115	0.112	0.109	0.104	0.096	0.083	0.063
0.6	0.107	0.107	0.106	0.105	0.103	0.100	0.096	0.088	0.077	0.058
0.5	0.097	0.096	0.096	0.095	0.093	0.090	0.086	0.080	0.070	0.053
0.4	0.84	0.083	0.083	0.082	0.081	0.079	0.075	0.070	0.062	0.048
0.3	0.69	0.068	0.068	0.068	0.067	0.065	0.063	0.059	0.052	0.040
0.2	0.51	0.051	0.050	0.050	0.049	0.048	0.047	0.045	0.040	0.032
0.1	0.028	0.028	0.028	0.028	0.028	0.028	0.027	0.026	0.024	0.021
0.09	0.026	0.026	0.026	0.026	0.025	0.025	0.025	0.024	0.022	0.019
0.08	0.023	0.023	0.023	0.023	0.023	0.023	0.022	0.022	0.020	0.017
0.07	0.021	0.021	0.021	0.021	0.020	0.020	0.020	0.019	0.018	0.016
0.06	0.018	0.018	0.018	0.018	0.018	0.017	0.017	0.017	0.016	0.014
0.05	0.015	0.015	0.015	0.015	0.015	0.015	0.015	0.014	0.014	0.013
0.04	0.012	0.012	0.012	0.012	0.012	0.012	0.012	0.012	0.011	0.010
0.03	0.009	0.009	0.009	0.009	0.009	0.009	0.009	0.009	0.009	0.008
0.02	0.006	0.006	0.006	0.006	0.006	0.006	0.006	0.006	0.006	0.006
0.01	0.003	0.003	0.003	0.003	0.003	0.003	0.003	0.003	0.003	0.003

will now be discussed which uses the additive property of the configuration factors and develops a useful concept of a reducing factor and the enclosing rectangle approximation.[4]

Most fires are considered to act as black-body radiators and this assumption does not vary that much from reality since most common fires have yellowish flames which have been shown by experiment to approximate the behaviour of black-body radiators.

WORKED EXAMPLE

Estimate the minimum separation between a school building in Fig. 3.16 and two mobile classrooms of wooden construction located side by side.

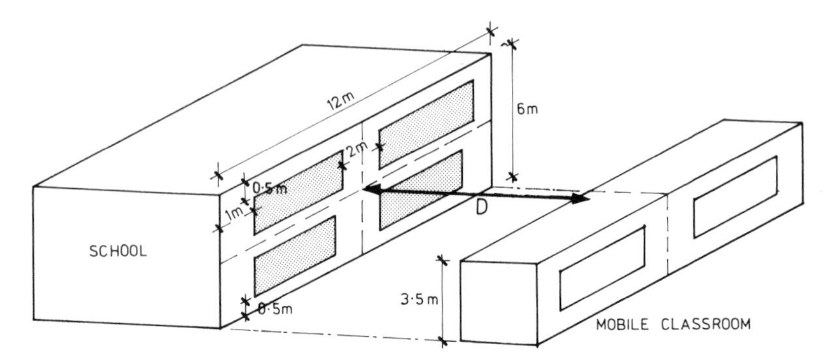

Fig. 3.16 Sketch of rectangular radiations surface for worked example 3.7

Assume that the critical level of radiation for pilot ignition is 24 kW m^{-2} and that the fire loading within the school is such that when on fire it will behave as a black body having a temperature of 1200 K.

Solution

In order to facilitate a simple solution it may be assumed that the glazed area that is enclosed by the rectangle, Fig. 3.17, can be replaced by a complete window having a fire of reduced intensity behind it. The actual reduction is given as the fraction of the original area of glazing of that of the enclosing rectangle,

$$\text{i.e.}\quad \text{reduction factor} = \frac{E_2}{E_1} = \frac{\text{reduced intensity}}{\text{original intensity}} = \frac{\text{area of glass}}{\text{enclosing area}} \qquad \ldots[15]$$

$$\text{reduction factor} = \frac{32}{50}$$

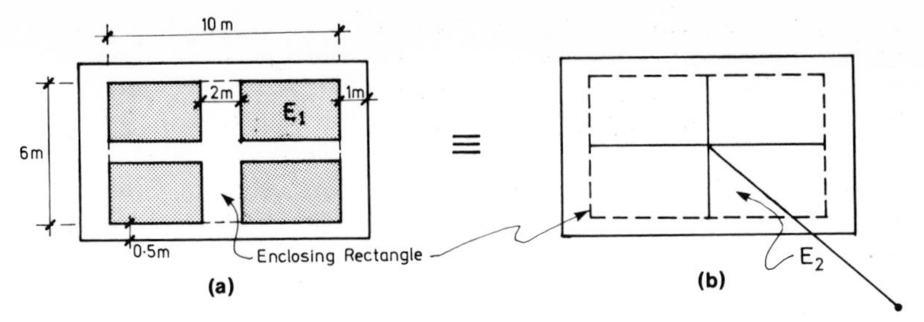

Fig. 3.17 Actual radiation surfaces replaced by a more easily analysed arrangement of surfaces at a reduced temperature

$$\therefore \qquad E_2 = \frac{32}{50} \times E_1 = \frac{32}{50} \times 5.67 \times 10^{-8} \times 1.2^4 \times 10^{12}$$

$$E_2 = 75.2\ \mathrm{kW\ m^{-2}}$$

Referring to Fig. 3.17 and using the law of addition for the configuration factors then:

$$\phi_{\mathrm{TOT}} = 4 \times \phi = \frac{24}{75.2} = \frac{\text{maximum allowed intensity}}{\text{incident intensity from fire}}$$

$$\therefore\ \phi = 0.079$$

Using Table 3.2 and the fact that $S = y/x = 2.5/5 = 0.5$, the corresponding α when $S = 0.5$ and $\phi = 0.079$ is determined directly or by extrapolation. In this case $\alpha = 0.4$ since $\alpha = xy/d^2 = (5 \times 2.5)/d^2 = 0.4$.

$$\text{Then} \quad d^2 = \frac{12.5}{0.4} = 62.5$$

$$\therefore \qquad d = 7.9\ \mathrm{m}.$$

Thus if the separation distance for this school façade and mobile classrooms is greater than 7.9 m, the probability of pilot ignition is greatly reduced in the event of a fire.

3.7 CONVECTION

The convection process plays a very important role in the spread of fire throughout a building since approximately 76–80 per cent of energy released from a fire is by this process. In simple terms when a heat source is introduced to a fluid as shown in Fig. 3.18, the fluid layer closest to the hot surface warms up and becomes more buoyant compared to the rest of the fluid. This fluid takes away some of the thermal energy from this region and is replaced by

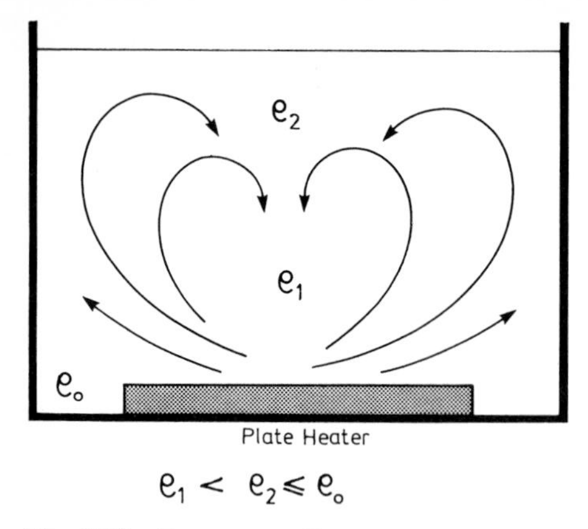

Fig. 3.18 Buoyancy effect causes a current to move and by a mixing process the cooler fluid is warmed up

cooler, denser fluid. The efficiency at which this occurs depends on factors such as:

1. Dynamic viscosity of the fluid
2. Temperature of the fluid
3. Velocity at which the fluid moves in relation to the hot surface.

It has been shown experimentally that forced convection is more efficient at transferring energy by this means than natural convection. Forced convection occurs when some external agency such as a stirrer helps to move the fluid over the hot surface. The energy that leaves the hot surface by convection is shared within the bulk of the fluid. This process is repeated until eventually the fluid reaches its boiling point.

3.8 FIRE INDUCED CONVECTION

In the fire situation the hot gases that are produced as a result of the combustion process move away due to the effect of buoyancy forces to be replaced by cooler gases. The hot gases usually remain close to the ceiling of an enclosure and act themselves as radiant sources to the surrounding surfaces as shown in Figs. 3.19 and 3.20.

If there are doors or other openings in the fire enclosure walling, the smoke and toxic gases can be forced out to make their way to other parts of the building. Later in the text, methods of controlling smoke movement will be discussed in the context of means of escape.

Fire by its nature produces large changes in the density of the

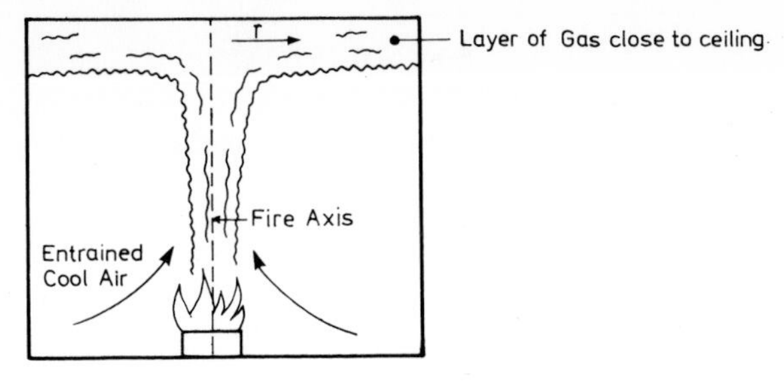

Fig. 3.19 Hot layer of gas forming at ceiling level

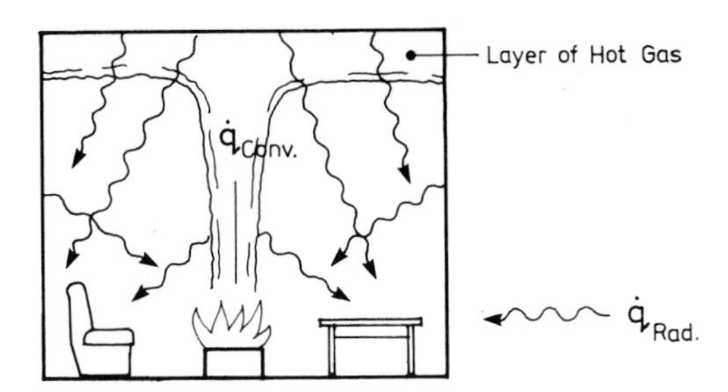

Fig. 3.20 Radiation feedback from hot gas layer at ceiling to combustible materials below

entrained air within the fire zone and plume. This induced buoyancy forces the air and combustion products towards the ceiling. This is illustrated in Fig. 3.19 where the gas flow impingement at the ceiling causes it to flow close to the ceiling for a considerable distance r measured from the fire axis.

Thus a further knowledge of this process would help the reader to understand more completely the behaviour of fire detectors and where they must be located at ceiling level.

It has been shown experimentally[5] and theoretically[6] that the variation of the gas temperatures and velocities on and near the fire axis is approximately Gaussian in shape for any height above the fire bed. It has been shown that this Gaussian or 'top hat' distribution for the maximum temperature θ_{max} and gas velocity V_{max} can be described mathematically in terms of:

(a) the ceiling height h above the fire zone
(b) the rate of heat energy release from the fire zone $\dot{q}_f$
(c) the radial distance r from the fire axis,

i.e. $$\theta_{max} - \theta_{amb} = \Delta\theta = \frac{K_1 \dot{q}_f^{2/3}}{h^{\frac{5}{3}}} \qquad \text{for } r \leqslant 0.18h \qquad \ldots [16]$$

and $$\Delta\theta = \frac{K_2(\dot{q}_f/r)^{\frac{2}{3}}}{h} \qquad \text{for } r > 0.18\text{ h} \qquad \ldots[17]$$

Here $\Delta\theta$ is the rise in temperature of the fire gases above the surrounding ambient temperature.

Similarly for the maximum gas velocity

$$V_{max} = K_3\left(\frac{\dot{q}_f}{\text{h}}\right)^{\frac{1}{3}} \qquad \text{for } r \leqslant 0.18\text{ h} \qquad \ldots[18]$$

$$V_{max} = \frac{K_4\dot{q}_f^{1/2}h^{\frac{1}{2}}}{r^{\frac{5}{6}}} \qquad \text{for } r > 0.18\text{ h} \qquad \ldots[19]$$

The variation in V_{max} and θ_{max} are shown for different heights h and distances r from the fire axis for the 18 MW fire (Fig. 3.21 and Fig. 3.22).

A sprinkler head or heat detector will be activated by the absorption of heat energy from the fire plume, and its response will depend on the rate at which the heat energy is transferred from the gases to the head. This rate of heat energy transfer $\dot{q}$ can be related to $\Delta\theta$ and V as follows:

$$\dot{q} = \text{const } \Delta\theta \, V^{\frac{1}{2}}$$

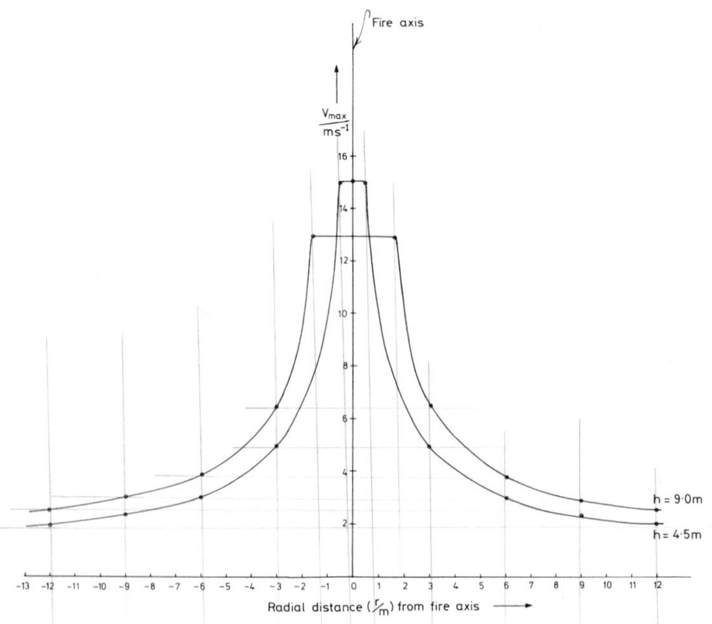

Fig. 3.21 Variation in maximum gas velocity (V_{max}) with radial distance from fire axis

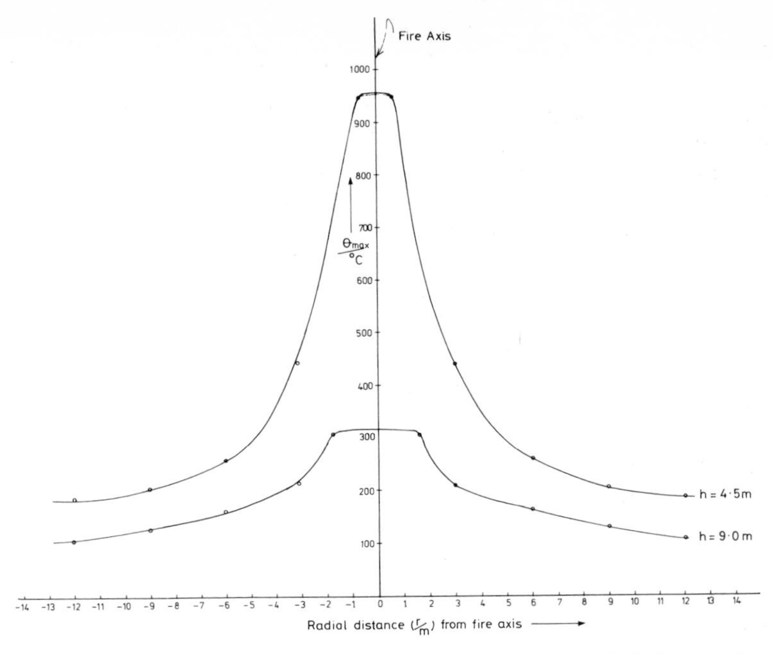

Fig. 3.22 Variation of maximum temperature θ_{max} with radial distance from fire axis

where $\Delta\theta = (\theta_g - \theta_d)$
θ_g = gas temperature
θ_d = detector temperature
V = gas flow temperature over detector.

For a maximum rate, $\dot{q}_{max}$,

i.e. $\dot{q}_{max} = \text{const}\ \Delta\theta_{max}\ V_{max}^{1/2}$

Table 3.3 gives the maximum gas temperatures and gas velocities expected at different heights h along the fire axis above fires of differing heat energy output rates.

Thus for even a small fire, the gas velocities may be large enough

Table 3.3 Fire type

	0.5 mW		1.25 mW		5.0 mW	
h	V_{max}	θ_{max}	V_{max}	θ_{max}	V_{max}	θ_{max}
3 m	5.1 m s^{-1}	200 °C	7.0 m s^{-1}	350 °C	11.0 m s^{-1}	845 °C
5 m	4.4 m s^{-1}	106 °C	6.0 m s^{-1}	175 °C	9.4 m s^{-1}	400 °C
6 m	4.1 m s^{-1}	81 °C	5.5 m s^{-1}	130 °C	8.8 m s^{-1}	280 °C

for the gas flow to be non-laminar in the vicinity of the detector head. If this is the case, 'boundary flow theory' shows that this turbulent-flow pattern is more effective regarding the transfer of energy from fluid to detector than that occurring in a streamlined flow situation.

The change from streamlined flow to turbulent flow can be evaluated using 'Reynolds number', which is a dimensionless grouping of the physical quantities that control the nature of the fluid flow.

Reynolds' number is defined as:

$$Re = \frac{\rho \cdot \ell \cdot v}{\eta}$$

which is the ratio of the momentum forces to viscous forces within the fluid,

where ρ = density of fluid ($\mathrm{kg\,m^{-3}}$)
ℓ = a characteristic dimension of the surface over which the fluid moves (m)
v = velocity of fluid over surface ($\mathrm{m\,s^{-1}}$)
η = dynamic viscosity of fluid ($\mathrm{N \cdot s \cdot m^{-2}}$).

A small Reynolds number is associated with the well-ordered streamlined or laminar flow case where the viscous forces or liquid friction within the fluid arises from the movement of fluid molecules between fluid layers moving at different velocities (see Fig. 3.23).

The turbulent-flow pattern occurs when the Reynolds number exceeds 2,300 approximately and is accompanied by the generation of large friction forces within the fluid due to the random and chaotic transfer of large globules of fluid between the liquid layers (see Fig. 3.24).

As mentioned earlier, the response of a detector will depend on the energy transfer rate from fluid to detector. This is important when it is necessary to consider the location of sprinkler heads and heat detectors to protect a zone from fire. Considering the likely

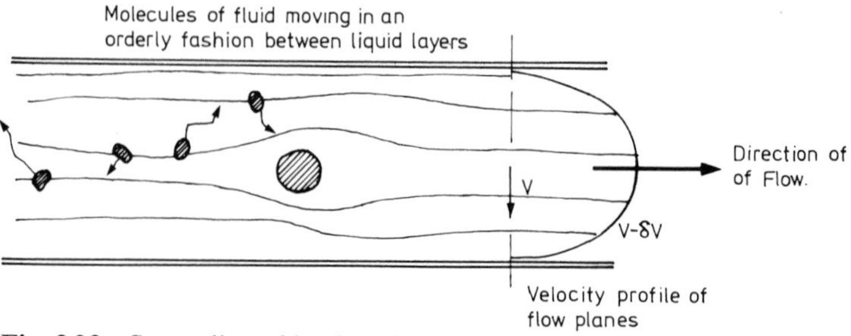

Fig. 3.23 Streamline of laminar flow

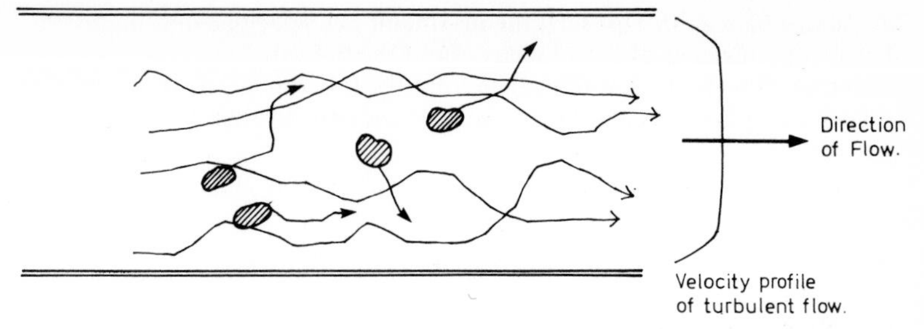

Fig. 3.24 Globules of fluid molecules moving in a chaotic manner between flow lines moving at random down the pipe

thermal output $\dot{q}_f$ from a fire load, the spacing of the heads at ceiling level must be such that even if the head closest to the fire axis only undergoes a streamlined heat transfer mechanism it must be sensitive enough to react.

The normal head distribution is based on a square-grid pattern array, the spacing of which is calculated so that in the event of a fire the sprinkler will respond either to control the fire or in the case of a heat detector to give adequate warning to the occupancy and the fire fighters.

WORKED EXAMPLE

Estimate whether or not the gas flow associated with a small fire (0.5 mW) in a room having a ceiling height of 3 m is turbulent or streamlined for a detector of 24 mm diameter located close to a ceiling:

(i) directly over fire zone
(ii) some distance from fire zone.

Solution

Using Table 3.3 it can be assumed that θ_{max} is 473 K; $V_{max} = 5.0\,\text{m s}^{-1}$. It is now necessary to get an estimate of the Reynolds number for each case:

(i) $$Re = \frac{\rho \cdot \ell \cdot v}{\eta} \qquad \ldots [20]$$

assuming at 473 K, $\eta = 3.0 \times 10^{-5}\,\text{N s}^{-1}\,\text{M}^{-2}$

and $$\rho_{473\,\text{K}} = \rho_{294\,\text{K}} \times \frac{294}{473} = 0.65\,\text{kg m}^{-3}$$

$\therefore$ $$Re = \frac{0.65 \times 2.4 \times 10^{-2} \times 5.0}{3.0 \times 10^{-5}} \simeq 2{,}700$$

In this case the flow is turbulent.

(ii) As can be seen in Fig. 3.21, the maximum gas velocity drops quite sharply at a distance of 2 or 3 metres from the fire axis.

Assume that the V_{max} at this range from the fire axis is 2.5 m s^{-1} then, *Re* will reduce to 1,350 which shows that the flow over the detector is now streamlined.

3.9 CONCLUSIONS

It should now be clear that each of these thermal-energy transfer mechanisms have an important part to play in the spread of fire and its effects within a building. In future chapters these thermal-energy transfer processes will be referred to either directly or indirectly when describing the effects and consequences of fires.

REFERENCES

1. Roberts J K and Miller A K, *Heat and Thermodynamics*. Blackie and Sons, 1960.
2. Cornwell K, 'The flow of heat', in *S.I. Units*, Van Nostrand Reinhold, 1978.
3. Hottel H C and Sarofin A F, *Radiative Transfer*, McGraw-Hill, 1967.
4. Law M, '*Heat Radiation from Fires and Building Separation*', Fire Research Technical Paper No. 5, HMSO, 1963.
5. Alpert R L, 'Calculation of response time of a ceiling-mounted fire detector', *Fire Technology*, **8**, pp. 181–95, 1972.
6. Lee S L and Emmons H W, 'A study of natural convection above a live fire', *J. Fluid Mech.*, **11**(3), pp. 353–68, 1961.

CHAPTER 4

Chemistry of fire

4.1 COMBUSTION: INTRODUCTION

The term combustion or burning used in this text refers to a rapid oxidation of combustible substances. The oxidation process is an exothermic chemical reaction.

This process is rapid and is considered to be practically adiabatic. Thus the temperatures reached by the reacting species are high, because the energy transferring mechanisms, i.e. conduction, convection and radiation are not capable of quickly dispersing the released thermal energy. The condition necessary for combustion to occur is known to be one where the fuel and oxygen are present in the correct proportions in the presence of an ignition source. This condition is simply represented by the fire triangle (Fig. 4.1). Usually some preheating of the fuel is necessary before the required volatiles are released to take part in the combustion process. Once the combustion process is initiated, it is necessary that fuel and sufficient thermal energy be available to allow the process to be self-sustaining. This energy can be achieved by thermal feedback, i.e. convection and radiation, as shown in Fig. 4.2.

① Here the thermal radiation and convected energy is fed back to the original fuel bed to release more volatiles
② This radiation component also preheats the virgin fuel causing it to pyrolyse, thus encouraging fire spread.

The ignition mechanism may be considered as that process which produces an exothermic reaction characterised by an increase in temperature greatly above ambient.

It is possible to identify two types of ignition:

1. Pilot ignition
2. Spontaneous ignition.

In pilot ignition, flaming is created in the flammable vapour–air

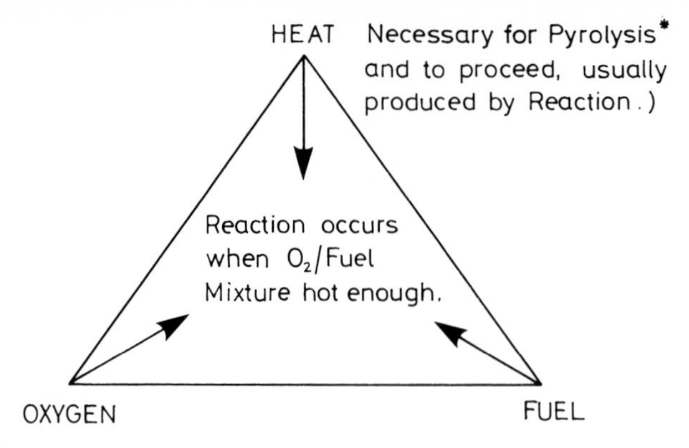

Fig. 4.1 Fire triangle

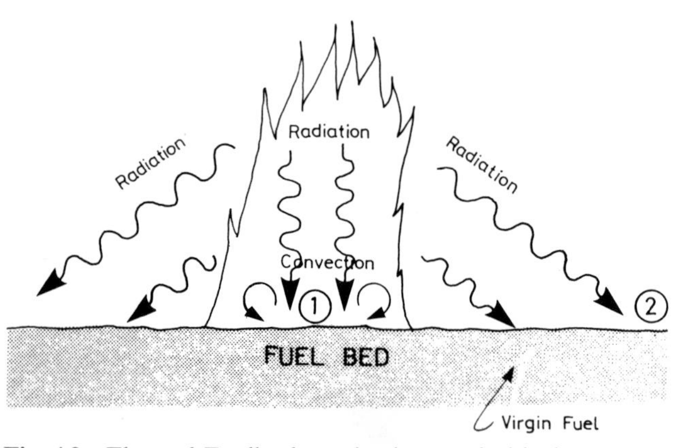

Fig. 4.2 Thermal Feedback mechanisms to fuel bed

mixture by a pilot source, e.g. flaming match or spark. With spontaneous ignition, however, flaming occurs without the application or presence of a pilot ignition source, if the environmental temperatures are sufficiently elevated. Pilot and spontaneous ignition will be discussed later in this chapter.

In simple terms the combustion process can be considered as follows:

Fuel + Oxidant → Combustion products

Consider the complete oxidation of methane the simplest of all the hydrocarbon fuels:

$$CH_4 + 2O_2 \rightarrow CO_2 + 2H_2O \qquad \ldots[1]$$

This reaction is one involving a 'stoichiometric mixture', i.e. 'A mixture in which the ratio of fuel to oxidant is theoretically correct for complete oxidation.'

Most combustion processes occur in air which is a chemical mixture of approximately 21 per cent oxygen and 79 per cent nitrogen, thus when air is used in eqn [1] describing a stoichiometric mixture it must be rewritten as follows (in molar terms):

$$CH_4 + 2(O_2 + \tfrac{79}{21}N_2) \rightarrow CO_2 + 2H_2O + 2\cdot\tfrac{79}{21}N_2$$

Fuel + (Oxidant + Dilutent → Combustion products + Dilutent ...[2]

Here the dilutent plays no part in the chemical process, but participates in a physical process which helps to dissipate some of the thermal energy produced by the combustion process.

Equation [2] describes a complete combustion process in air. However, when there is not enough fuel or oxygen an incomplete reaction may occur. Consider the following incomplete oxidation process, which is short of oxidant, so it is considered as *fuel rich*.

$$CH_4 + 1\tfrac{1}{2}(O_2 + \tfrac{79}{21}N_2) \rightarrow CO + 2H_2O + (\tfrac{3}{2}\cdot\tfrac{79}{21})N_2 \qquad \ldots[3]$$

Alternatively, a *lean mixture* reaction can now be considered as follows:

$$CH_4 + (2+a)(O_2 + \tfrac{79}{21}N_2) \rightarrow CO_2 + 2H_2O + (2+a)\tfrac{79}{21}N_2 + aO_2$$

Fuel Air Combustion products Dilutents ...[4]

Here the unreacting oxygen becomes a dilutent and helps to absorb some of the combustion energy.

4.2 COMBUSTION TYPES

There are essentially two types of combustion:

1. Flaming combustion
2. Smouldering combustion.

Flaming combustion is characterised by premixed or diffusion flames. Smouldering combustion is a non-flaming self-propagating combustion process. Only those materials which form a solid carbonaceous char on heating can exhibit smouldering combustion, e.g. paper, cellulosic fabrics, sawdust, latex rubber and fibreboard.

4.3 PREMIXED FLAMES

Figure 4.3 depicts the combustion of a premixed gas in a pipe. The pre-combustion interaction is one that modifies the air/fuel mixture that exists before the ignition and combustion occur, and is endothermic requiring energy to be fed to it from the combustion zone. This pre-combustion interaction does influence the combustion

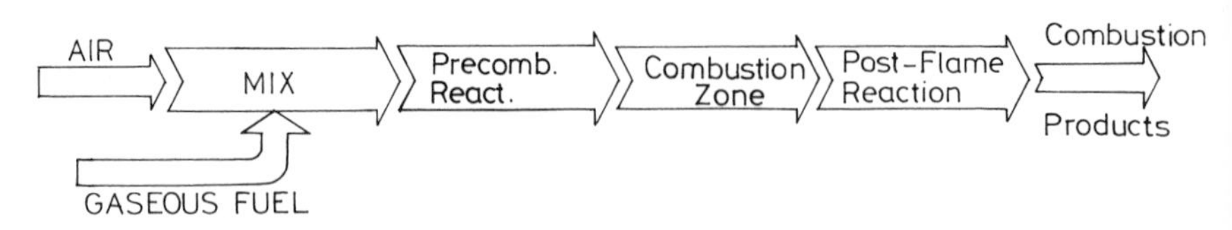

Fig. 4.3 Premixed combustion process within a pipe

process and the nature of the emissions from this zone. The temperature variation through the pipe is as shown in Fig. 4.4. The temperature of the flame can be estimated by appealing to elementary thermodynamics.[1] It has been shown that this adiabatic flame temperature depends on the fuel/air mixture and that every fuel oxidant mixture has a lower limit of flammability below which the combustion process will not occur. For every lower limit there is a lower adiabatic flame temperature, which for most hydrocarbon fuels is approximately 1400 K.

Thus the premixed flame combustion process is well ordered and the rate of reaction is controlled by chemical kinetics.

An important characteristic of premixed flames is their fundamental burning velocity.[2] This may be defined as the maximum burning velocity observed for a given fuel in air in metres per second. It is important to note that the burning velocity of premixed

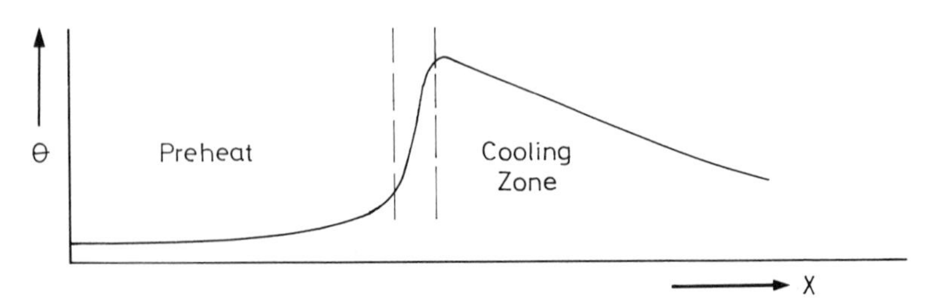

Fig. 4.4 Temperature variation through pipe when premixed combustion is occurring

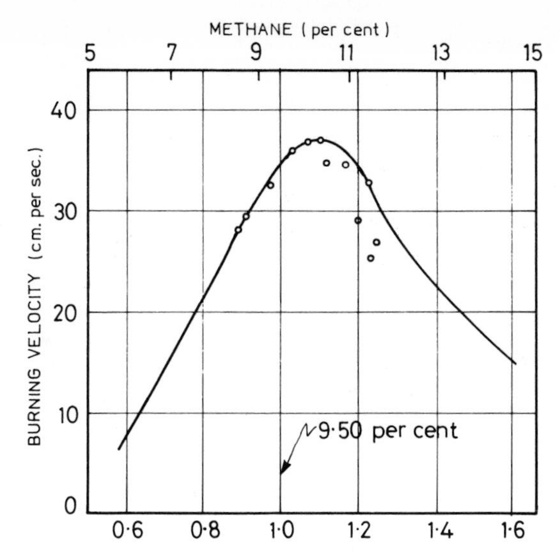

Fig. 4.5 Methane fraction of stoichiometric mixture

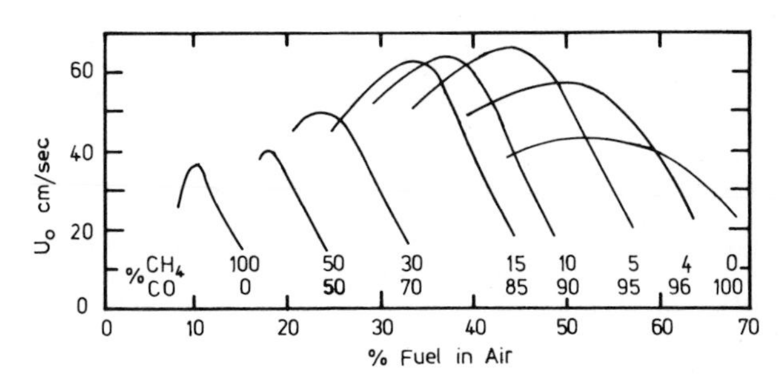

Fig. 4.6 Flame speed of $CH_4 + CO$ mixtures burning in air

flames will vary with:

(a) mixture composition
(b) temperature
(c) pressure
(d) presence of chemical inhibitors
(e) turbulence

and the influence of some of these factors are illustrated in Figs. 4.5, 4.6 and 4.7.

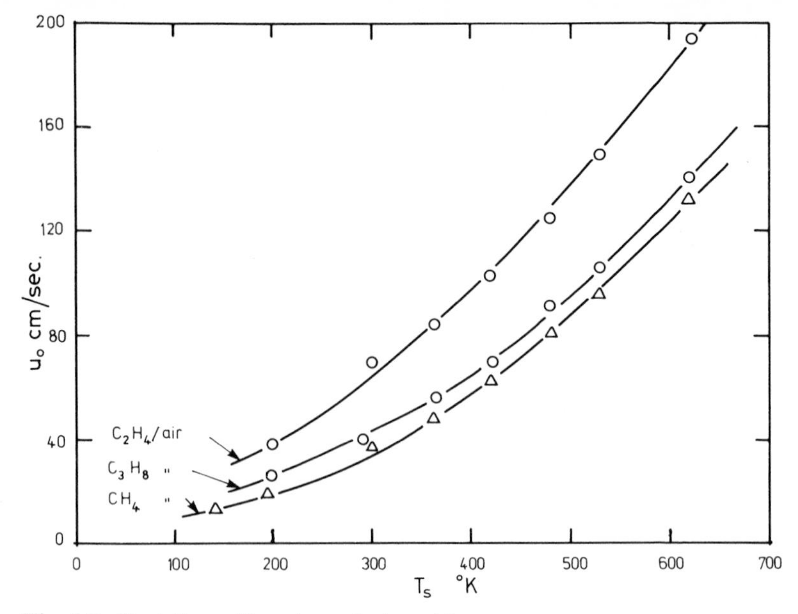

Fig. 4.7 Variation of burning velocity with temperature

4.4 FLAMMABILITY LIMITS

A flammable mixture is one through which a flame will propagate. Fortunately not all mixtures of a flammable gas and air are flammable.

The limits of flammability are defined by Zabetakis[3] as follows:

Lower flammability limit:

$$L = \tfrac{1}{2}(C_w + C_r)$$

where C_w = greatest concentration of fuel in air which is non-flammable
C_r = least concentration of fuel in air which is flammable.

Upper flammability limit:

$$U = \tfrac{1}{2}(C_{w_1} + C_{r_1})$$

where C_{w_1} = least concentration of fuel in air which is non-flammable
C_{r_1} = greatest concentration of fuel in air which is flammable.

The flammability limits are generally expressed as a volume percentage of fuel in air and the flammability envelope for a methane–oxygen–nitrogen system is illustrated in Fig. 4.8. The lower flammability limit is also a thermal limit which connects it to the

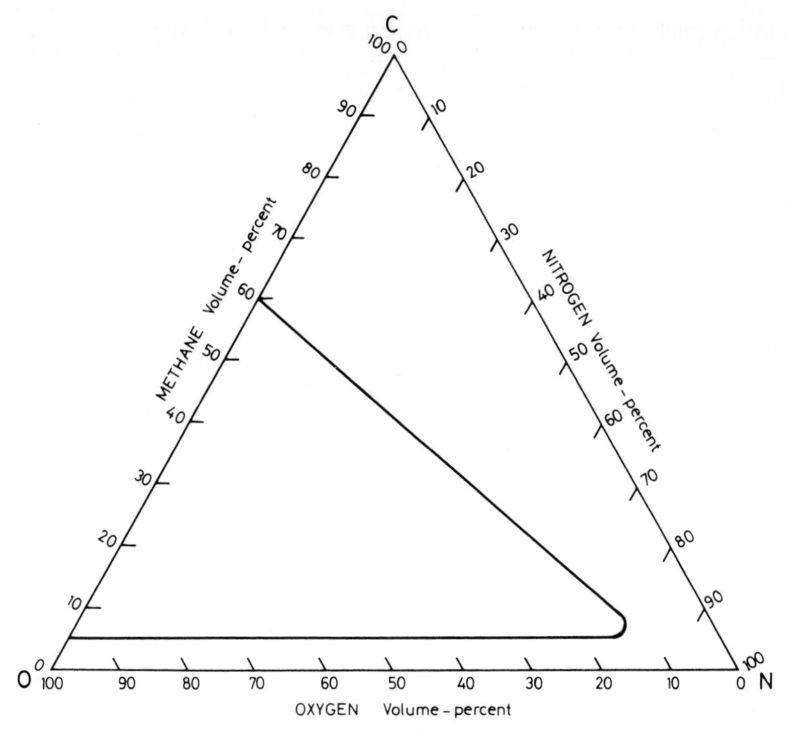

Fig. 4.8 Flammability diagram for methane, oxygen and nitrogen at STP

limiting adiabatic flame temperature. Also the lower flammability limit approximates to 0.55 of the stoichiometric concentration, i.e. for alkanes 45–50 mg litre^{-1}.

4.5 STOICHIOMETRIC CONCENTRATION

The stoichiometric concentration may be termed the ideal concentration of fuel in air, i.e. complete combustion of the fuel is achieved and all the oxygen is consumed (Fig. 4.5).

4.6 LIMITING ADIABATIC FLAME TEMPERATURE

Adiabatic means no heat gains or losses from the system. The limiting adiabatic flame temperature is the temperature of the flame which must be maintained to support combustion. If the temperature of premixed flames drops below this critical temperature the flame cannot be supported and will be extinguished.

4.7 DIFFUSION FLAMES

With diffusion flames the rate of combustion is determined by the rate of mixing of the fuel vapour with the air/oxidant and is essentially a physical process. Diffusion flames are those most often encountered and are not characterised by such factors as burning velocity and mixture strength.

In order to study diffusion flame phenomena and assist with mathematical modelling certain assumptions are made:

1. Reaction zone in infinitesimally thin
2. Reaction zone corresponds to the stoichiometric surface
3. Rate of diffusion determines the rate of burning
4. Diffusion co-efficient is constant.

A diffusion flame is the manifestation of a combustion process where the gaseous fuel mixes with the oxygen by the process of molecular and/or turbulent diffusion. If the diffusion process is slow the mixture of gaseous fuel and oxygen will not occur within the flammability limits until much higher up the buoyant plume, and consequently flame height will be greater.

A knowledge of this phenomena is very important to our understanding of fire spread in buildings. The rising buoyant plume will impinge upon ceilings and induced turbulence will produce gaseous fuel/oxygen mixtures within the flammability limits at considerable distances from the area of first contact between the rising plume and ceiling, thus increasing flame length even on non-combustible ceilings. The physical structure of a diffusion flame is not symmetrical (Fig. 4.9). It relies on the mixture composition being correct before the flame can develop. This involves the diffusion process referred to earlier. Some preheating plus some chemical excitation occurs as the fuel and oxygen carriers break down in the preheating zone.

Consider a Bunsen burner in which the fuel stream issues from the jet and mixing occurs immediately at the edges of the outlet so that close to the rim a premixed situation is almost created and is sufficient to anchor the flame. As the mixture progresses upwards in the buoyant plume the diffusion process occurs and consequently flaming, which varies in thickness and stability. In each combustion zone a state of equilibrium is achieved, i.e. further oxygen diffuses to the combustion zone to replenish that already consumed. However, if

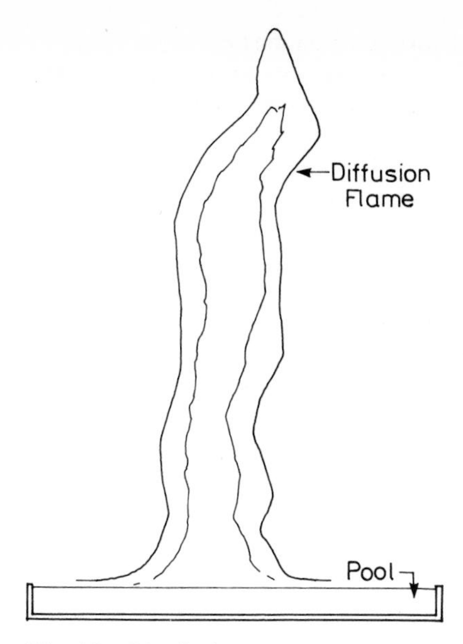

Fig. 4.9 Physical structure of a diffusion flame

for example the pressure were to drop, molecular activity would increase, i.e. the diffusion process would increase and although the chemical reaction rate may slow a little in the combustion zone, a situation close to the premixed case could be achieved.

4.8 POOL BURNING

Features common to all fires may be conveniently examined with a burning pool which is in essence a simple geometrical form. These are:

1. Flame height
2. Ignition characteristics
3. Burning rate.

4.8.1 Flame height

Burke and Schumann[4] consider the effect on flame height of variables such as flow rate, oxygen requirement of the fuel, and tube diameter. It can be shown for small pools that if the volumetric flow rate of the fuel remains constant the flame height is independent of the burner diameter, d; the mean rate of consumption of fuel per unit area of flame thus varies directly as ℓ/d (ℓ = flame height). If the fuel is changed the height varies inversely with the diffusion coefficient.

It is worth mentioning here that the burning of a liquid or solid differs from that of a gas in that the fuel vapour must be made available by the feedback of heat from the combustion zone to the condensed phase. Thus the burning rate of a liquid is equal to the rate of vaporisation, which in turn depends upon the rate of heat transfer to the liquid.

4.8.2 Ignition characteristics

If an ignition source is introduced to the vapour of a liquid at its flash point, combustion will occur momentarily.

Flash point may be defined as the lowest temperature at which there is enough evaporation to reach the lower flammability limit. As the temperature is increased so the rate of evaporation is increased.

A higher temperature, termed fire point, must be achieved in order to ensure that once flaming has commenced it is self-propagating, i.e. the rate of vaporisation is sufficient to maintain combustion. Fire point may be defined as the lowest temperature at which there is enough evaporation to produce sustained combustion. Fuel in a wick may be readily ignited because:

1. Liquid currents are inhibited
2. Small amount of fuel is involved
3. Wick acts as insulant
4. Ignition energy is concentrated.

4.8.3 Burning rate

The burning rate of liquid fuels are dependent on pool diameter and for small diameter pool fires the thermal conductivity and thickness of the container walls. For small pools a heat sink effect is created inhibiting flame spread. Other factors which can effect the burning rates are:

1. Induction period; during this period the burning rate increases steadily to reach a constant value
2. Thermal conductance of the container
3. Fuel temperature
4. Air velocity.

4.9 SOLID-FUEL COMBUSTION

The process whereby fire consumes solids is called pyrolysis (Fig. 4.10). As the solid fuel is heated a decomposition process is initiated, latent moisture in the fuel is first vaporised and then combustible vapours are produced. It is the latter which actually burn. Most solid organic compounds such as wood or plastics do not burn, they pyrolyse. It is the combustible products of their pyrolytic decomposition that burn.

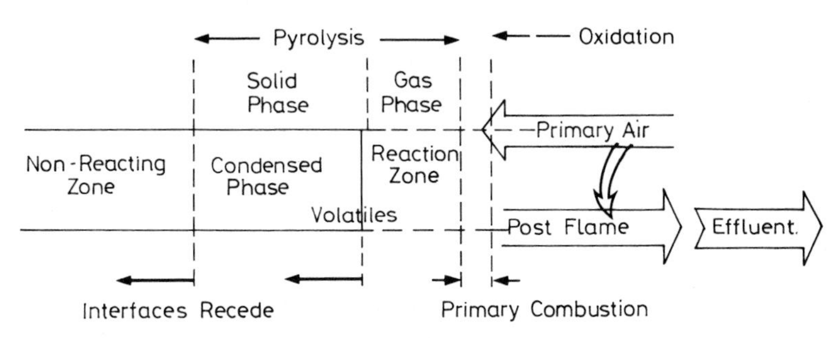

Fig. 4.10 Combustion of solid fuel

Consider for example a common building material such as wood, one of the major constituents of which is cellulose. As the temperature of the wood is increased the cellulose begins to decompose, i.e. it is pyrolysed. The first vapour produced will be water vapour since a large percentage of the volume of wood is actually water. As the heating process continues carbon monoxide, carbon dioxide and hydrogen vapours will be distilled until finally all that remains is carbon in the form of charcoal (Fig. 4.11). Softwoods contain large quantities of resins and when exposed to heat these resins vaporise rapidly. Softwoods have a larger cellular structure than hardwoods, and the combustible vapours are distilled from softwoods at lower temperatures than from hardwoods. This is one of the factors which accounts for the difference in the ignition temperatures of various woods.

Pyrolysis can occur at temperatures well below the ignition temperature of wood. Timber components in timber-framed buildings in close proximity to flue and heating pipes[5] have been found to become charred over a period of time as the heat has gradually produced pyrolytic decomposition.

Thus the diffusion flame process is the one associated with the naturally-burning fire.

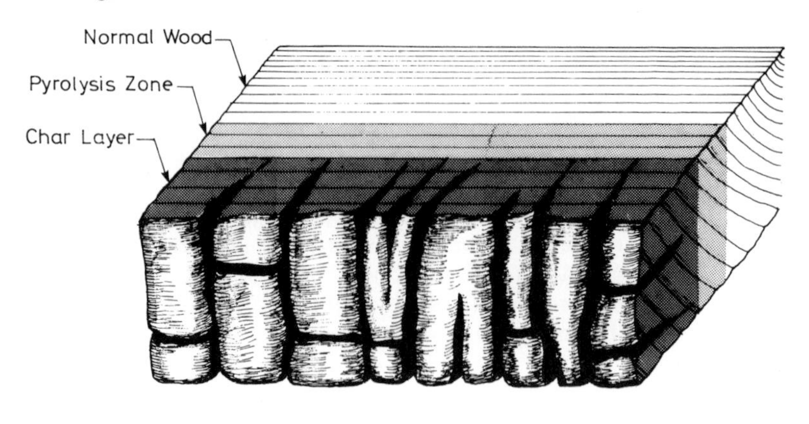

Fig. 4.11 Pyrolysis of wood

4.10 CHEMISTRY OF THE FLAME

The chemical processes that occur within the combustion zone do so very quickly and at very high temperatures. A great deal of work has been directed to this subject of combustion chemistry.[6] From the mechanical viewpoint these reactions involve two- or three-body collision processes, the probability of two-body collision being much greater than for a three-body collision.

Consider the reactions possible for the H* and OH* radicals as they collide within the combustion zone to form water as indicated earlier. The hydrogen ion reacts with molecular oxygen to form an oxygen ion plus a hydroxial ion, i.e.

$$H^* + O_2 \rightarrow {}^*O^* + {}^*OH$$

whereas the OH* reacts with molecular hydrogen to yield water and a hydrogen radical, i.e.

$$OH^* + H_2 \rightarrow H_2O + H^*$$

It is now obvious that the combustion of hydrogen is not a single direct reaction but a series of separate steps.

Consider the following chain of reactions in the combustion of hydrogen:

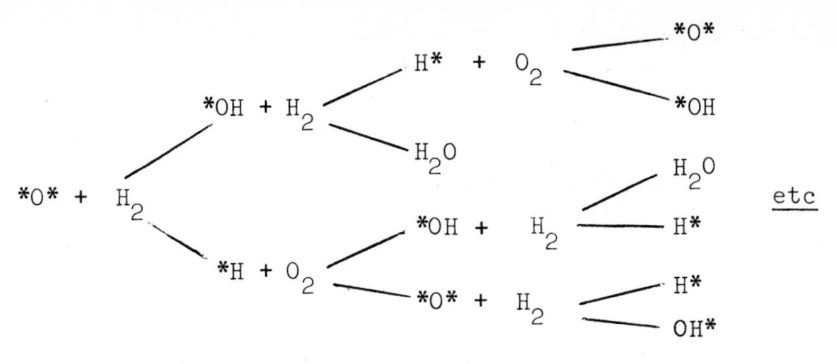

Fig. 4.12 Branching chain combustion of hydrogen

In this chain or branching chain reaction it is not certain whether all these reactions are instantaneous or whether some could be metastable and more complex than suggested in Fig. 4.12. However, it has been shown experimentally and from analysis that the energetic H* radical plays a leading role in the combustion processes of materials. Thus if the concentration of these H* radicals can be reduced to molecular hydrogen then the combustion process may be retarded, i.e.

$$H^* + H^* \rightarrow H_2$$

These combinations can be enhanced by extinguishing agents which encourage similar processes to occur in the combustion zone. The action of chemicals suitable as extinguishing agents will be considered in a later chapter.

4.11 SPONTANEOUS COMBUSTION

This combustion process does not require any flame or external heat or radiation source. It occurs within materials which are separate or of a composite nature by an exothermic chemical or biological reaction generating heat internally. Thus when this process starts and the cooling is inadequate, the temperature of the substance will rise (particularly so when it is well insulated) to the lowest self-ignition temperature for the composite as a whole or for one of its components. Materials exhibiting this tendency are referred to as 'active'. The mechanism of heat build-up can be easily understood by reference to Fig. 4.13, curve 1, which shows the rate of heat generation as an exponentially increasing function depending on the temperature of the material.

When the temperature of the material equals that of the ambient

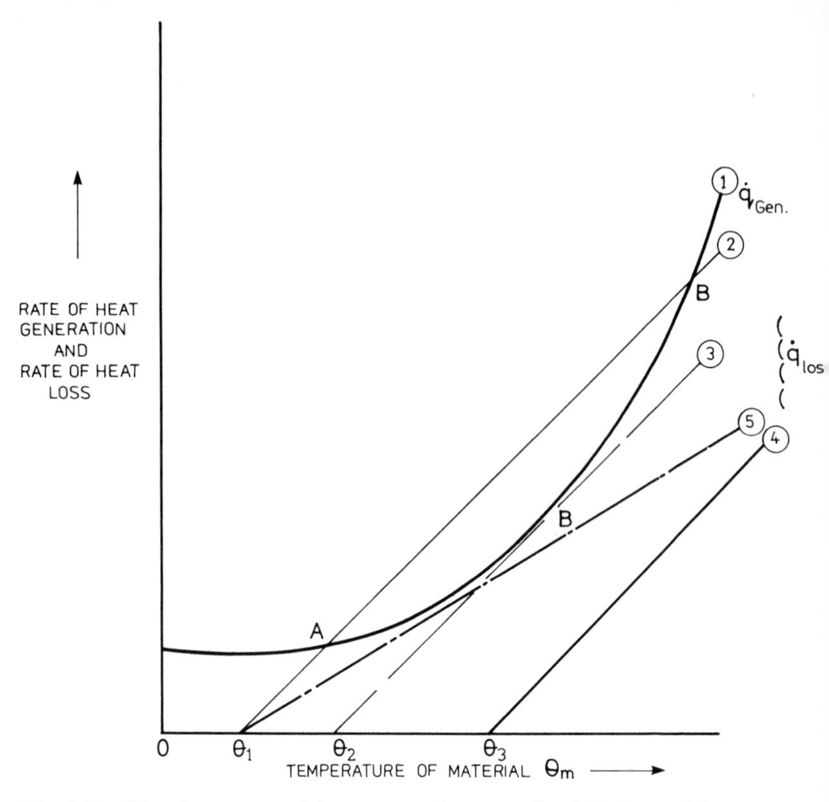

Fig. 4.13 Heat loss rate and heat generation rates for differing ambient conditions

temperature of the surrounding air the heat loss is zero according to the basic physical laws (see Fig. 4.14). Thus if the ambient temperature is θ_1 the rate of heat loss from the body will follow curve 2, Fig. 4.13.

$$\dot{q}_{\text{loss}_1} = C_1(\theta_m - \theta_1)$$

At the intersection of these two lines A, equilibrium is reached and a stable situation results, i.e. thermal runaway cannot occur.

However, if the ambient temperature is increased from θ_1 to θ_2 the heat loss will follow curve 3 which is the limiting case as indicated by point B, i.e. the tangent to curve 1, Fig. 4.13.

If the temperature θ_2 referred to hereafter as the **Critical temperature** is exceeded, say to θ_3, then the heat loss will follow curve 4 and thermal runaway will occur.

In the situation where more material or better insulated material (Fig. 4.15) is located in the same initial surrounding air temperature

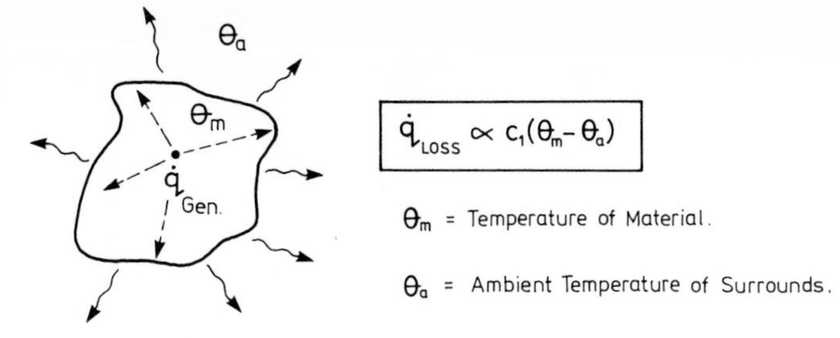

Fig. 4.14 Heat loss from a hot body to the surrounds

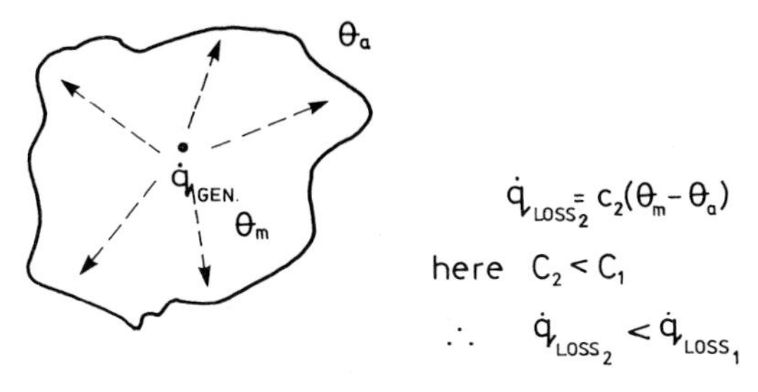

Fig. 4.15 Bigger quantity of material or better insulated system

θ_1, the heat loss rate $\dot{q}_{\text{loss}_2} = C_2(\theta_m - \theta_1)$ will follow 5, which indicates a much lower critical temperature, where C_1 and C_2 represent the gradients of the respective curves and in this case $C_2 < C_1$.

It should be noted from this that extreme caution should be exercised in the storage of excessive amounts of active materials.

Spontaneous combustion has occurred in a variety of circumstances, e.g.

1. Heating oil leaking into intrinsically[7] inert lagging
2. Rags soaked with vegetable oils in contact with a heat source or stored in large quantities.

In Table 4.1 a list of a few common active materials is given as an aid to highlighting the hazard presented when such materials have to be stored.

Table 4.1 Active materials

Bonemeal	Peat
Fats/oils	Soap powder
Fertilisers	Straw
Lagging	

4.12 PILOT IGNITION

Here an auxiliary source of heat in the form of a flame or spark is necessary to initiate the combustion process.

A piece of paper or pile of wood shavings is much easier to set alight than the corresponding book or block of wood. Thus a match flame can set off a piece of paper very easily since the radiation and convection energy exchange processes are able to raise the temperature of the paper to a temperature at which the necessary volatiles are released and the combustion process can occur.

If the combustible material is already being bombarded with radiation the pilot ignition of the material is very easy. Thus combustible surfaces must always be located at sufficient distances from possible sources of radiation. This is the basis upon which the concept of space separation is formulated and prescribed in building regulations.

4.13 SPONTANEOUS IGNITION

Here the instant spark, i.e. source of ignition, is not available so that the combustion process must use thermal mechanisms to initiate the burning process. For spontaneous ignition to occur a higher level of radiation intensity is needed compared to that required for pilot ignition (Fig. 4.16).

However, if combustible surfaces of adjacent buildings are located such that in the event of a fire the levels of radiation necessary for spontaneous ignition are reached and maintained for long periods of time, fire spread between buildings will be encouraged. For example, a light wooden fence sufficiently close to a burning building may experience a large enough intensity of radiation to undergo spontaneous ignition notwithstanding the possibility of pilot ignition caused by flying burning brands.

A series of experimental tests[8] were carried out on 510 mm × 510 mm × 19 mm thick samples of whitewood to determine the variation of ignition time with incident radiation intensity for both pilot and spontaneous ignition. These results of these tests are as shown in Fig. 4.16. Using these results it was deduced that the critical intensities for pilot and spontaneous ignition were 23 kW m^{-2} and 36 kW m^{-2}, respectively.

It was also found that for intensities of radiation greater than these critical values the ignition time could be related to the intensity

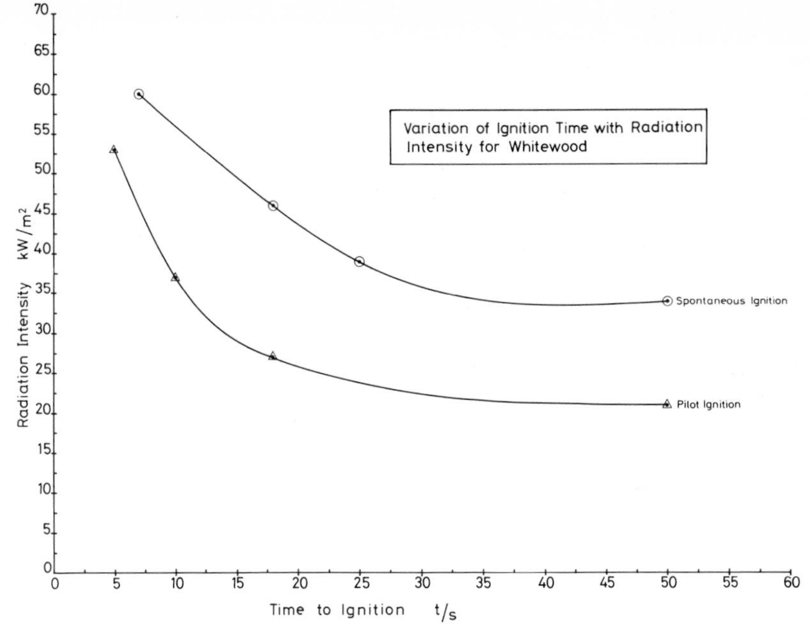

Fig. 4.16 Variation of ignition time with radiation intensity for whitewood

of radiation by the following expressions:

$$(I - I_{cp})t^{\frac{2}{3}} = A \qquad \text{for pilot ignition}$$

and $$(I - I_{cs})t^{\frac{4}{5}} = B \qquad \text{for spontaneous ignition}$$

and where A and B were directly proportioned to the thermal inertia $(\lambda \rho c)$ of the wooden samples.

It was also found that the critical intensity for pilot ignition I_{cp} is approximately similar for most wooden species. This was also found to be the case for spontaneous ignition.

Pilot and spontaneous ignition temperatures of materials will also vary with the mode of heating, i.e. convection and radiation.

REFERENCES

1. Goodger E M, '*Combustion Calculations*', *Theory*, *Worked Examples and Problems*, MacMillan Press, 1977.
2. Lewis B and vol Elbe G, *Combustion*, *Flames*, *Explosions in Gases*, Academic Press, 1962.
3. Zabetakis M G, *Flammability Characteristics of Combustible Gases and Vapors*, US Bureau of Mines Bulletin 627, 1965.

4. Burke S P and Schumann T E W, Symp. (Int), *Combustion, 1st and 2nd*, Combustion Institute, Pittsburgh, 1965, pp. 2–11.
5. Franklin P, Symposium, *Fire Safety in the Design and Construction of Dwellings*, Ulster Polytechnic, 1984.
6. Gaydon A G and Wolfhard H G, *Flames, their Structure Radiation and Temperature*, Chapman & Hall, 1979.
7. Beever P F, *Spontaneous Combustion*, BRE Information Paper, 1982.
8. Lawson D I and Sims D L, 'The ignition of wood by radiation', *British Journal of Applied Physics*, pp. 288–292, **3**, 1952.

CHAPTER 5

Development and growth of fires in enclosures

5.1 INTRODUCTION

The growth and development of fire has been shown to be dependent to a large extent on the geometry and ventilation of the enclosure containing the fire.

Friedman[1] reported the differences observed between fires burning in the open and those burning in enclosures. Figure 5.1 shows the effect of an enclosure on the burning rate of a square slab of polymethyl methacrylate which illustrates the difference between a material burning in the open and the same material burning under a hood or roof. In the former case, except for the heat required to produce the volatiles from the fuel bed, all of the remaining heat energy is lost to the atmosphere. In the latter case, however, the roof plays a significant role in that the loss of heat energy is considerably reduced and an energy feedback mechanism is created which significantly increases the pyrolisation and hence the burning rate.

The rapidity with which fire so tragically engulfed the Bradford City Football Stadium in 1985[2] clearly demonstrates the latter.

5.2 FACTORS AFFECTING FIRE DEVELOPMENT

A fire usually starts because a material is ignited by a heat source. The development of the fire within the compartment depends on many factors, namely:

1. The item first ignited is sufficiently flammable to allow flame spread over its surfaces
2. The heat flux from the first fuel package is sufficient to irradiate adjacent fuel packages which in turn will begin to burn

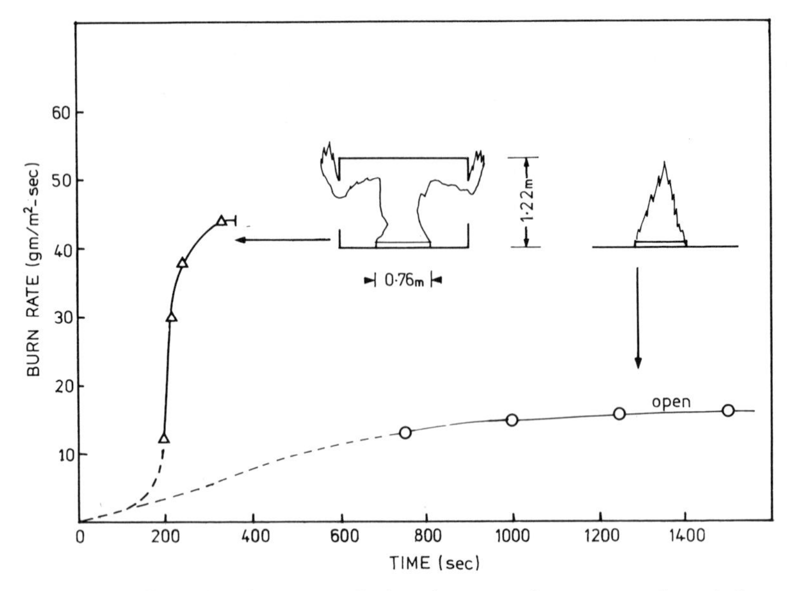

Fig. 5.1 Effect of enclosure on the burning rate of a square polymethyl methacrylate slab

3. Sufficient fuel exists within the compartment otherwise the fire may simply burn itself out
4. The fire may burn very slowly because of a restricted oxygen supply, e.g. in a well-sealed compartment the fire may eventually smother itself
5. Providing that there is sufficient fuel and oxygen available the fire may totally involve the compartment.

5.3 FLASHOVER

Flashover has been defined as the rapid involvement of a compartment's combustible contents as they ignite almost simultaneously. Unfortunately such a definition tends to regard flashover as an event. A better definition of flashover would be 'the time when flames cease to be localised and flaming can be observed throughout the whole compartment volume', i.e. the burning activity changes from being a surface phenomenon to a volume process. Flashover is, in fact, the transition from the growth period to the fully-developed stage in fire development. It is used as the demarcation point between two stages of a compartment fire, i.e. pre-flashover and post-flashover.

In model studies designed to study the effects of the variation of room and window geometries on flashover in residential sized rooms Waterman[3] chose a heat flux–time criterion to indicate flashover. A low level heat flux of 2 W cm^{-2} was used in these experiments.

Hagglund *et al.*[4] in attempting to develop a technique for predicting probable flashover conducted several room experiments using different types of furnishings. Visual observations of flames emerging from the windows were found to correlate with an average room-gas-temperature of 600 °C at the ceiling. This 600 °C 'flashover criterion' was then applied to a further series of experiments where the size of crib fire and window opening were varied. This gave an expression for peak burning rate required to cause flashover (Fig. 5.2).

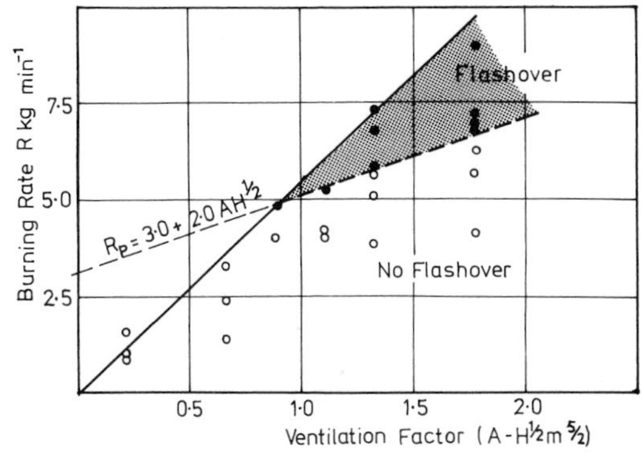

Fig. 5.2 Flashover, ventilation and burning rate

The expression so generated was:

$R_{\text{peak}} = 3.0 + 2.0AH^{\frac{1}{2}}$ when $AH^{\frac{1}{2}} \geqslant 0.9$

R_{peak} = 'peak burning rate' kg min^{-1}
(i.e. minimum burning rate necessary for flashover to occur)
A = window area (m^2)
H = window height (m)

Although this work was limited to fixed room sizes and boundary constructions, it does offer a useful approximation of the flashover potential of various rooms and their furnishings. It also offers an insight into the complex phenomenon of flashover and the factors which influence its occurrence. Some of these factors are: burning rate; compartment geometry; and ventilation.

Thermal properties of the enclosure construction will be discussed further, later in this Chapter.

5.4 FIRE DEVELOPMENT IN COMPARTMENTS

The time versus temperature development of a fire is shown in Fig. 5.3, the curve representing the average temperature determined under test conditions.

The period A–B is known as the growth period. It is essentially the pre-flashover period during which the temperatures in the compartment remain relatively low and the chance of escape are relatively high. At B the fire progresses rapidly through flashover to the fully-developed stage and it can be seen that flashover in essence is the transition from the growth period to the fully-developed period. During this period all the combustibles in the compartment are burning and the temperature within the enclosure increases sharply.

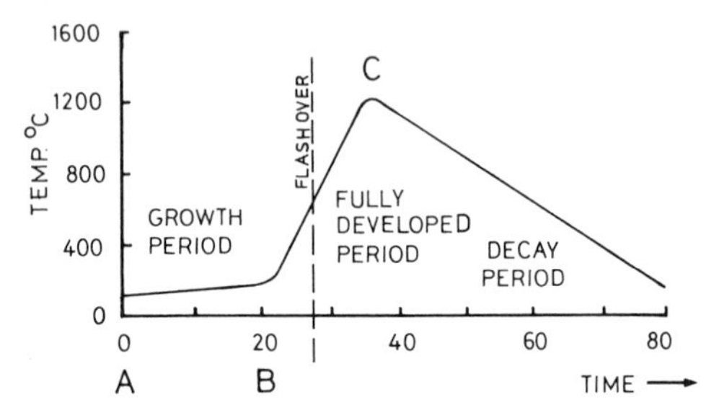

Fig. 5.3 Time/temperature fire profile

At C the burning period ends, the temperature begins to fall and the decay period begins. During the decay period the temperatures are such that for a period of time the direct threat to other spaces remains because of the risk of propagation by radiation or penetration of constructional components. In the model heat balance described above, the temperature within the compartment is assumed to be uniform and the properties of materials forming the boundary construction are assumed constant. In practice, however, the temperature within the compartment during a fire will vary, not only with time but also with location, and the properties of the materials forming the enclosing construction will also vary.

Much of the interest in modelling fire development focuses on the flashover transition and full compartment involvement in fire so that the pre-flashover situation may be overlooked. This would be a mistake. As can be seen from Fig. 5.3 the temperatures during the growth period are low and consequently can be ignored. It is the duration of the growth period, however, which is very important as it determines the time available for escape and for the effective operation of the emergency services.

During the growth period, heat from the fire causes materials in the compartment, e.g. wall lining, to evolve gases and vapours. If the rate of vapour production is sufficiently high a vapour–air mixture will be formed within the flammability limits[5] which may be ignited by flames from the already burning materials. It follows that the easier a material is to ignite and the greater the rate of heat production, the faster the growth of the fire. The time required to cause ignition of a material and its subsequent rate of heat production is dependent not only on the nature and dimensions of the physical characteristics of the material, but also on the heat flux transferred to the material itself.

Large expanses of combustible materials, such as wall and ceiling linings, can contribute significantly to the rapid growth of a fire. The radiation from large areas of burning surfaces, exponential rate of flame spread over vertical surfaces and relatively low ceilings interact to promote the rapid development of fire within a compartment.

Drysdale[6] categorised the factors affecting the rate of flame spread over combustible solids (see Table 5.1).

Table 5.1 Factors affecting the rate of flame spread on combustible solids

Material		Environmental
Chemical	**Physical**	
Composition of fuel	Initial temperature	Composition of the
	Surface orientation	atmosphere, temperature
Presence of retardants	Direction of propagation	Imposed heat flux
	Thickness of specimen	Initial pressure
	Thermal conductivity	Air velocity
	Density	
Geometry		

As well as the factors listed in Table 5.1, other factors that affect the duration of the growth period are:

1. Spacing of combustible fuel packages within the compartment
2. Mass and surface area of the combustible materials dispersed within the room
3. Size and location of ignition sources
4. Size and location of the openings in the compartment boundaries
5. Geometry of the compartment.

5.5 BURNING REGIMES

Two identifiable regimes of burning will now be considered. These are:

1. Ventilation-controlled regime
2. Fuel-controlled regime.

5.5.1 Ventilation-controlled regime

In the ventilation regime it is assumed that the rate of burning (R), i.e. the rate of mass loss, is strictly controlled by the rate of ingress of air. The underlying assumption is that the rate of air supply is exactly sufficient to burn all the volatiles which are produced at rate R. Figure 5.4 indicates the limiting ventilation rate between the

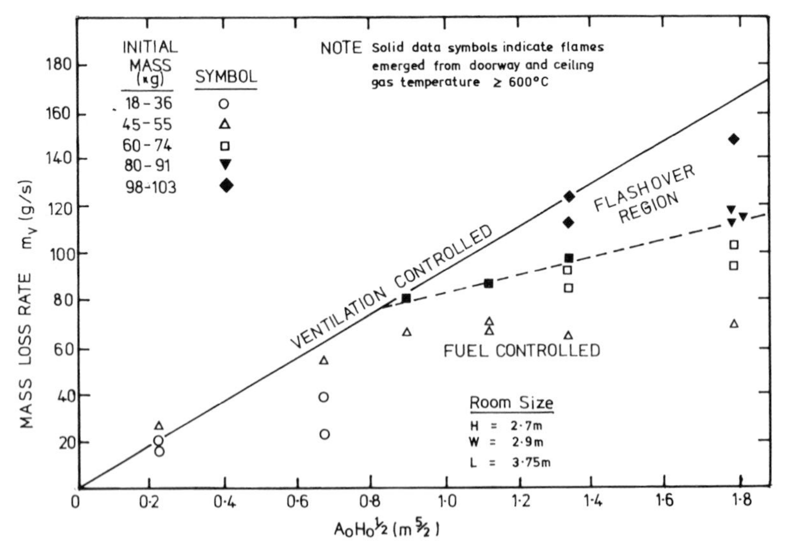

Fig. 5.4 Limiting ventilation rates between ventilation and fuel-controlled burning

ventilation- and fuel-controlled regimes. Figure 5.5 shows the relationship between temperatures attained in fires and ventilation enclosures which have similar fire loadings.

If compartment (a) represents a well-sealed compartment with no windows, the temperature profile in the compartment will be low

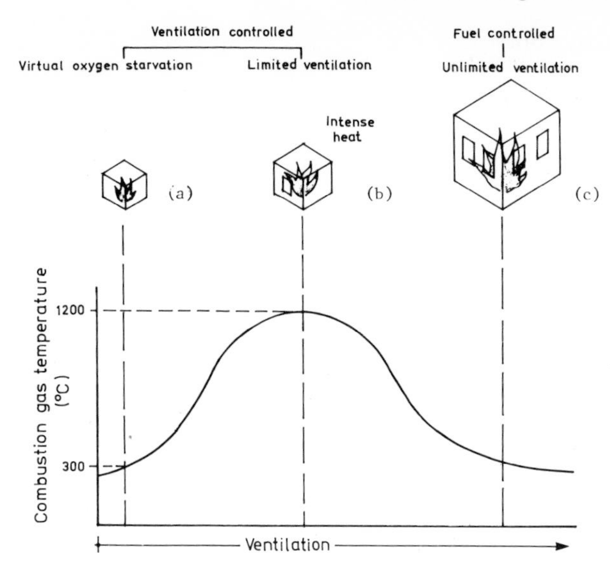

Fig. 5.5 Variation in combustion gas temperature with ventilation

and the fire would eventually self-extinguish due to oxygen starvation. With compartment (b) there is sufficient oxygen available to allow all the fuel to be consumed and the relationships between ventilation rates, fire loading compartment geometry and the thermal characteristics of the compartment-boundary construction are such that maximum fire temperatures are achieved. In compartment (c) there is too much ventilation and the size of the compartment reduces the effect of the boundary construction in fire development. The available excess air entrained into the fire process actually has a cooling effect and the fire burns as if it were in the open. Thus it can be seen that the more severe fires occur in the ventilation regime.

From Kawogoe's[7] work for a ventilation-controlled regime

$$R = 5.5AH^{\frac{1}{2}} \text{ kg min}$$

It should, however, be noted that the ratio $R/AH^{\frac{1}{2}}$ is not constant, i.e. does not equal 5.5 over a wide range of conditions. It is dependent to some extent on the geometry of the enclosure especially the ratio of width/depth. The relationship is also empirical in that strictly speaking it applies only to wood crib fires which although are useful in exploratory experimental work to represent a standard fire may not represent a real fire in an enclosure. This is particularly so for a fire in a room containing modern furniture and materials. In an enclosure with more than one opening:

$$A = A_1 + A_2 + A_3 + \cdots$$

and H is calculated from:

$$H = \sum_{1}^{n} A_i H_i / A_i$$

From the example which follows the dependence on window height should be noted.

WORKED EXAMPLE

Compare the rates of burning to be expected in two enclosures A and B, identical in all respects except that one has four tall narrow windows 0.5 m wide × 1.5 m high and the other has four windows 1.5 m wide × 0.5 m high. In each case the window lintels are the same distance from the ceiling.

Solution

Assume a ventilation-controlled fire:

$$R_A = 5.5 A_A \sqrt{H_A}$$

$$R_B = 5.5 A_B \sqrt{H_B}$$

But $A_A = A_B$

$$\therefore \quad \frac{R_A}{R_B} = \sqrt{\frac{H_A}{H_B}}$$

$$= \sqrt{3}$$

$$\therefore \quad R_A = \sqrt{3} R_B$$

5.5.2 Fuel-controlled regime

Thomas[8] has suggested a burning rate equal to the 'charring rate' of wood, $R = 6 \times 10^{-3} A_s \text{ kg s}^{-1}$ (for cellulosic materials), where A_s is the surface area of the combustible material.

However R/A_s is known to be function of stick size in crib fires and consequently the relationship is rather contrived. Any significant changes in the heat feedback mechanisms could influence the burning rate as expressed by the relationship above, which is empirical and does not hold for non-cellulosic fires.

Figure 5.6 further illustrates the conditions necessary for the transition from the fuel-controlled regime to the ventilation-controlled regime or vice versa.

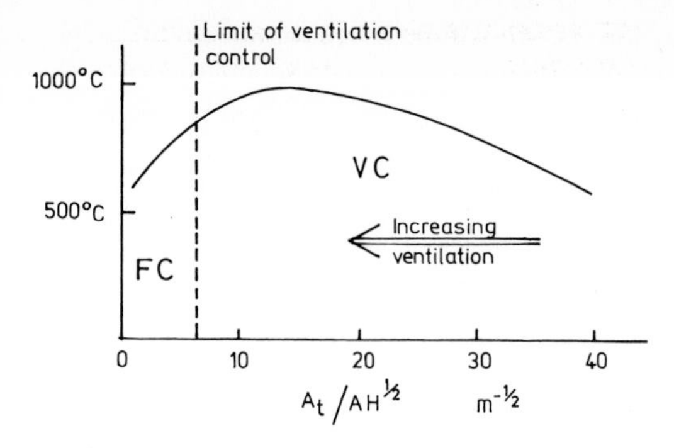

Fig. 5.6 Maximum temperatures in experimental fires correlated with ventilation

5.6 HEAT BALANCE FOR AN ENCLOSURE

A heat energy balance for an enclosure will assist in developing a fuller understanding of fire growth and development. This energy balance can be expressed as:

$$\dot{Q}_T = \dot{Q}_E + \dot{Q}_B + \dot{Q}_L$$

i.e. the heat release in the enclosure should be equal to the sum of all heat losses, Fig. 5.7, where:

$\dot{Q}_T$ = the quantity of heat released per unit time by combustion of the combustibles in the enclosure

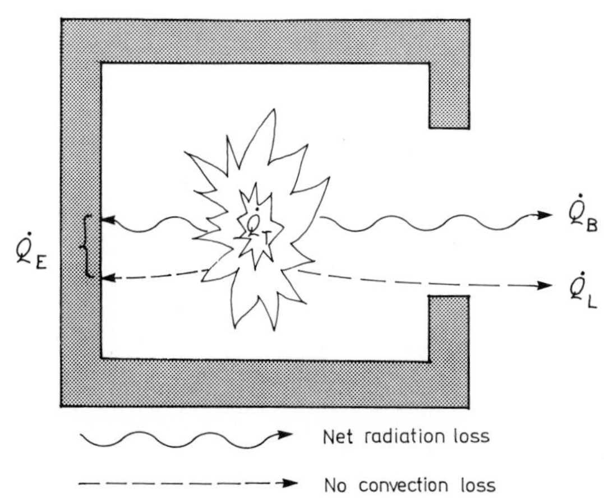

Fig. 5.7 Energy balance for an enclosure

$\dot{Q}_E$ = the quantity of heat lost per unit time by radiation and convection to the boundary construction of the enclosure
$\dot{Q}_B$ = the quantity of heat lost per unit time by radiation through the openings
$\dot{Q}_L$ = the quantity of heat carried away per unit time by the combustion gas.

5.6.1 Calculation of Q_T

The heat release Q_T is calculated by the formula:

$$\dot{Q}_T = R \cdot q$$

where:

q = calorific value of wood (kJ kg^{-1})
R = the rate of burning (kg min^{-1})

It should be noted here that $\dot{Q}_T$ in a real fire will vary with time.

Kawagoe[7] in experimental work involving the burning of wooden cribs in enclosures within the ventilation-controlled regime found that the rate of burning was essentially independent of the amount of fuel, but increased with the size of the ventilation opening. Correlating the rate of weight loss with the ventilation opening he obtained:

$$R = 5.5 A_w \sqrt{H} \text{ kg/min}$$

Figure 5.8 illustrates this relationship where experimental results are shown which verified the theoretical determination of R.

The results of Kawagoe and others are summarised by Thomas[9] who suggests the relationship for ventilation-controlled fires to be:

$$R/AH^{\frac{1}{2}} = 6 \text{ kg} \cdot \text{min}^{-1} \text{ m}^{-\frac{5}{2}}$$

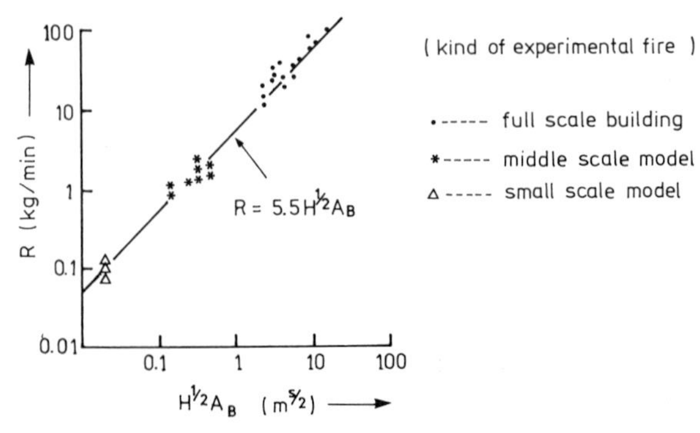

Fig. 5.8 Variation of burning rate with opening factor

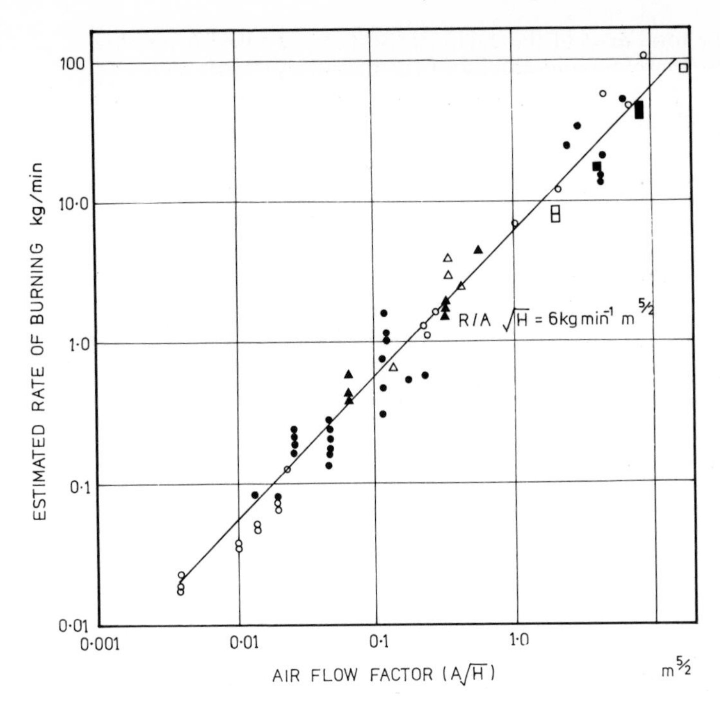

Fig. 5.9 Burning rate and air flow factor

where R = burning rate

A = window area

H = window height

Thomas's correlation is shown in Fig. 5.9.

5.6.2 Calculation of $\dot{Q}_E$

The heat loss to the enclosure boundary $\dot{Q}_E$ is obtained from the equation:

$$\dot{Q}_E = \dot{q}_R + \dot{q}_C$$

$$\dot{Q}_E = A_E \cdot \lambda \cdot \frac{d\theta}{dx}$$

where:

q_R = the net heat flux radiation from the flame to the inside surfaces of the enclosure

q_C = the net heat transferred by convection from the flame to the inside surfaces of the enclosure

A_E = the total area of the enclosure surfaces exposed to the fire
λ = the thermal conductivity of the boundary construction
$\frac{d\theta}{dx}$ = temperature gradient in the boundary construction.

In this equation q_R and q_C are expressed as:

$$q_R = A_E \varepsilon \sigma (T_g^4 - T_c^4) \qquad \text{and} \qquad \dot{q}_C = hA_E(T_g - T_c)$$

where:

ε = emissivity of the enclosure surfaces
h = co-efficient of heat transfer by convection
T_g = temperature of the flame
T_c = temperature of the enclosure surfaces.

In the equation: $\dot{Q}_E = A_E \cdot \lambda \cdot \frac{d\theta}{dx}$

$A_E \cdot \lambda \cdot d\theta/dx$ is the term for the heat transfer by conduction through the walls and can be calculated by a graphical or numerical techniques.

WORKED EXAMPLE

Using the constructional element described in Section 3.5 estimate:

1. The temperature variation through the sample after a period of one hour, and
2. The variation in the heat flux entering the element

if one side suffers a temperature–time variation expressed by $\theta_{(t)} = \theta_0 + kt$ where θ_0 is initial ambient temperature of slab and k is the rate of rise of temperature taken for this example as $10\,°\mathrm{C\,min}^{-1}$.

Solution

1. The temperature variation can be estimated graphically as shown in Fig. 5.10.
   Assuming $\Delta x = 1.4 \times 10^{-2}$ m and $\Delta t = 600$ seconds or 10 minutes, or using a tabular method which is to be preferred in most cases, Table 5.2.
2. The flux entering the slab of insulation $\dot{q}_{w(t)} = \lambda(\Delta\theta/\Delta x)_t$ is estimated using the $\Delta\theta/\Delta x$ between layer (1) and (2), the results being given in Table 5.3 and Fig. 5.11.

5.6.3 Calculation of $\dot{Q}_B$

$\dot{Q}_B$ is the net heat lost by radiation through the openings in the enclosure walls to the external environment and is obtained from the

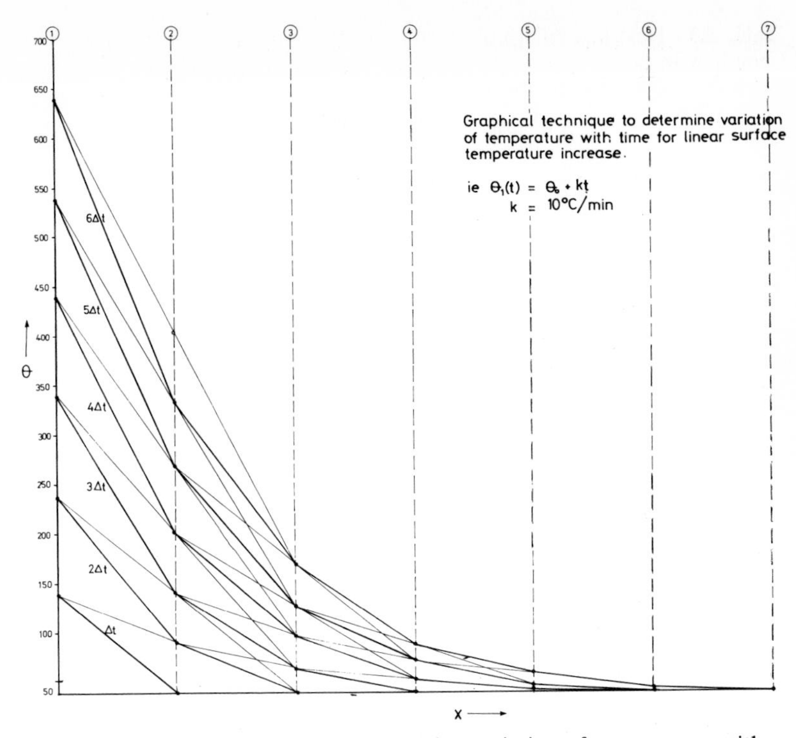

Fig. 5.10 Graphical technique to determine variation of temperature with time

Table 5.2 Numerical method of determining temperature variation with time

t (min)	Interfacing surface number						
	1 θ_1	2 θ_2	3 θ_3	4 θ_4	5 θ_5	6 θ_6	7 θ_7
0–0	38	38	38	38	38	38	38
$10 = \Delta t$	138	38	38	38	38	38	38
$20 = 2\,\Delta t$	238	88	38	38	38	38	38
$30 = 3\,\Delta t$	338	138	63	38	38	38	38
$40 = 4\,\Delta t$	438	200	88	51	38	38	38
$50 = 5\,\Delta t$	538	263	126	63	44	38	38
$60 = 6\,\Delta t$	638	332	163	85	51	41	38
$70 = 7\,\Delta t$	738	400	208	107	63	45	–

Table 5.3 Heat flux entering sample with time

t (min)	$\dot{q}_w$ (kW/m²)
0	0
10	1.07
20	1.60
30	2.14
40	2.55
50	2.95
60	3.28
70	3.62

where $\dot{q}_w = \lambda \cdot \dfrac{(\theta_1 - \theta_2)}{\Delta x}(t)$

$\lambda = 0.15 \text{ W m}^{-1} \text{ K}^{-1}$

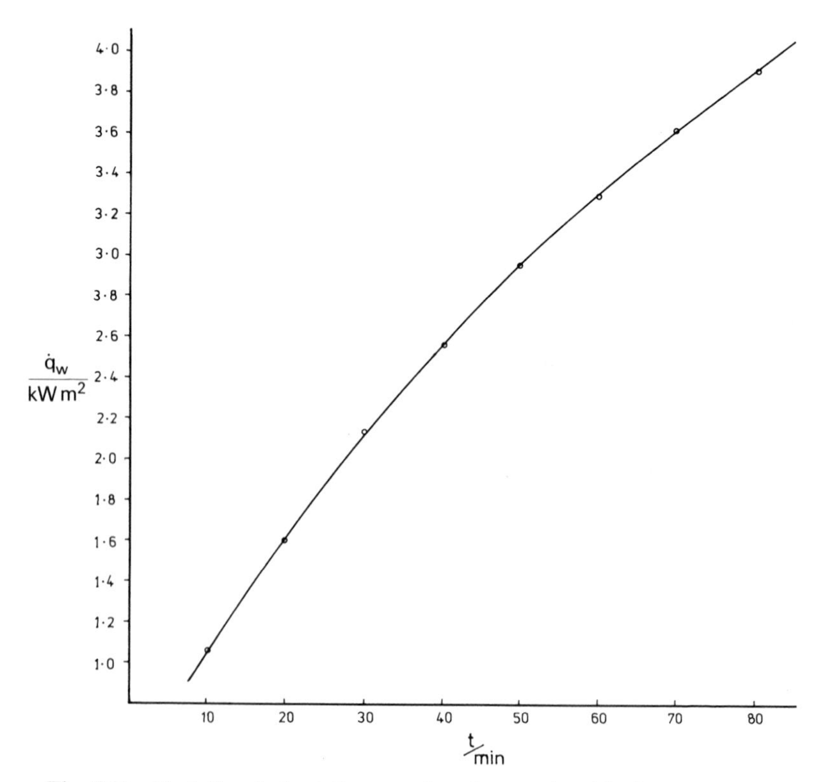

Fig. 5.11 Variation in heat flux entering the sample with time

formula

$\dot{Q}_B = A_B \varepsilon_f \sigma (T_g^4 - T_0^4)$

where:

A_B = area of the opening in the walls
ε_f = emissivity of the flame, assumed to be unity.

5.6.4 Calculation of $\dot{Q}_L$

$\dot{Q}_L$ is the quantity of heat lost by replacing hot gases by cold air. Assuming equal inflow and outflow air and gases then:

$\dot{Q}_L = \dot{M}_{out} C_p (T_g - T_0)$

where $\dot{M}_{out} = \frac{2}{3} CB(H_A^{2/3})[2g \cdot \rho_g/(\rho_0 - \rho_g)]^{\frac{1}{2}}$ kg s^{-1}

where:

C = discharge coefficient
B = width of the opening (m)
H_A = height of the opening (m)
ρ_0 = density of external air (kg/m^{-3})
ρ_g = density of fire gases (kg m^{-3})
V = velocity of the gases
V_m = mean velocity of air
g = gravitational constant

Figure 5.12 illustrates the quantities used in the above expression.

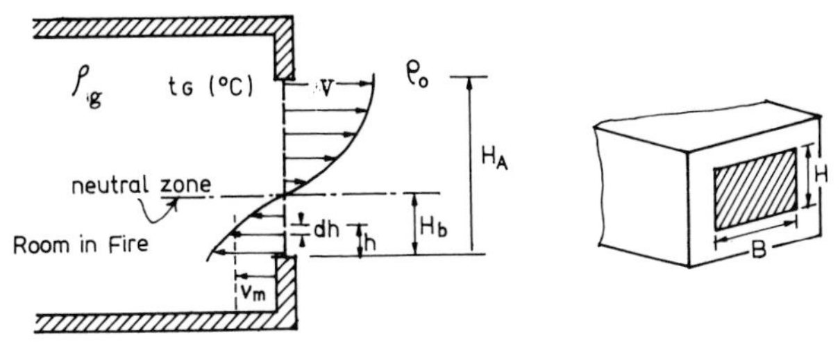

Fig. 5.12 Gas and air flow in a room with one opening

Simplified:

$\dot{M}_{out} = \rho_g H_A^{3/2} B(V_{out})$

where:

$$V_{out} = \frac{2}{3} C \sqrt{\frac{2g \cdot \rho_g}{\rho_0 - \rho_g}}$$

Thus the equation of heat balance can now be rewritten as follows:

$$R\cdot q = A_E\varepsilon\sigma(T_g^4 - T_0^4) + A_E h(T_g - T_c) + A_B\varepsilon\sigma(T_g - T_c)^4 + \dot{M}_{out}C_p(T_g - T_0)$$

$$\dot{Q}_T = [\quad \dot{Q}_W \quad] + [\quad \dot{Q}_B \quad] + [\quad \dot{Q}_L \quad]$$

where A_B = window and opening area
A_E = enclosure area.

Thus:

$$q\frac{R}{A_E} - \frac{\dot{M}_{out}}{A_E}\cdot C_p(T_g - T_0) = \varepsilon\sigma(T_g^4 - T_c^4) + h(T_g - T_c) + \frac{A_B}{A_E}\varepsilon\sigma(T_g^4 - T_c^4)$$

$$q\frac{5.5\sqrt{H}\cdot A_B}{A_E} - \frac{\dot{M}_{out}}{A_E}\cdot C_p(T_g - T_0) = \varepsilon_E\sigma(T_g^4 - T_c^4) + \frac{A_B}{A_E}\varepsilon\sigma(T_g^4 - T_c^4) + h(T_g - T_c) \quad [1]$$

and from the wall flux input:

$$A_E\varepsilon\sigma(T_g^4 - T_c^4) + A_E h(T_g - T_c) = A_E\lambda\frac{d\theta}{dx} \quad \ldots[2]$$

In this equation $d\theta/dx$ is the thermal gradient across the first layer of thickness Δx which can be evaluated graphically or by tabular means. To remind the reader of this method the worked example 3.6 should be re-read as the knowledge of $\lambda(d\theta/dx) = \lambda_0[(T_c - T_1)/x]n\,\Delta t$ is essential to the solution of these equations to evaluate T_g and T_c with passage of time.

Method of solution

It may be assumed that at $t = 0$; $T_c = T_0 = T_1$

where T_1 = temperature at first interface.

Step 1: Evaluation $T_{g(0)}$ using eqn [1].
Step 2: Substitute this value of $T_{g(0)}$ into eqn [2] to obtain a value for $T_c(1\cdot\Delta t)$, i.e. the wall surface temperature $(1\cdot\Delta t)$ minutes later.
Step 3: Using this value $T_c(1\cdot\Delta t)$, obtain (using graphical or tabular method given in worked example 3.6) $T_1(2\cdot\Delta t)$, Fig. 5.13(a), which will be needed to obtain $T_c(2\cdot\Delta t)$ via eqn [2].
Step 4: Now substitute $T_c(1\cdot\Delta t)$ into eqn [1] to get $T_g(1\cdot\Delta t)$.
Step 5: Substitute $T_g(1\cdot\Delta t)$ into eqn [2] and using value of $T_1(2\cdot\Delta t)$ previously evaluated in Step 3 determine a value for $T_c(2\cdot\Delta t)$.

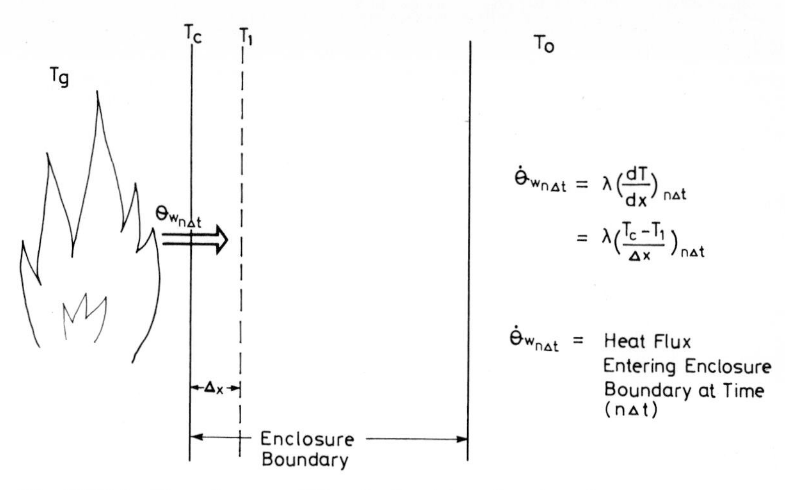

Fig. 5.13(a) Boundary condition for heat transfer at surface

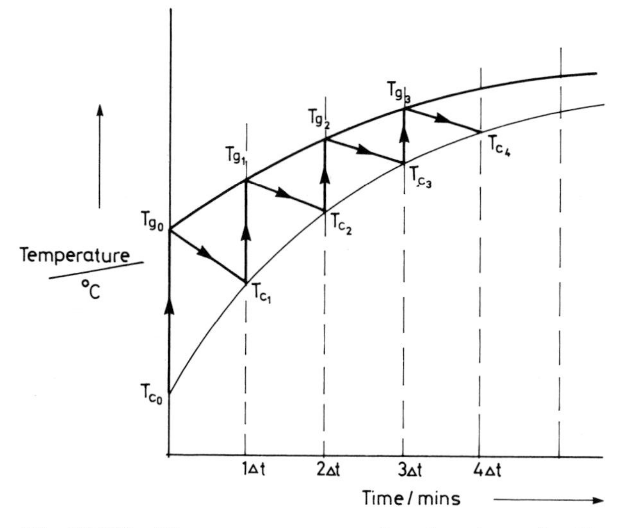

Fig. 5.13(b) Time step process employed to determine Tg and Tc

This process is then repeated as many times as necessary to obtain values for T_g and T_c for the time interval $0 \rightarrow n \cdot \Delta t$. These steps as described above can be expressed pictorially as shown in Fig. 5.13(b).

This method was employed by Kagowe[15] to obtain:

1. Variation with time of fire temperature T_g with opening factor (Fig. 5.14)

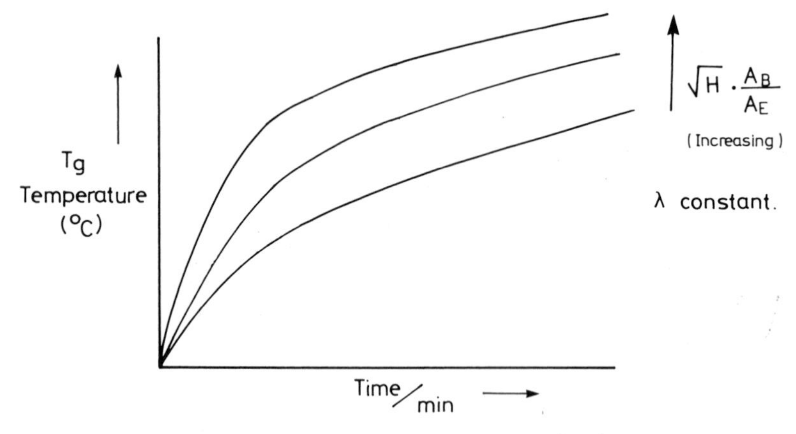

Fig. 5.14 Variation of fire temperature Tg with opening factor

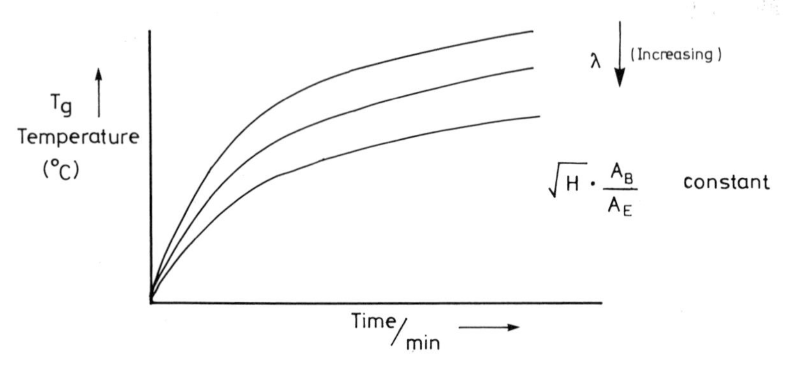

Fig. 5.15 The influence of the thermal conductivity of wall lining materials on fire temperature

2. Influence of the thermal conductivity of the wall materials on the fire temperature curves (Fig. 5.15).

Reference will be made later in Chapter 8 to these findings.

5.7 FIRE SPREAD

The spread of fire may be influenced by essentially three factors:

1. Interspatial interactions
2. Fire loading
3. Mode of heat transfer.

Items 1 and 2 have been discussed earlier.

5.8 INTERSPATIAL INTERACTIONS

The determination of interspatial interactions is an integral part of design, which demands as a prerequisite the establishment of building functional requirements related to dedicated space. Primarily the functional requirements of different parts of a building will require the provision of separating constructional components. These components can then be utilised by increasing their fire endurance characteristics thus providing an initial barrier to fire spread.

Buildings are designed and constructed to accommodate people and processes, among other things. Consequently the flow of people, materials and products must be incorporated in such a way so as to achieve a pre-determined level of fire safety. The minimum level of fire safety to be achieved may be prescribed by building legislation or be determined by insurance requirements or other influences.

Figure 5.16 shows the plan of simple factory building and Fig. 5.17 shows the longitudinal elevation.

The three-storey section is devoted to administration processes and the single-storey section is concerned with production processes. Clearly there is a general division of function although communications between both parts must obviously be provided.

Within the production-oriented section, various sub-divisions are also apparent, related to the activity carried on in particular spaces, e.g. storage. For an apparently simple building such as the one illustrated, the building designer to satisfy minimum legislative requirements will perhaps unconsciously carry out a notional hazard analysis. The sequential analysis process can be outlined as follows:

1. Problem definition; establish clearly the need
2. Choice of objectives; a definition of physical needs and of the criteria within which they must be met
3. System synthesis; the creation of possible alternative systems
4. Systems analysis; analysis of alternative systems against defined objectives
5. Systems selection; selection of the most promising alternative
6. Implementation; test the preferred solution
7. Systems engineering; monitor, modification and information feedback.

Having determined the nature and scale of the risk, the designer has several options which may be conceived as follows:

(a) isolation
(b) containment – compartmentation
(c) segregation.

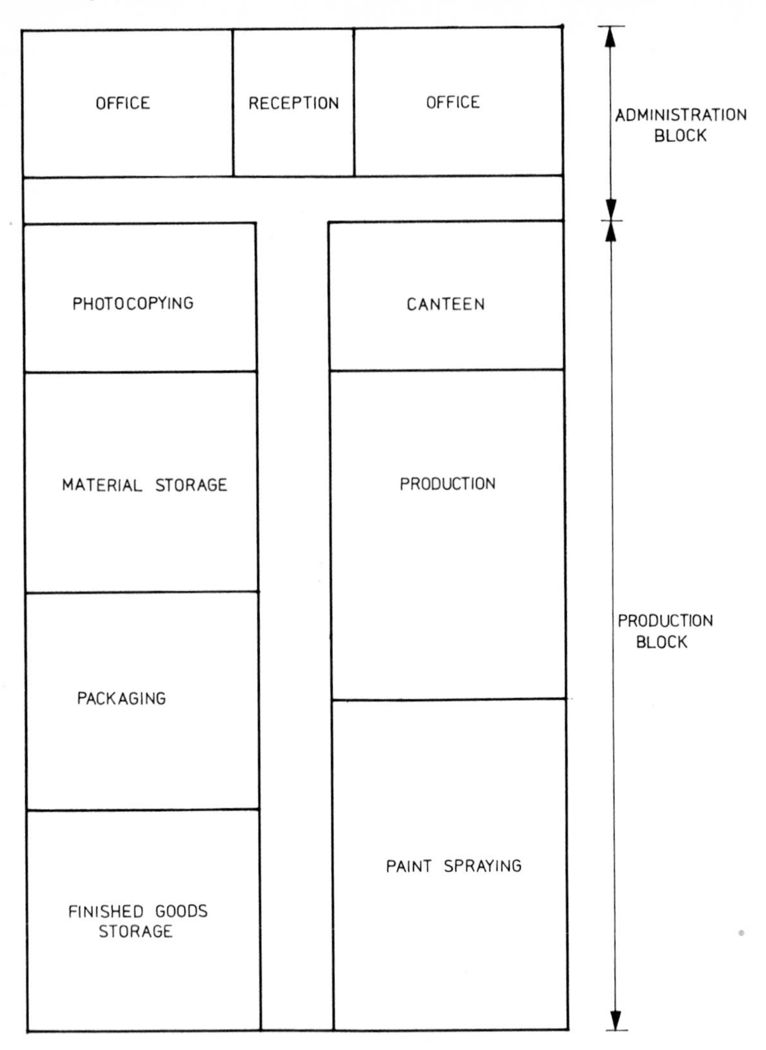

Fig. 5.16 Plan of factory building

5.8.1 Isolation

If the risk is such that the threat to the building and its occupants cannot be accepted then the risk may be removed and located on another part of the same site. This is particularly the case with industrial processes where the risk of explosion exists.

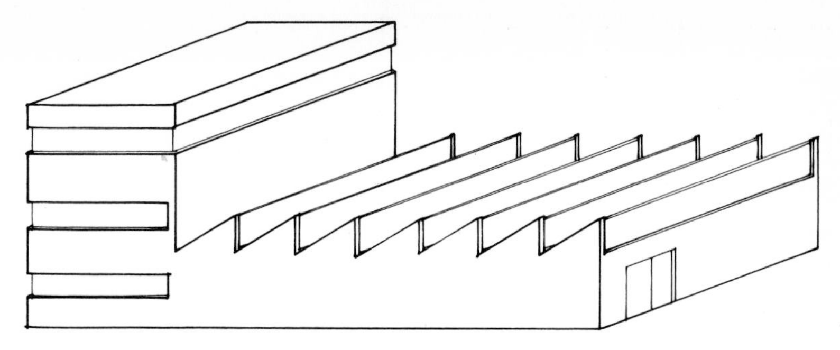

Fig. 5.17 Longitudinal elevation of building

5.8.2 Compartmentation

The factors which generally determine the need for compartmentation are floor area and cubic capacity of the spaces under consideration. Thus building legislation relates these two criteria to risk and it follows that if floor areas and cubic capacities of predetermined dimensions are suitably protected by fire-resisting boundary-constructional components the probability of fire spread beyond the compartment of origin is diminished. Figure 5.18 illustrates an uncompartmented building with an open plan floor. A fire in room X can spread uncontrolled laterally and vertically. However, if the same building is compartmented as illustrated in Fig. 5.19, the fire originating in room X cannot spread so rapidly throughout the building. Consequently the fire hazard is mitigated and fire safety is enhanced. Compartmentation as a component of fire safety in building regulations will be discussed more fully in Chapter 7.

5.8.3 Segregation

Segregation of risk elements of areas requires a hazard analysis study. The risks of potential threats to the building and its occupants are identified and classified. Risks of similar character and nature can then be collectively accommodated in spaces especially provided for that specific purpose.

5.9 FIRE LOADING

The concept of fire loading has its origins in the *Post-War Building Studies*[10] as a means of estimating the potential demand hazard. The

Fig. 5.18 Elevation – uncompartmented Building 'A'

term fire load is used to describe the heat energy which could be released per square metre of floor area of a compartment or storey by the combustion of the contents of the building and any combustible parts of the superstructure itself.

$$\text{Fire load} = \frac{M \times C}{A} \quad \text{kJ m}^{-2}$$

where M = mass of combustible materials in the compartment or storey (kg)
C = calorific value of the materials (kJ/kg^{-1})
A = floor area in metres (m^2)

For example, if a building contained 6,000 kg of combustible material of calorific value 20,000 kJ kg^{-1} over an area of 3,000 m^2, the fire load would be:

$$\frac{6{,}000 \times 20{,}000}{3{,}000} = 40{,}000 \text{ kJ m}^{-2}$$

Building regulations have adopted a purpose grouping system

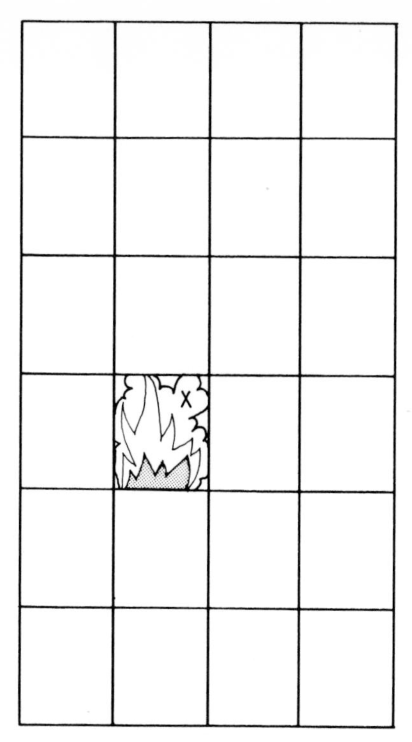

Fig. 5.19 Elevation – compartmented Building 'B'

which in reality is a grading of occupancies based on assumed fire loadings. The purpose grouping of buildings is then used as a determinant in establishing the desirable fire-resisting characteristics of the elements of the structure of the building. Thus it can be shown that the concept of fire loading attempts to relate the combustible contents of a building to the potential severity of a fire in that building and consequently to the fire-resisting capabilities of the elements of the structure. It is, therefore, assumed that the building will remain structurally intact, during a complete burnout of all the combustible materials.

The relationships mentioned above are rather crude and are based upon several assumptions:

1. That the combustibles are uniformly distributed throughout the building
2. That all the combustibles will be involved in a fire
3. That combustion of the combustibles will be complete
4. That non-cellulosics will behave in the same manner as cellulosics, i.e. the rate of heat production will be the same and

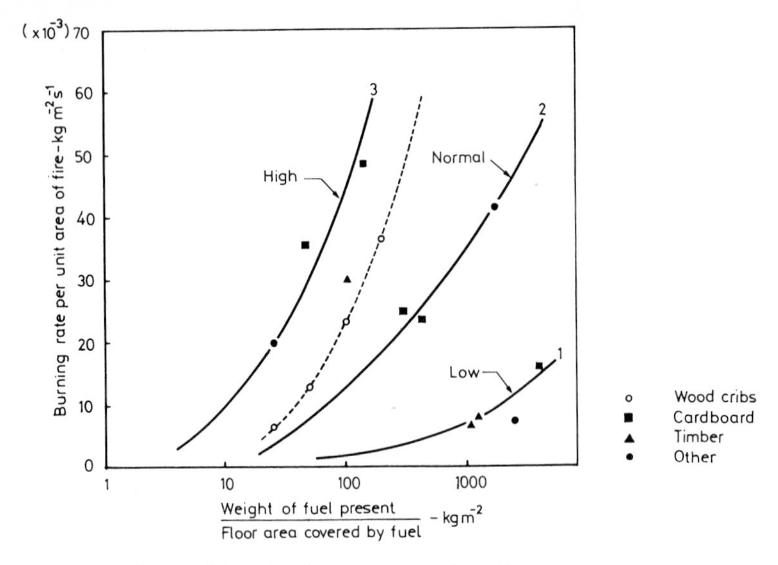

Fig. 5.20 Rates of burning for fuels in various forms

can be treated as wood equivalents

5. That fire load per unit area can be related to the fire resistance of the elements of the structure.

Thomas *et al.*[11] shows that the nature and disposition of a fuel could have a considerable effect on the rate of burning. In Fig. 5.20. Curve 1 refers to materials such as books and reels of cardboard which offer a comparatively small surface area for combustion. Curve 2 represents the same type of material in a different configuration, e.g. cardboard cartons stored folded flat, and Curve 3 represents stacks of cardboard cartons.

Thus it can be seen that whilst fire load is a necessary component of fire growth it is like the blood in a person's veins, necessary for life, but not a measure of the whole person.

The nature, disposition and height of the fuel will contribute also to fire growth and must be considered for each building, not just each building type. For example, two buildings may be used for storage purposes, both storing cardboard. One building uses high rack storage for cartons and the other low rack storage for folded cartons. In the latter it would be reasonable to expect a slow burning deep-seated fire which a sprinkler system may control but not extinguish, whereas in the former it would be reasonable to expect a very rapid fast growing fire with flames at ceiling level in the very early stages of development, which the sprinkler system may not be able to control.

5.10 HEAT TRANSFER

Heat may be transmitted in three distinct ways in any situation or by a combination of all three ways.

1. Conduction: transfer of heat through material from a high temperature zone to a lower temperature zone. Building regulations require hearths to be non-combustible and a minimum thickness of 150 mm so as to prevent the transfer of heat by conduction (Fig. 5.21(a). Heat conducted through a masonry wall can easily ignite combustibles stored against the wall at the face opposite to the fire face (Fig. 5.21(b).

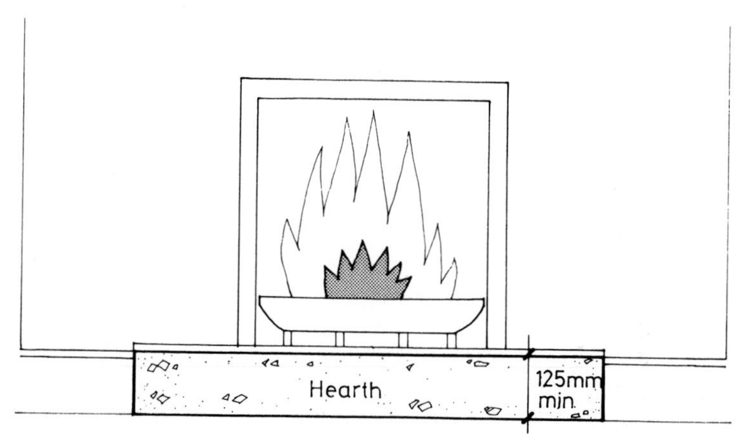

Fig. 5.21(a) Heat transfer by conduction

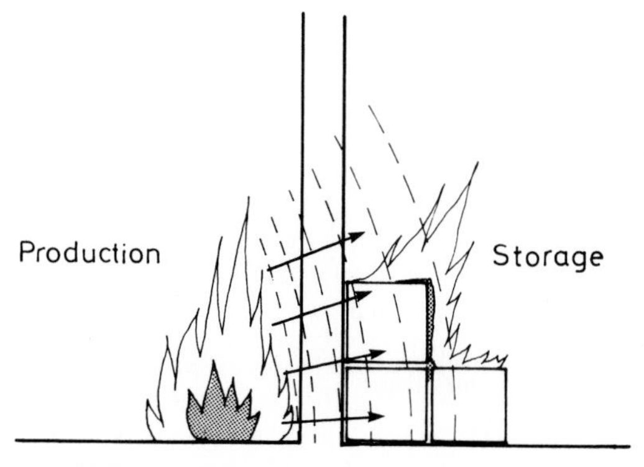

Fig. 5.21(b) Heat transfer by conduction

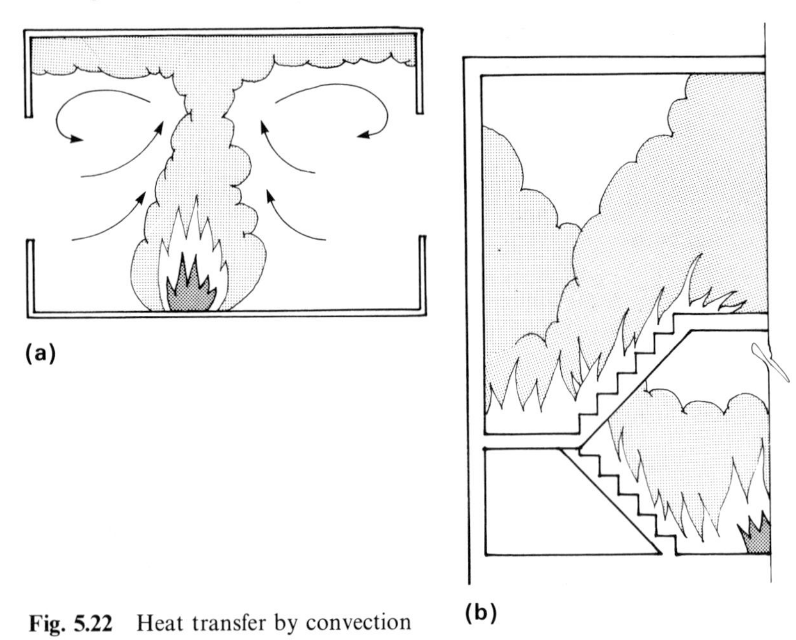

Fig. 5.22 Heat transfer by convection

2. Convection: transfer of heat through or by a fluid by means of induced convective currents within the fluid. When a fire burns in a room a buoyant plume of hot smoke rises from the fuel bed, contacts the ceiling and spreads across the ceiling progressively heating up the surface and the entire volume until a point is reached when all the combustibles within the room can be easily ignited (Fig. 5.22(a)). In the simple design of domestic flues, etc., reliance is placed on convection to remove the smoke and toxic gases. Within a building a fire may develop on one floor and progress upwardly through the building by means of convective forces. In buildings, staircases, lift shafts and ducts provide ready-made artificial flues which will channel fire rapidly upwards through the building until every floor above the fire floor is engulfed in fire (Fig. 5.22(b)).
3. Radiation: transfer of heat through a fluid or vacuum by means of electromagnetic waves. In many fire situations radiation may be the dominant factor controlling spread and growth. As the fire develops within an enclosure all the surfaces gradually (sometimes very quickly) heat up. The surfaces themselves begin to radiate heat energy (Fig. 5.23) significantly influencing:
   (a) temperature rise in the enclosure
   (b) the burning rate of the combustibles
   (c) fire development and growth.

So far we have concentrated on fire spread within buildings, but

Fig. 5.23 Heat transfer by radiation

radiation is also very important when considering fire spread between buildings. Building regulations consider unprotected areas in the façades of buildings as potential radiators in a fire situation and require that buildings be sited sufficiently distant from respective site boundaries so as not to present a fire hazard to other buildings on adjoining or within the same site. The distances quoted in building regulations are derived in relation to the amount of unprotected areas in the external walling of the building. Unprotected areas are defined as:

1. Areas of walling not having the required degree of fire resistance
2. Windows, doors or other openings
3. Any part of the external wall which has combustible material more than 1 mm thick attached or applied to its external face whether for cladding or any other purpose.

Figure 5.24 shows the elevation of a dwelling. If it is assumed that the unprotected areas act as black-body radiators at 1,000 °C, it is possible to calculate the radiant intensity at various distances (m) from the burning building. Knowing the radiant intensity at various

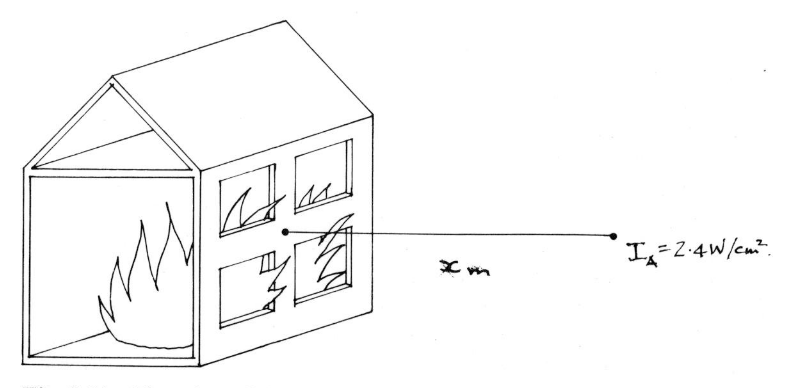

Fig. 5.24 Elevation of dwelling as a potential source of radiation

distances, buildings can be sited so as to avoid spontaneous ignition on their exposed façades. Alternatively having prescribed a level of radiant intensity, e.g. 2.4 W cm^{-2} (pilot ignition), necessary for ignition it is possible, making some basic assumptions, to determine the distance at which the prescribed heat flux will not occur and site other buildings accordingly. This aspect was discussed more fully in Chapter 3.

5.11 FIRE SEVERITY

Reference to Chapter 1 will indicate that the concept of resisting the impact of fire developed as a response to an identifiable need and in the total absence of any notion of fire severity. Henceforth fire resistance has been coupled with fire severity.

Fire severity may be simply defined as the destructive potential of a compartment fire, i.e. the potential impact that a fire in a given compartment will have upon the structural and constructional components which form the compartment and the contents of the compartment. Inevitably fire severity has been linked with structural performance in terms of a component's fire-resisting capabilities. Ingberg[12] established a direct relationship between fire-load density and fire severity. This concept has been widely utilised in building legislation for determining the fire resistance requirements of building components within buildings of various purpose groups. An equal-area concept (Fig. 5.25) was developed on the basis of this work by calculating the area under the average temperature curve and equating it to the area under the standard temperature–time curve used in fire-resistance furnace tests.[13]

Fire severity is not a function of a single parameter such as fire-load density, but depends on other factors such as ventilation rate, burning rate, fire duration and the thermal properties of the enclosure construction. Fujita[14] and Kawagoe[15] recognised that the flow of air into a compartment was an important factor in the severity of fully-developed fires. Butcher[16] using purpose-built brickwork rooms with controlled ventilation showed the effects of ventilation and fire load on fire-temperature profiles for a standard compartment (Fig. 5.26). Subsequent work (Heselden)[17] was carried out using the fire load, ventilation and thermal properties of the enclosure boundary construction as variable parameters to derive expressions for the potential severity of compartment fires. The information has tended to be expressed in terms of fire-temperature histories particularly during the fully-developed period of the fire. Heselden[17] reporting the results of an international co-operative

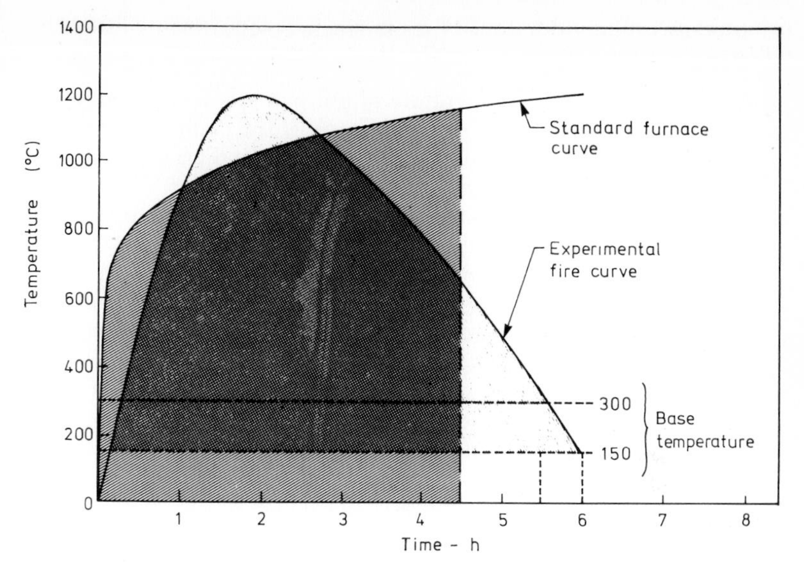

Fig. 5.25 Equal-area concept to relate fire severity

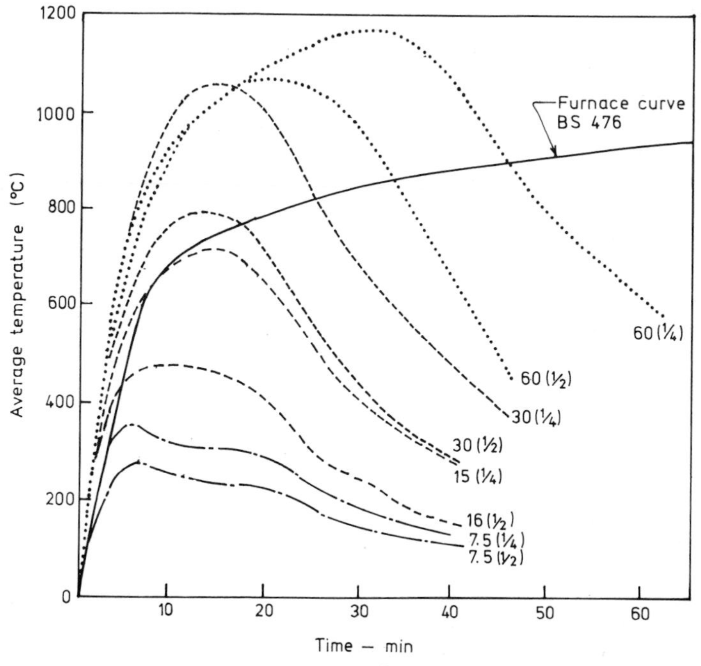

[60 (½) means fire load 60kg/m² floor area and ventilation 50% of one wall]

Fig. 5.26 Effect of fire-load density and ventilation on fire temperatures

programme on fully-developed fires in single compartments listed the following conclusions:

1. Rate of burning and temperature of a fire in a compartment are dependent mainly on the size and shape of the compartment and the size and shape of its ventilation opening, and also to some extent on fire-load density
2. Although burning rate changes widely with scale, the burning rate per unit floor area for fires with full ventilation openings and the values of $R/A_wH^{\frac{1}{2}}$ for $\frac{1}{4}$ openings vary only slightly with scale. With large openings, temperature is independent of scale, but with $\frac{1}{4}$ openings temperature increases slightly with increasing scale
3. Burning rate and temperature depend to a small extent on the thermal properties of the ceiling and wall material
4. Intensity of radiation from the ventilation opening can be related to the rate of burning, the area of the opening and the temperature within a given compartment
5. Intensity of radiation from the flame can be related to the rate of burning, the intensity of radiation at the ventilation opening and the dimensions of the compartment
6. Changes in the exposure hazard caused by wind deflection of the emergency flames are likely to be more important than changes caused in the way the fire burns in the original compartment.

The experimental conditions used in this programme of work are summarised as follows:

Shape	relative dimensions of width/depth/height of 211, 121, 221 and 441 (Fig. 5.27)

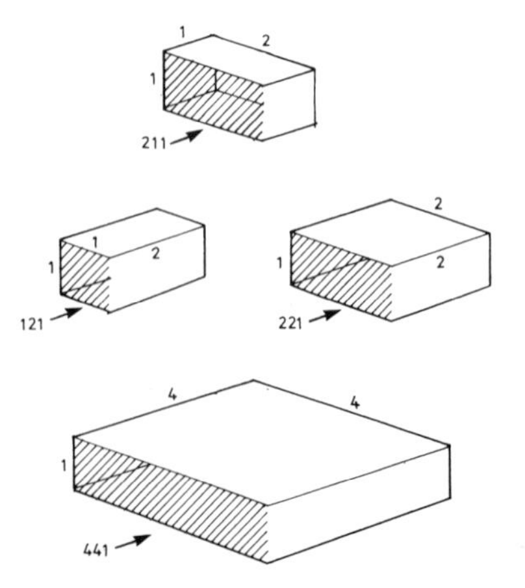

Fig. 5.27 Ratio of width/depth/height

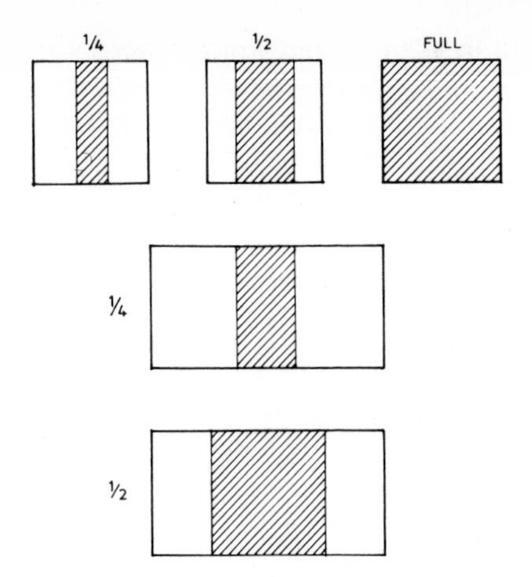

Fig. 5.28 Ratio of ventilation opening to front wall

Scale (compartment height)	$\frac{1}{2}$, 1 and $1\frac{1}{2}$ m
Ventilation opening	$\frac{1}{4}$, $1\frac{1}{2}$ or whole of front wall (Fig. 5.28)
Fire-load density (fire load/floor area)	20, 30 and 40 kg m^{-2}. The fuel consisted of standardised wooden cribs
Fuel thickness	10, 20 and 40 mm
Fuel spacing	$\frac{1}{3}$, 1 or 3 stick thickness (Fib. 5.29)

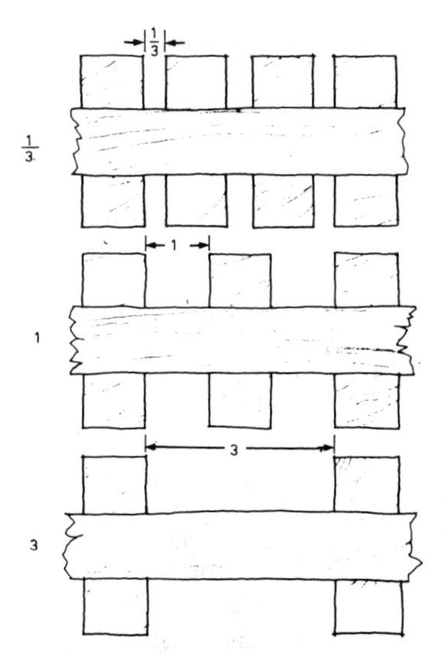

Fig. 5.29 Fuel spacing in terms of stick thickness

Winds	winds speeds of 5 and 7 m s^{-1} were used blowing perpendicularly to the plane of the ventilation opening and at 60 °C to this direction for $\frac{1}{2}$ m scale compartments having shapes 111, 211, 121 and 221.

Harmathy[18] introduced three fire-severity parameters. These are:

1. The duration of fully-developed fire
2. The overall penetration flux, $\bar{q}$, i.e. the heat flux incident on the inside surface of the compartment boundaries, averaged spatially over the boundary surfaces and temporarily over the period of full-fire development
3. The average temperature of the compartment gases, $\bar{T}_g$, averaged temporarily and spatially over the compartment volume.

Figure 5.30 illustrates an attempt to relate the components, processes and factors which must be considered in order to quantify the potential severity of a fire in a compartment.

5.12 CONCLUSIONS

From the foregoing it can be concluded that variation in the components of fire severity can create fires that have a widely-varying impact on the enclosures containing them.

Maximum fire duration is one component of fire severity, not a total measure. If a fire in a compartment burns slowly, for a long time, to consume all the combustibles it will have relatively little impact upon the constructional components of the compartment.

Fire-temperature profiles have been shown to be dependent upon fuel loading and ventilation rates. In some fires the thermal shock imposed upon the constructional components during the early stages of a fire may be more important than maximum fire duration. The internal surface linings of a compartment can also influence fire severity. For example, consider two compartments identical in all respects except that Compartment A is lined with a high thermal conductivity (λ) material and Compartment B is lined with a material having a low thermal conductivity. The temperature in the Compartment B will reach higher values than that in Compartment A, thus the fire in Compartment B will appear to be the more severe. However, the rate of heat transfer into the boundary construction of Compartment B will be lower due to the low conductivity of the surface linings, and thus the damage caused to the constructional components could be much less than the damage suffered in Compartment A.

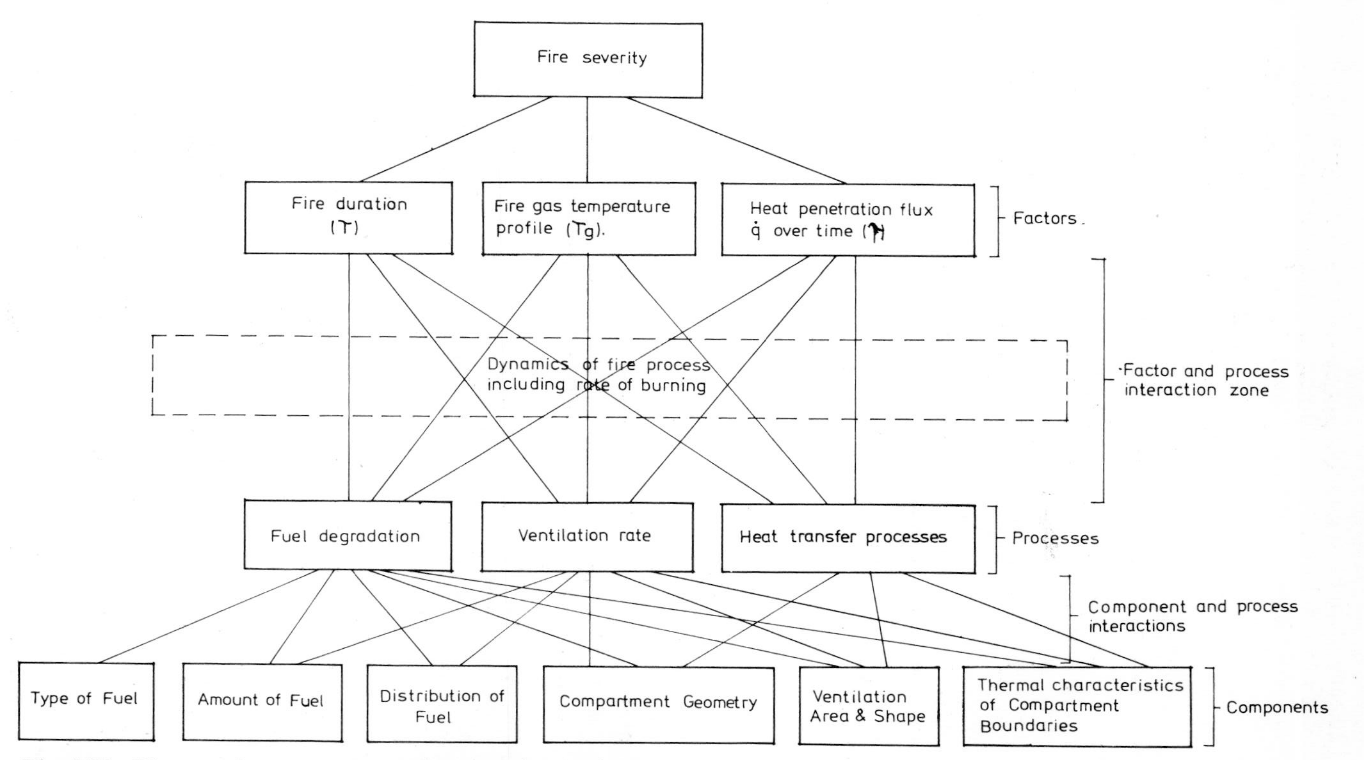

Fig. 5.30 Fire severity component and process interactions

Thus any attempt to quantify fire severity can now be seen to be a complex process. For all practical purposes the approach suggested by Harmathy warrants consideration.

REFERENCES

1. Friedman R, 'Behaviour of fires in compartments', *Fuel Symposium Fire Safety of Combustible Material*, Edinburgh University 1975.
2. *Fire tragedy at football stadium*, *Fire Prevention No. 181*, *July/August*, Fire Protection Association, London 1985.
3. Waterman T E, *Fire Technology* **8**(4), Nov. 1972.
4. Hagglund B, Janson R and Ornemark B, *Fire development in residential rooms after ignition from nuclear explosion.*, R.O.A. Dept. C. 20016 – DG (A3), Sweden 1974.
5. Zabetakis M G, '*Flammability characteristics of combustible gases and vapours*', US Bureau of Mines, Bulletin 627, 1965.
6. Drysdale D D, '*An Introduction to Fire Dynamics*', Wiley 1984.
7. Kawagoe K and Sekine T, '*Estimation of temperature–time curves in rooms*', B.R.I. Occasional Report No. 11. Building Research Institute, Tokyo, 1963.
8. Thomas P H and Hoselden A J M, *Fire Research Note 983*, 1972.
9. Thomas P H, 'Studies of fires in buildings using models', *Research*, Feb–March 1964.
10. Ministry of Works, *Post-War Building Studies*, *No. 20*, HMSO, London 1946.
11. Thomas P H and Theobald C R, '*The burning rates and duration of fires*', Fire Prevention Science and Technology, No. 17.
12. Ingberg S H, '*Fire Tests of Office Occupancies*', US National Fire Protection Association Quarterly 20, 1927.
13. Malhotra H L, '*Design of fire-resisting Structures*', Surrey University Press 1982.
14. Fujita K, '*Research report concerning characteristics of fire inside of non-combustible room and prevention of fire damage*', Report 2(n), Building Research Institute, Japan.
15. Kawagoe K, '*Fire behaviour in rooms*', Report No. 27, Building Research Institute, Jàan 1958.
16. Butcher E G, *et al.* '*The temperature attained by steel in building fires*', Fire Research Station Technical Paper No. 15, HMSO London, 1966.
17. Heselden A J M, 'Results of an international co-operative programme on fully-developed fires in single compartments', Symposium No. 5, *Fire Resistance Requirements for Buildings – A New Approach*, HMSO, London, 1971.
18. Harmathy T Z, *Fire Technology*, pp. 324–51 **8** 1972.

CHAPTER 6

The standard fire tests

6.1 INTRODUCTION

In Chapter 1 an account is given of how the technique of fire testing has been developed over many years and is subject to change as new materials, building components and methods of construction are developed.

Although no two fires are the same, it is necessary that the conditions used to determine responses are standardised to resemble closely certain features of one or more of the stages of fire behaviour shown graphically in Fig. 6.1. Consequently there are major differences between fire tests and real fires. This does not mean that fire tests are useless in themselves, but that it is necessary to be aware of their limitations and to interpret and use the data they provide in a technically sensible manner. It is also necessary that the development of fire tests follows some agreed format. Six basic steps have been outlined as essential procedural requirements in the development of fire tests.[1] These six steps are:

1. Preparation of a fire scenario
2. Selection of an appropriate part of the scenario
3. Preparation of a test method with appropriate physio-chemical simulation of the selected scenario
4. Establishing the appropriate levels of repeatability and reproductibility
5. Validation of the test method
6. Development of a data utilisation scheme.

Using the concept of the 'ideal fire' Fig. 6.1, Step 2 can be considered and represented as in Fig. 6.2.

There are, of course, many assumptions made in using the 'ideal fire' concept and these have been discussed in other chapters. Nevertheless, the use of the idealised curve is a convenient way of

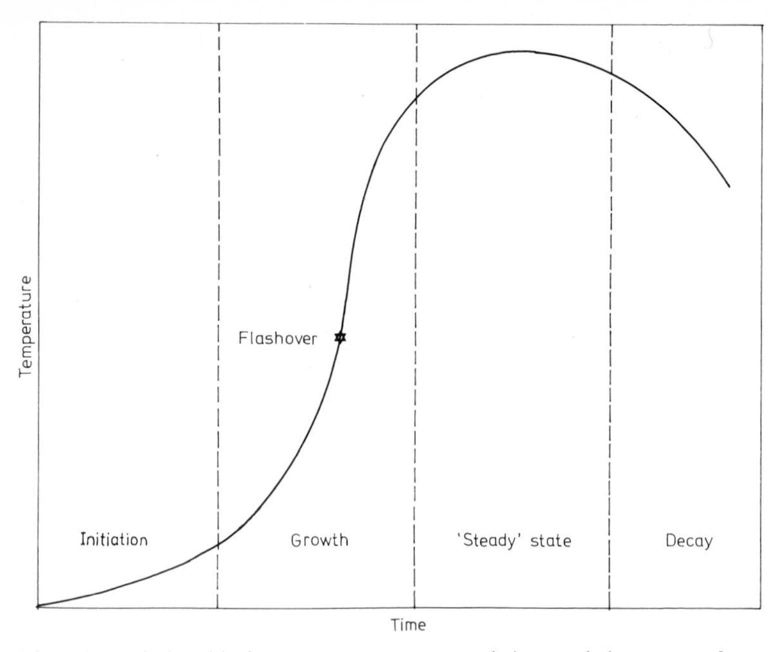

Fig. 6.1 Relationship between temperature and time and the stages of a typical uncontrolled fire in a compartment

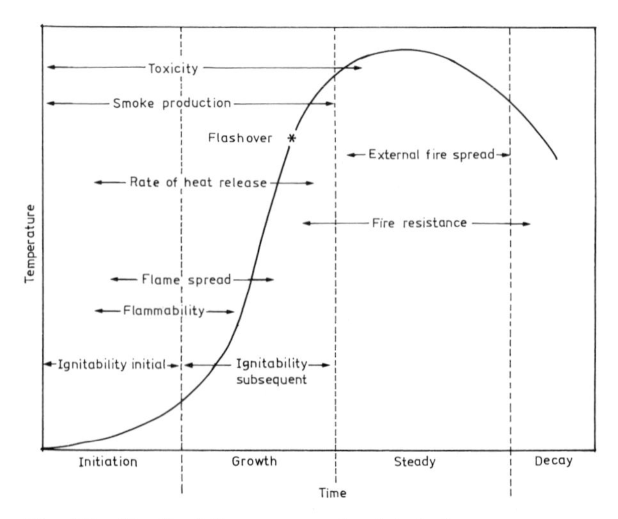

Fig. 6.2 Idealised fire curve and related phenomena

simplifying a complex process and associating occurrences of interest.

British Standard 476 contains the principal criteria for assessing the fire properties of building materials and elements of structure. These standard methods of test are referred to in building regulations and represent a body of knowledge developed over many years. Much research is currently being undertaken in the whole area of fire development and these tests consequently may be altered or new tests devised.

British Standard 476[2] is published in a number of parts, these being († indicates British Standards in course of preparation (1986)).

BS 476: Part 3:	*External fire exposure roof test*
BS 476: Part 4:	*Non-combustibility test for materials*
BS 476: Part 5:	*Method of test for ignitability*
BS 476: Part 6:	*Method of test for fire propagation for products*
BS 476: Part 7:	*Surface spread of flame tests for materials*
BS 476: Part 8:	*Test methods and criteria for the fire resistance of elements of building construction*
BS 476: Part 11:	*Method of assessing the heat emission from building materials*
†BS 476: Part 12:	*Method of measuring the ignitability of products using direct flame impingement*
†BS 476: Part 13:	*Method of measuring the ignitability of products subjected to thermal irradiance*
†BS 476: Part 14:	*Method of measuring the rate of flame spread on surface of products*
†BS 476: Part 15:	*Method of measuring the rate of heat release of products*
†BS 476: Part 16:	*Method of measuring the smoke release (obscuration) of products*
†BS 476: Part 20:	*General principles and requirements for the determination of the fire resistance of elements of building construction*
†BS 476: Part 21:	*Methods of the determination of the fire resistance of loadbearing elements of building construction*
†BS:476: Part 22:	*Methods for the determination of the fire resistance of non-loadbearing elements of building construction*
†BS 476: Part 23:	*Methods of the determination of the contribution provided by components and elements to the fire resistance of a structure*
†BS 476: Part 24:	*Methods for the determination of the fire resistance of elements of construction penetrated by building services*

†BS 476: Part 30: *Methods for measuring the performance of flat and sloping roofs exposed to an external fire*
BS 476: Part 31: *Methods for measuring smoke penetration through door sets and shutter assemblies*

Some of the tests are applicable to materials only and the others to components and elements of structure.

Figure 6.3 shows the part numbers of BS 476 located on the idealised fire curve referred to earlier.

To assist with the interpretation and use of fire test data the parts of BS 476 currently referred to in building regulations are considered briefly.

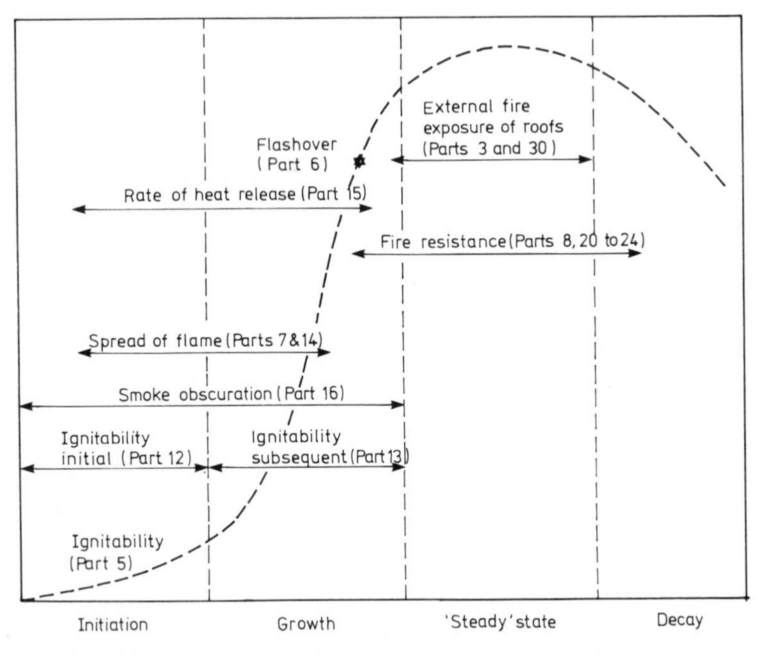

Fig. 6.3 Fire phenomena and the part numbers of BS 476 describing methods to examine them, related to the stages of an uncontrolled fire in the compartment of origin

6.2 BS 476: PART 3: 1975 – EXTERNAL FIRE EXPOSURE ROOF TEST

The external fire exposure roof test attempts to simulate conditions likely to arise in an actual fire situation by assuming a radiation

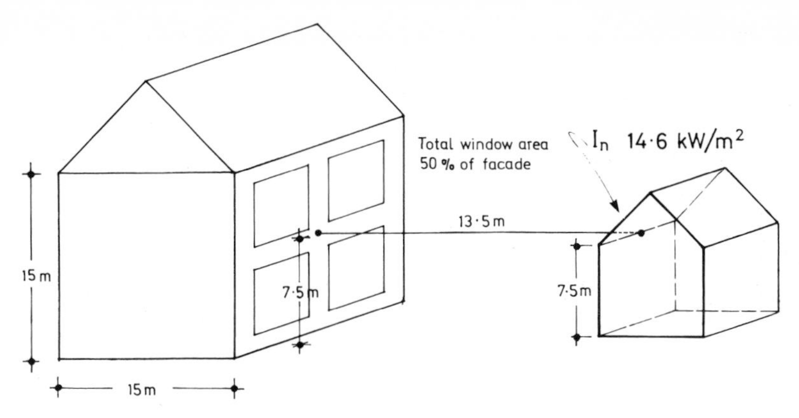

Fig. 6.4 Assumed geometrical relationships according to BS 476 pt (3)

intensity of 14.6 k W m^{-2} incident on a roof 7.5 m above ground level from a fire 13.5 m away in a building having a façade 15 m × 15 m with 50 per cent unprotected area Fig. 6.4.

Four samples measuring not less than 1.5 m × 1.2 m of the roof construction are required to be tested. These samples must first be pre-conditioned at a temperature of 10–20 °C and at 55 per cent to 65 per cent relative humidity before testing. This test is composed of three elements:

1. Tests to evaluate the ignitability of roof surfacing
2. Tests to determine the resistance of the roof construction to fire penetration
3. Tests to determine the surface ignition characteristics of the roof.

The external fire exposure roof tests should not be confused with or compared to BS 476: Part 8: Fire resistance test or BS 476: Part 7: Surface spread of flame tests. The test is not concerned with how the roof construction would behave when exposed to the effects of a fire within a building but rather how the roof construction would behave when exposed to the effects of a fire in an adjacent building, e.g. if the roof is subjected to external fire (burning fragments from adjacent buildings). It would be useful to know if:

1. The construction in the absence of radiant heat and ignited by a burning fragment would become completely involved in fire
2. The integrity of the construction when subjected to radiant heat can be maintained for a given time
3. The construction contributes to flame spread across its external surface.

They must also be tested within four hours of leaving the conditioned atmosphere.

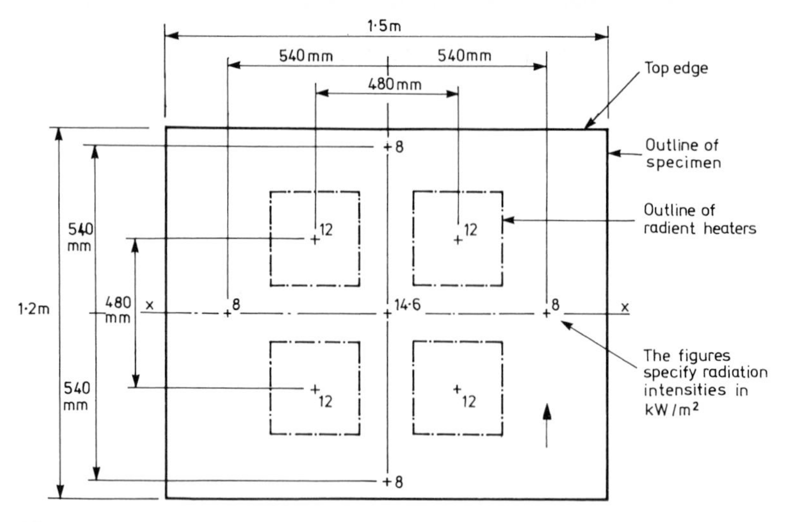

Fig. 6.5 Location of radiant heating panels in test apparatus

The testing apparatus, Fig. 6.5, consists of four radiant panels mounted on a steel framework so that their centres coincide with the corners of a square of 480 mm side. The sample is itself mounted on a steel framework in such a way that its thermal movement is not restricted. Both the radiant panel framework and the sample rig can be adjusted through 45° to simulate the pitch of the roof.

Figure 6.4 indicates the underlying assumption in this test. Figure 6.6 shows the test apparatus and it can be seen that the radiant panels follow the plane of the roof specimen to be tested.

However, Figs. 6.7(a)–(d) show situations where the intensity of radiation incident in reality on roofs or parts of roofs will not be 14.6 kW m^{-2} and may even be zero.

For example in Fig. 6.7(a), half of the roof will not be subject to radiant heating. In Fig. 6.7(b), in reality all of the roof will not be exposed to a radiant heat flux of 14.6 kW m^{-2}. In Fig. 6.7(c), a flat roof may well be subjected to a greater degree of convective heating than radiant heating. Figure 6.7(d) shows a mono-pitch roof orientation such that the intensity of radiation incident on this roof would be virtually zero.

In the preliminary ignition test a test flame is held 5–10 mm from the centre of the sample surface. The flame is removed after one minute and observations made for continued flaming on the upper surface and for fire penetration to the underside. If penetration occurs the subsequent tests are not carried out.

During this part of the test the specimen is not subjected to radiant heating flux of 14.6 kW m^{-2} at the centre of the sample and flame spread is measured in any direction (Fig. 6.8(a), (b), (c)).

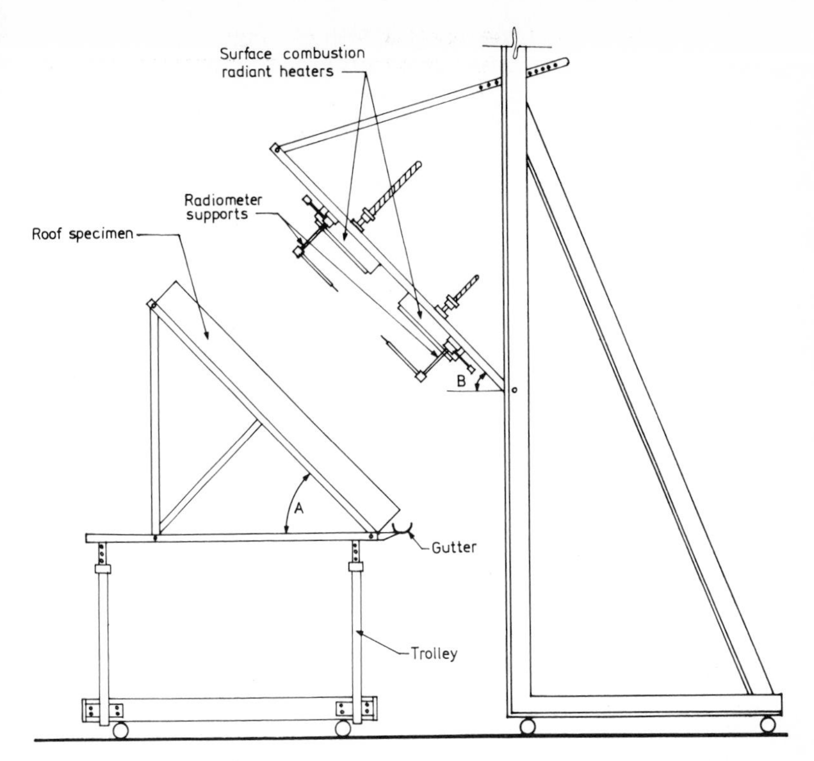

Fig. 6.6 Roof test rig

Obviously, the pitch of the roof will influence vertical flame spread due to convective preheating of the surface should ignition occur.

If the duration of flaming exceeds five minutes after the withdrawal of the test flame, or if the maximum distance of flaming in any direction exceeds 370 mm, the performance of the specimen is expressed by the letter *X*. However, if the duration of flaming is less than five minutes or the maximum distance of flaming in any direction is less than 370 mm, the performance of the specimen is expressed by the letter P.

The fire penetration and surface ignition test are combined and three samples must be tested. When the apparatus is calibrated, the sample is positioned for testing and during the test a flame is applied over the surface of the sample for a period of one minute at 5, 10, 15, 30, 45 and 75 minutes from the start of the test. The penetration time to the nearest minute is recorded for each specimen and if no penetration occurs the time is recorded as the maximum duration of the test. The extent of surface ignition is recorded at 60 minutes or at the time of penetration to the nearest 25 mm.

During the surface ignition part of the test, the specimen is

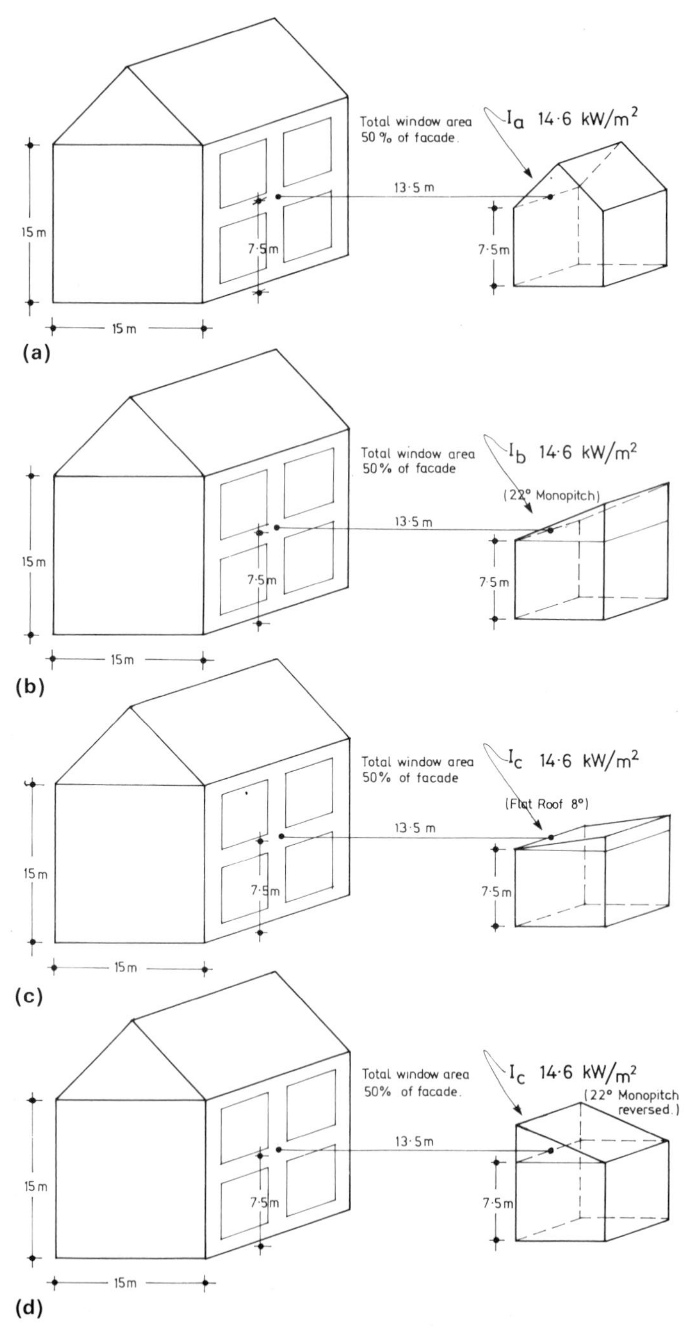

Fig. 6.7(a)(b)(c)(d) The geometrical relationships in roof design which could invalidate the assumption in the roof test

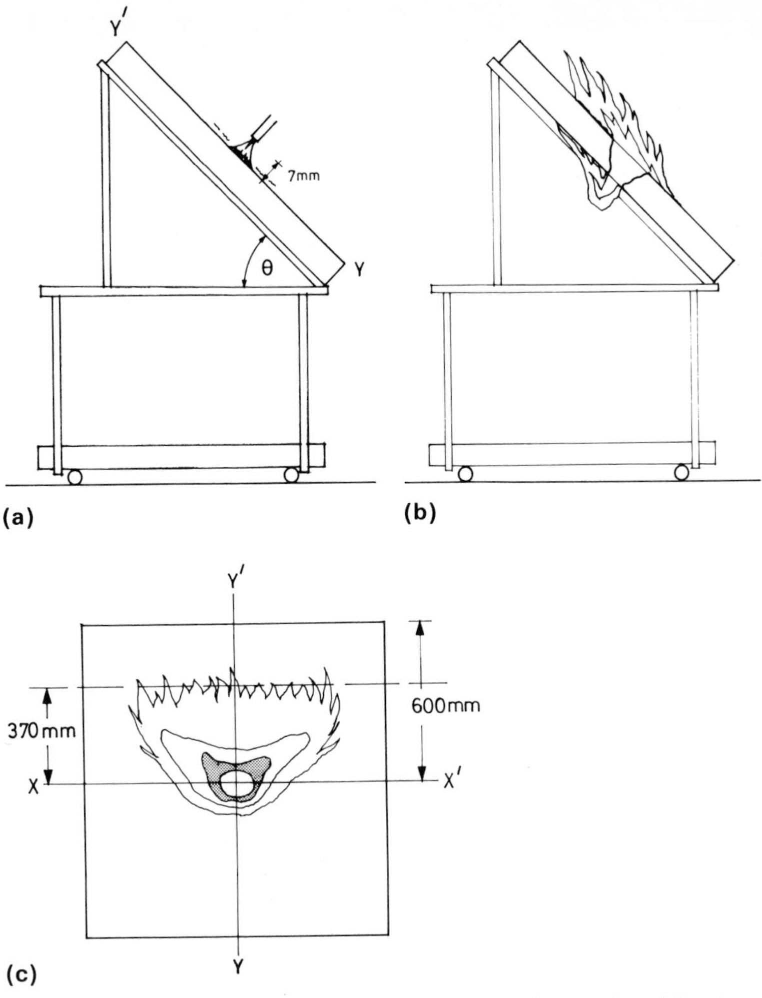

Fig. 6.8 (a), Pilot flame removed after one minute; (b), continued flaming and fire penetration for five minutes or more; (c), surface flaming exceeds 370 mm

subjected to a radiant heat flux of 14.6 kW m^{-2} and lateral flame spread only is measured. The influence of vertical spread is regulated by directing the pilot flame along the top edge of the specimen from one side, Fig. 6.9 and through the central area of the specimen parallel to the X–X axis to the opposite side.

This part of the test considers the lateral flame spread velocity vector as more realistic for comparability purposes.

During the test observations are made of:

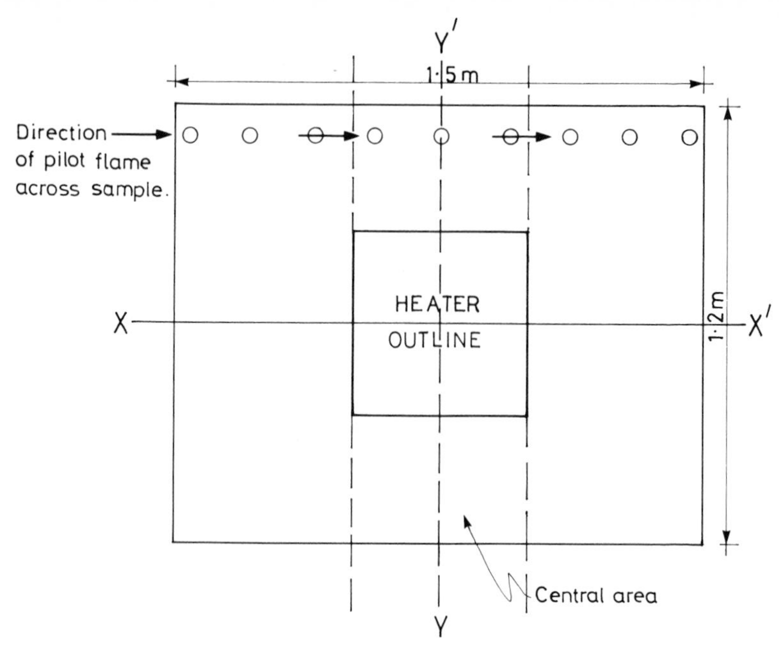

Fig. 6.9 Surface ignition of test specimen

1. Time and occurrence of any flaming on the upper surface and its duration
2. Time at which glowing or flaming appears on the underside
3. Time of development of holes or fissures or collapse
4. Visual changes in appearance
5. Any fall of molten materials and any flaming of molten materials
6. Behaviour of eaves when these are incorporated
7. Maximum distance of lateral flame spread on the upper surface which has accrued at any time during the test to the nearest 25 mm, as follows:
   (a) maximum spread in any direction in the preliminary ignition test
   (b) maximum spread measured horizontally through the centre of the sample.

BS 476: Part 3: 1975 is different from its predecessor BS 476: Part 3: 1958, particularly in expressing the results of tests. In the 1958 test, letters were used as designations expressing the performance of samples of materials. Thus designations AA, AB, AC, etc., were used. The first letter referred to the penetration times of the sample, where penetrations had not occurred for not less than one hour – A, for not less than 30 minutes – B, for less than 30 minutes – C and during the preliminary test – D.

Table 6.1 Method of expressing test results

BS 476: Part 3: 1958	BS 476: Part 3: 1975
AA, AB, AC	P60
BA, BB, BC	P30
AD, BD, CA	P15
CB, CC, CD	P15
Unclassifiable	P5

The 1975 Part 3 test uses the actual performance expressed as, e.g. P60 which means the sample passed the preliminary test and that fire penetration did not occur in less than one hour. Since the 1958 test is still quoted in building regulations, Table 6.1 shows the relationships between existing groupings and the 1975 method of expressing the performance of the sample.

From the foregoing the test assumes a radiant heat flux incident on a roof, which in reality may not be the case. Many roofs, e.g. mono pitch roofs, will be shielded from the potential radiation by a neighbouring building. The test is a crude attempt to predict the performance of roofs in a fire situation and a great deal depends upon the interpretation placed on the test results. Further, the methodology of expressing test results leaves a great deal to be desired, i.e. P60 indicates that the specimen has passed the preliminary ignition test and that penetration of the specimen has not occurred within the 60 minute period. The P60 classification conveys little information on the spread of flame over the specimen surface and those placing reliance upon this test must look to the actual test certificate for such detailed information.

6.3 BS 476: PART 4: 1970 – NON-COMBUSTIBILITY TEST FOR MATERIALS

In building design it is useful for the designer to know if the materials he proposes to use in his design support combustion or will add heat (significantly) to a fire. Materials are considered combustible if they are capable of undergoing combustion. This test determines whether materials are non-combustible or combustible by comparing the test results with predetermined criteria.

The test is carried out by placing a sample of the material to be tested 40 mm (± 2 mm) + 40 mm + 50 mm (± 3 mm) volume = 80 cm^3 ($\pm 5\ cm^2$) in a small calibrated electric furnace (Fig. 6.10) which is maintained at 750 °C for 20 minutes after the apparatus has been calibrated. The test apparatus is shown in Figs 6.10 and 6.11.

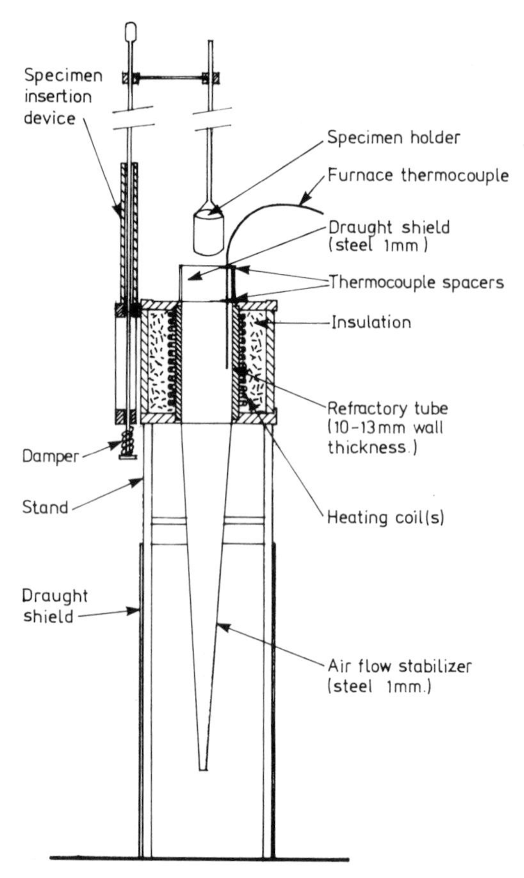

Fig. 6.10 Non-combustibility apparatus

Two thermocouples, one of which is inserted into the sample, are used to monitor temperatures and the occurrence and duration of any flaming that may occur is recorded.

Three samples are necessary for the test and the material is designated non-combustible if, during the test, none of the three specimens either:

1. Causes the temperature reading from either thermocouple to rise by 50 °C or more above the initial furnace temperature, or
2. Is observed to flame continuously for ten seconds or more inside the furnace (Fig. 6.12).

This test severely limits the amount of organic material present in the sample for it to be classed as non-combustible. Plasterboard under this test would be classified as combustible although it is a very useful material in that it can contribute to fire resistance. The

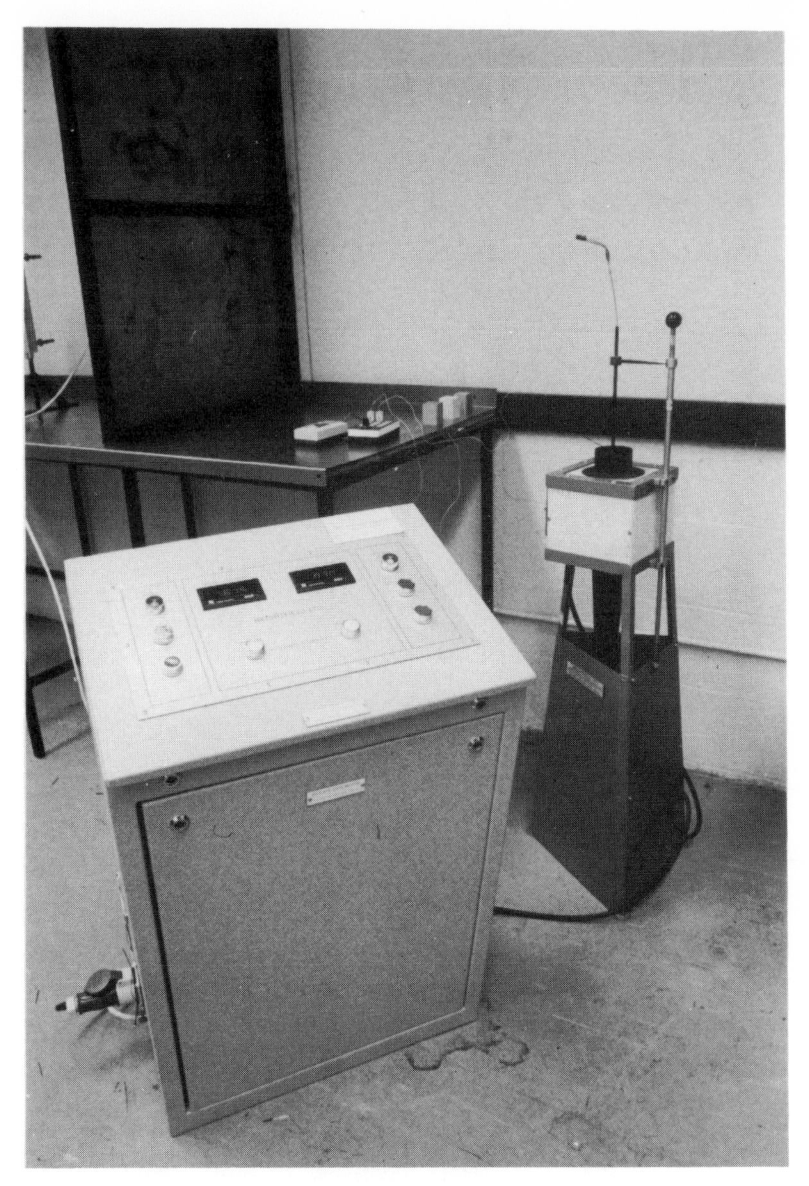

Fig. 6.11 Non-combustibility test apparatus

results of this test may determine the suitability of a material to be used as a wall or ceiling lining along an escape route.

As seen above, BS 476: Part 4 has serious shortcomings in that useful materials which contribute to the attainment of fire safety, e.g. plasterboard, may be restricted in their application and use. What is

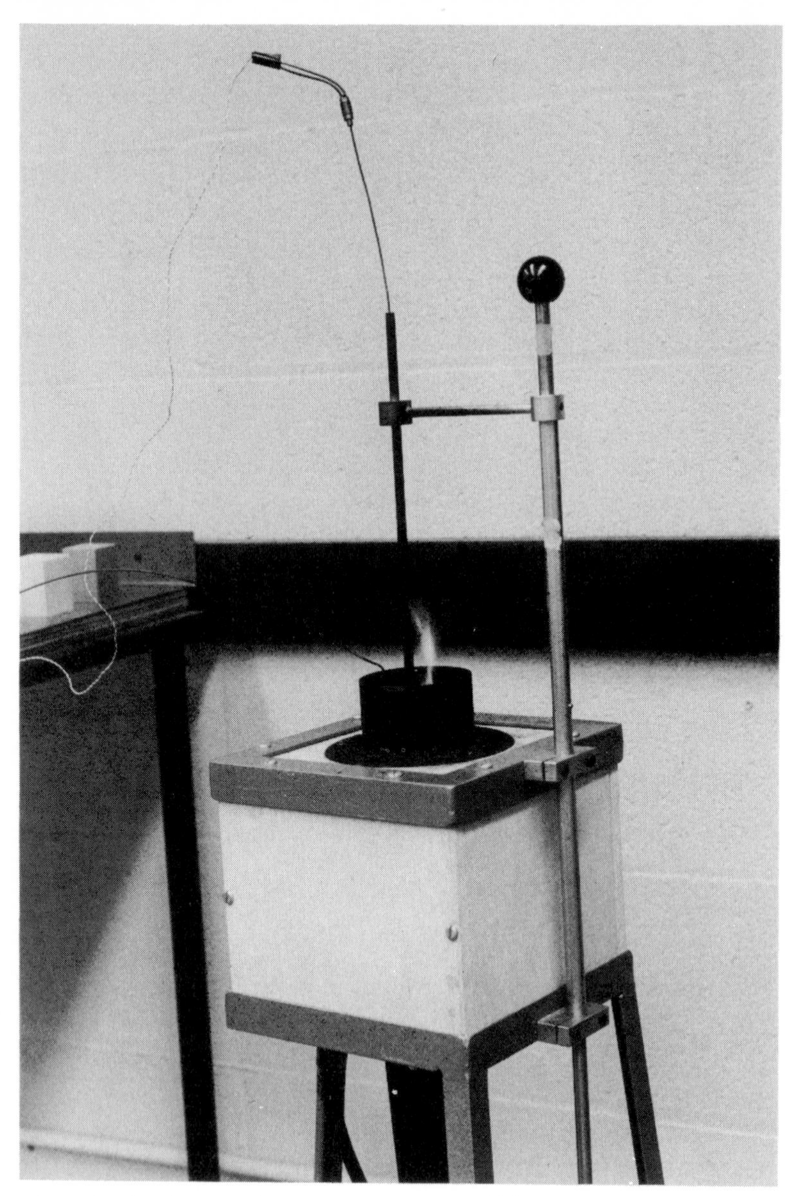

Fig. 6.12 Specimen under test in non-combustibility apparatus

desirable is a test which measures the heat release or rate of heat release from a material in a real fire situation.

It is worth pointing out here that there is no reference to the non-combustibility test in the idealised fire scenario figures contained in BS 476: Part 10: 1983[3] and reproduced here as Fig. 6.13.

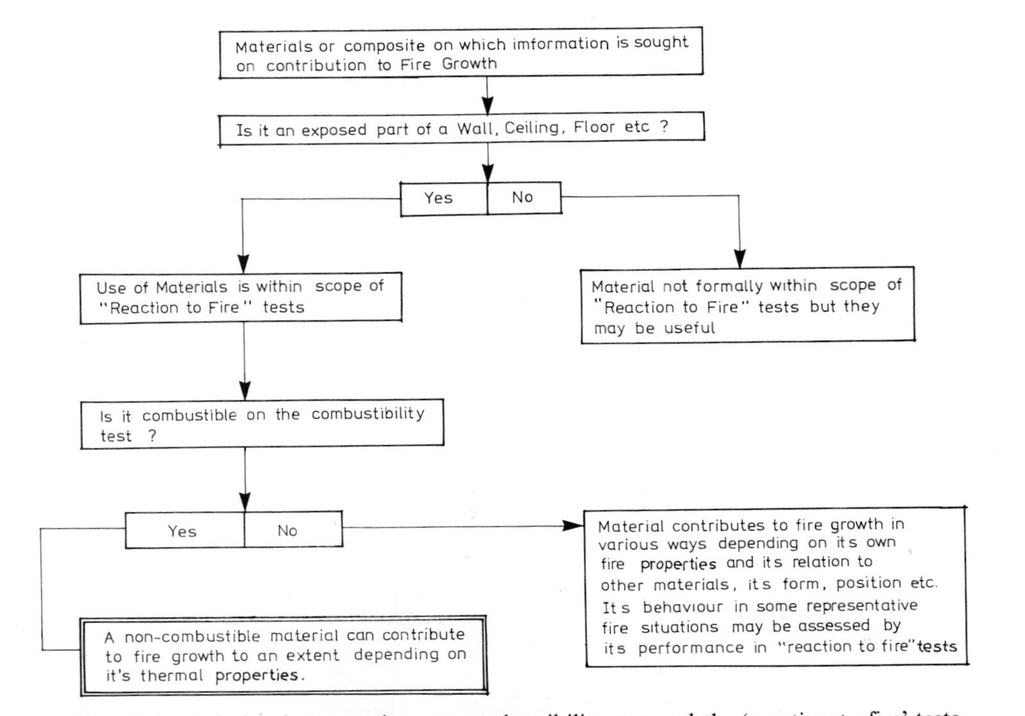

Fig. 6.13 Relation between the non-combustibility test and the 'reaction to fire' tests

In 1971 Malhotra[4] commented that although non-combustible materials have been traditionally regarded as safest to use because they do not add to the fuel load (and may even act as a heat sink), assist fire growth or produce hazardous by-products, investigations had shown that other factors besides non-combustibility of materials influence fire growth. Factors such as the thermal properties of compartment boundary constructions and ventilation rates could in some circumstances reduce the non-combustibility of materials to a subordinate role in the hierarchy of component contribution to fire growth and development. Consequently non-combustibility in itself is not a completely reliable guide to distinguishing between safe and unsafe materials.

The distinction between non-combustible and combustible materials is made on the basis of a 50° rise in temperature during the test and the fact that non-combustible materials can to some degree contribute to fire growth is recognised.

The development of 'reaction to fire' type tests as illustrated in Fig. 6.13 and the omission of the non-combustibility test therefrom gives some indication as to the future of this test method, particularly with the introduction of BS 476: Part 11: 1982, which is considered later.

Figure 6.13 shows the relationship between non-combustibility and 'reaction to fire' tests.[5] Malhotra[4] has concluded that the non-combustibility test has outlived its usefulness and the changes proposed in the building regulations for England and Wales support this view.

6.4 BS 476: PART 5: 1979 – METHOD OF TEST FOR IGNITABILITY

Construction materials in a fire can be hazardous for a number of reasons. They may give off toxic fumes and may be easily ignited. This test allows discrimination between materials based on their ease of ignition and those materials that fail the test. The latter type should not be used where ignition is a possibility.

In the test a square sample of 225 mm sides is held in a frame (Fig. 6.14) while a gas flame is allowed to play on it for ten seconds. If no specimen used flames for more than ten seconds after removal of the test flame, or if flaming does not extend to any edge during the application of the test flame or within a ten-second period following the removal of the test flame, the materials performed are designated P. Materials not satisfying the above criteria are designated X. The letter P indicates that the material has passed the test whereas the X designation indicates failure.

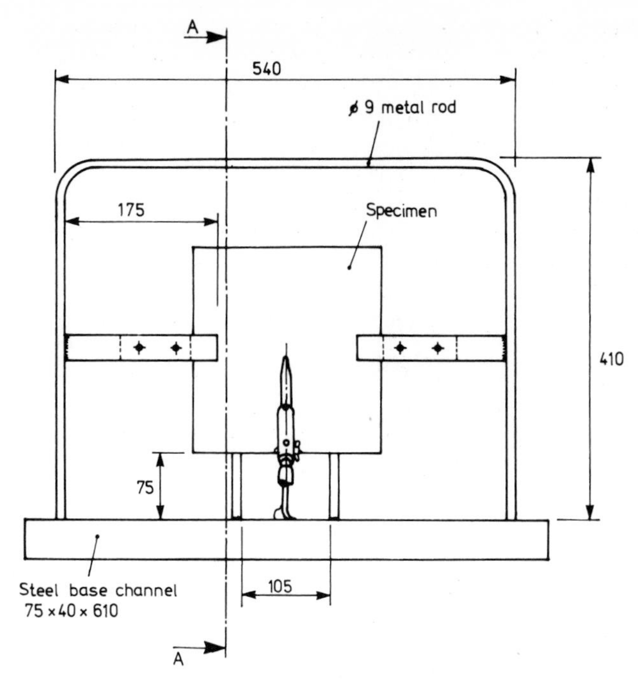

Fig. 6.14 Ignitability test apparatus

6.5 BS 476: PART 6: 1968 – FIRE PROPAGATION TEST FOR MATERIALS

With the development of fire testing it has become apparent that many properties of materials can contribute significantly to the growth and progress of a fire in its early stages. Some of the relevant factors are ignitability, combustibility, surface spread of flame, smoke emission and the rate of heat release of a material when subjected to fire. The test apparatus is shown in Fig. 6.15.

The fire propagation test has been developed in order that account be taken in the grading of materials of the amount and rate of heat evolved by such materials when subjected to a fire situation. In effect the test compares the time–temperature curve of a standard material with that of the specimen under consideration. The rear face of the well-ventilated combustion chamber is removable. After calibration, Fig. 6.16, the specimen replaces the standard material on the removal face of the combustion chamber. The time–temperature curve of the specimen is then compared with that of the chamber when all the internal faces are lined with standard material (Fig. 6.17).

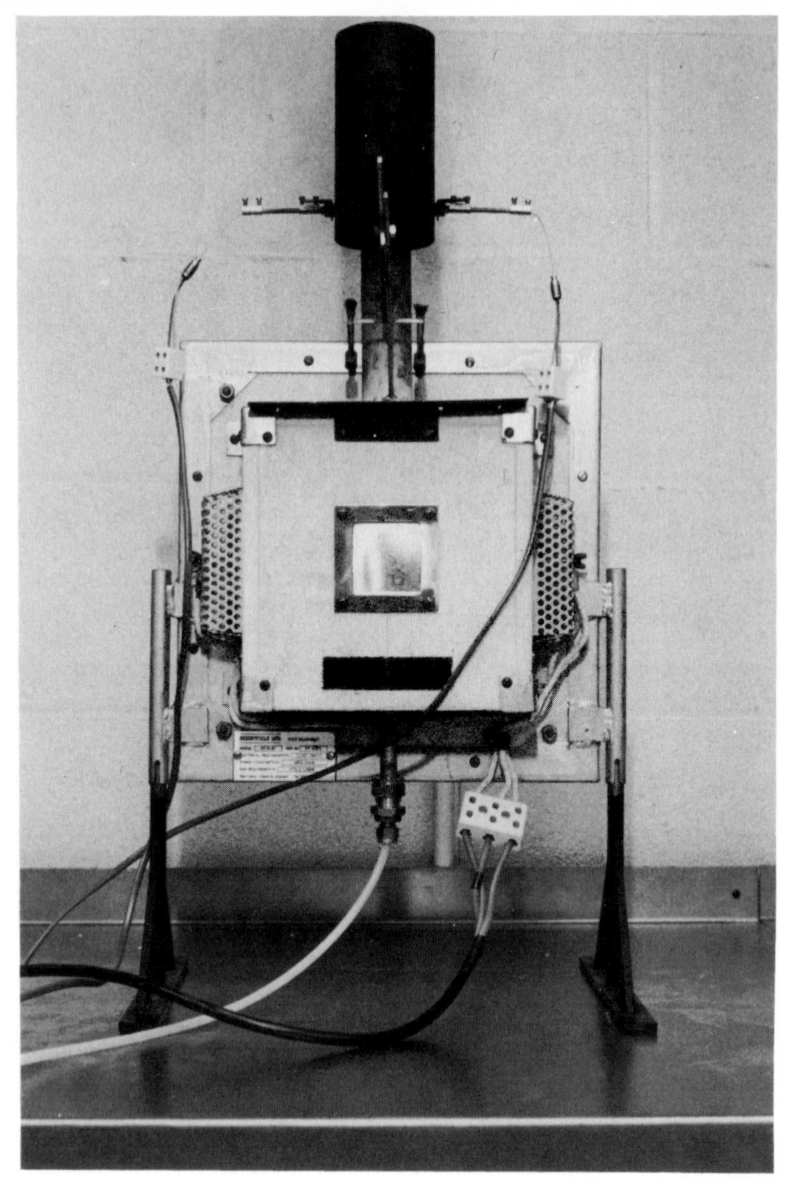

Fig. 6.15 Fire propagation test apparatus

By calculating the difference between the temperature rise obtained with the specimen and that with the standard calibration board, related to time, it is possible to obtain an assessment of the amount and rate of heat release from the material being tested.

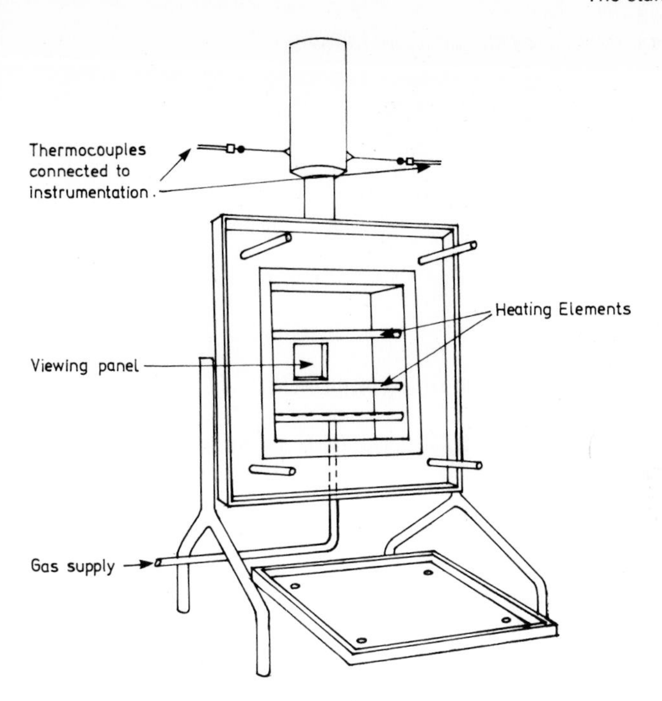

Fig. 6.16 View of fire propagation test apparatus

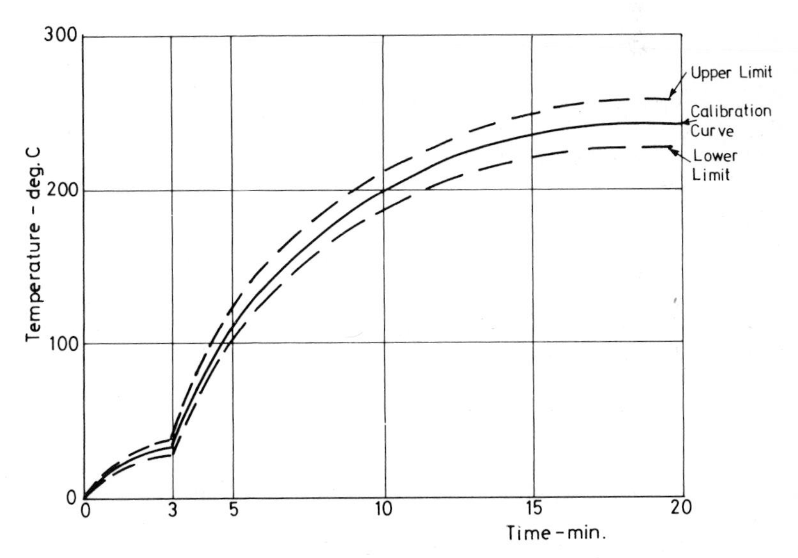

Fig. 6.17 Calibration curve

Performance indices are calculated as follows:

$$I = \underbrace{\sum_{\frac{1}{2}}^{3} \frac{(\theta_m - \theta_c)}{10t}}_{\substack{i_1 \\ \text{at } \frac{1}{2} \text{ minute intervals}}} + \underbrace{\sum_{4}^{10} \frac{(\theta_m - \theta_c)}{10t}}_{\substack{i_2 \\ \text{at 1 minute intervals}}} + \underbrace{\sum_{12}^{20} \frac{(\theta_m - \theta_c)}{10t}}_{\substack{i_3 \\ \text{at 2 minute intervals}}}$$

where:

$I =$ = index of performance

i_1, i_2, i_3 = sub-indices for the three time components

θ_m = temperature rise recorded for the material at time t

θ_c = temperature rise recorded for the non-combustible standard at time t

t = time in minutes from the beginning of the test.

Thus from the Fire Propagation Test a performance index I is calculated which provides a measure in comparative terms of the contribution of a material to the build-up of heat and potential fire spread. The individual use of the three sub-indices may be unnecessary, but the initial component i_1 may be useful as an indication of the ignitability and flammability of materials.

WORKED EXAMPLE

During a fire propagation test undertaken within the University of Ulster's Fire Engineering Laboratory, the following results for two specimens (1) and (2) were obtained (Table 6.2).

Using these results:

1. Plot a graph comparing both specimen (1) and (2) with the calibration curve
2. Evaluate the sub-indices of performance and the total index of performance for each specimen
3. Analyse the results and indicate which of the specimens is more hazardous in a fire situation.

Solution

1. As shown from graph Fig. 6.18, specimen (2) has a much bigger deviation from the calibration curve over the first three minutes of test compared with specimen (1). This means that sub-index i, will be larger for specimen (2) than for specimen (1).

Table 6.2 Specimen temperature profiles

Specimen temperature T_{s_1}	Specimen temperature T_{s_2}	Calibration temperature T_c
21.00	63.50	19.00
35.50	121.00	21.00
45.50	152.50	22.00
66.50	168.50	24.00
91.50	183.50	26.00
122.50	188.00	29.00
254.00	290.50	67.50
316.00	329.50	108.00
356.00	373.50	131.00
397.00	399.50	149.50
425.00	418.50	165.50
440.00	423.50	176.50
456.00	432.50	186.50
460.50	430.50	199.50
456.00	415.50	211.50
453.50	433.00	220.00
455.00	414.00	232.00
466.00	383.50	232.50

2. The sub-components are evaluated according to the equation above and are given in Tables 6.3 and 6.4 for specimens (1) and (2), respectively.
3. (a) The total index of performance for specimen (2) is greater than for specimen (1).
   (b) The sub-index i, for specimen (2) > for specimen (1). This suggests that specimen (1) will contribute a lot to the propagation of a fire particularly during the early stages, indicating that it is a more hazardous material than specimen (1) during a fire situation.

Building regulations make use of this test by prescribing maximum limits for both index I and sub-index i_1 averaged over three specimens. Consequently the general surface spread of flame test is augmented, consolidated and anomalies removed.

The fire propagation test was originally devised because the Class 1 grading in the Part 7 surface spread of flame test was too broad and further definition was required. Unfortunately the Part 6 test is often referred to as a rate of heat release test whereas in effect the test is really concerned with ignitability of a specimen of given area subject to a given heat flux over a given period of time.

It is important to know which materials will contribute to fire propagation at an early stage in fire development and consequently these materials are heavily penalised in this test.

For the purpose of building regulations, materials which under the test achieve an index $I = 12$, and a sub-index $i_i = 6$, are rated as Class 0 materials and consequently may be used in situations where maximum

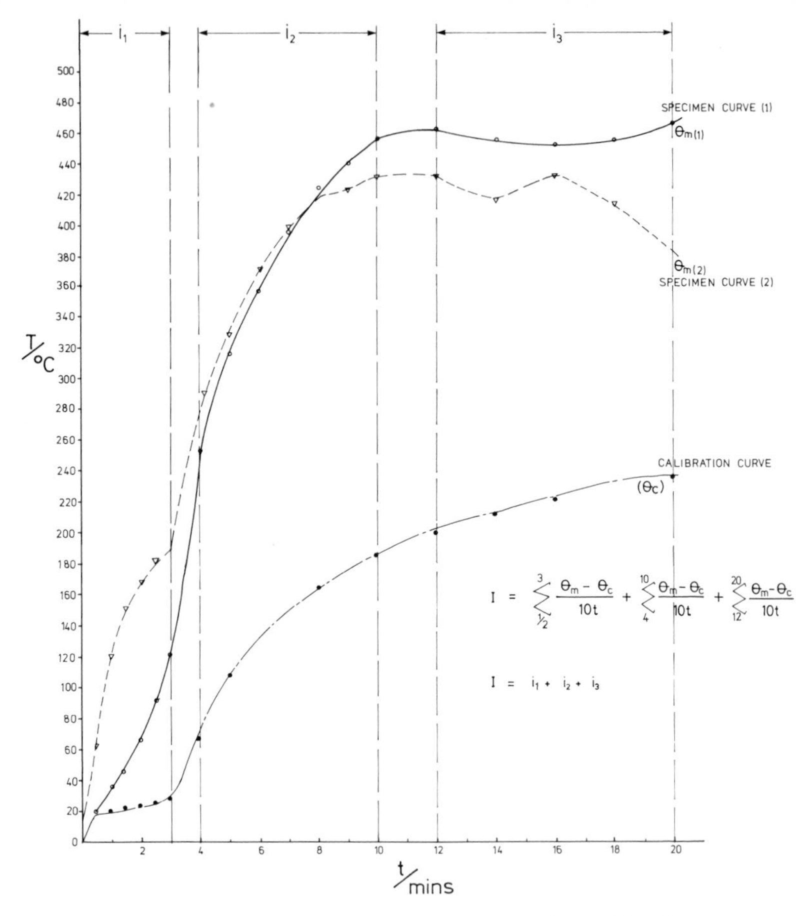

Fig. 6.18 Graphs of specimens (1) and (2) compared to calibration curve

protection is required. Figures 6.19, 6.20, 6.21 and 6.22 show a louvered timber ceiling boarding treated with an intumescent coating after being subjected to the Part 6 test. The boarding satisfied the criterion for Class 0 and the plates show how the substrate charred underneath the intumescent coating.

6.6 BS 476: PART 7: 1971 – SURFACE SPREAD OF FLAME TESTS FOR MATERIALS

Materials used for walls and ceilings have a significant effect on the way in which fire spreads. In particular it is essential that flame

Table 6.3 Computation of index I and sub-indices i_1, i_2 and i_3 for specimen (1)

t	$\theta_s - \theta_c$	$10t$	$\theta_s - \theta_c$
0.5	2.0	5	0.40
1.0	14.50	10	1.45
1.5	23.0	15	1.56
2.0	42.5	20	2.12
2.5	65.60	25	2.62
3.0	93.50	30	3.11
			$i_1 = 11.26$
4.0	186.5	40	4.66
5.0	208.0	50	4.16
6.0	225.0	60	3.75
7.0	247.5	70	3.53
8.0	259.5	80	3.24
9.0	263.5	90	2.92
10.0	269.5	100	2.69
			$i_2 = 24.95$
12.0	261.0	120	2.17
14.0	244.5	140	1.74
16.0	233.5	160	1.45
18.0	223.0	180	1.23
20.0	233.5	200	1.16
			$i_3 = 7.75$

$$I = i_1 + i_2 + i_3 = \underline{43.96}$$

spread be negligible in evacuation spaces and means of escape. The surface spread of flame test was devised to determine the burning characteristics of materials by considering the flame spread over their surfaces. BS 476: Part 7: 1971 specifies two tests:

1. Large-scale test
2. Small-scale test.

The latter is conducted with the apparatus reduced in scale by one-third of that used in the large-scale test and may be used in the development of products to predict the possible performance of the large-scale test. It must be pointed out that there is no direct correlation between the large-scale test and the small-scale test. Therefore, only the large-scale test will be described here.

The apparatus (Fig. 6.23) consists of four radiant panels enclosed in a refractory-concrete frame with two hinged wings mounted

Table 6.4 Computation of index i and sub-indices i_1, i_2 and i_3 for specimen (2)

t	$\theta_s - \theta_c$	$10t$	$\theta_s - \theta^c$
0.5	44.5	5	8.90
1.0	100.0	10	10.00
1.5	130.5	15	8.70
2.0	144.5	20	7.23
2.5	157.5	25	6.30
3.0	169.0	30	5.63
			$i_1 = 46.76$
4.0	223.0	40	5.57
5.0	221.5	50	4.43
6.0	242.5	60	4.04
7.0	250.0	70	3.57
8.0	253.0	80	3.16
9.0	247.0	90	2.74
10.0	246.0	100	2.46
			$i_2 = 25.97$
12.0	231.0	120	1.92
14.0	204.0	140	1.46
16.0	213.0	160	1.33
18.0	182.0	180	1.01
20.0	151.0	200	0.75
			$i_3 = 6.47$

$$I = i_1 + i_2 + i_3 = \underline{79.2}$$

centrally on the panel's height. These wings should be capable of taking samples measuring 900 mm × 225 mm. The samples are preconditioned, the apparatus being calibrated by means of thermocouples located at 75 mm centres along the wings to achieve the necessary temperature gradient across the sample. When the test commences, a gas flame is also applied at the end of the sample for one minute and then removed. The distance the flame has spread is recorded at the end of $1\frac{1}{2}$ minutes and measurements of spread against time are continued up to a total time of ten minutes, unless the flame has reached the far end of the specimen in less than ten minutes. The material is then classified in one of the categories as shown in Table 6.5.

Some materials may be unclassifiable under this test, e.g. some plastics, and other tests will have to be used. Building regulations make use of this test by prescribing Class 1, Class 2 and Class 3, surface flame spread characteristics for lining materials for walls and

Fig. 6.19 Louvered timber ceiling board with intumescent coating which achieved class O in Part 6 test

Fig. 6.20

Fig. 6.21

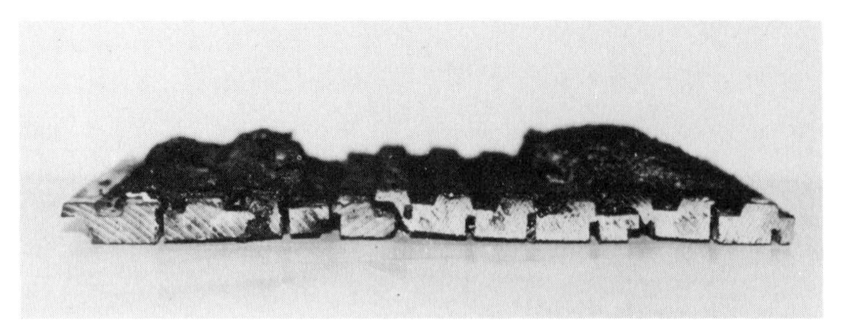

Fig. 6.22

ceilings in different locations in buildings. Class 1 materials have the lowest rate of flame spread and Class 4 has the highest. Building regulations do not permit the use of a Class 4 material. For example, the data contained in Table 6.6 obtained from a test would show that the material tested has a Class 1 surface spread of flame.

It is possible using impregnation and painting treatments to upgrade a material, e.g. from a Class 3 to a Class 1 material. However, it is just as equally possible inadvertently to downgrade a material, e.g. from a Class 3 to Class 4. Consider untreated timber,

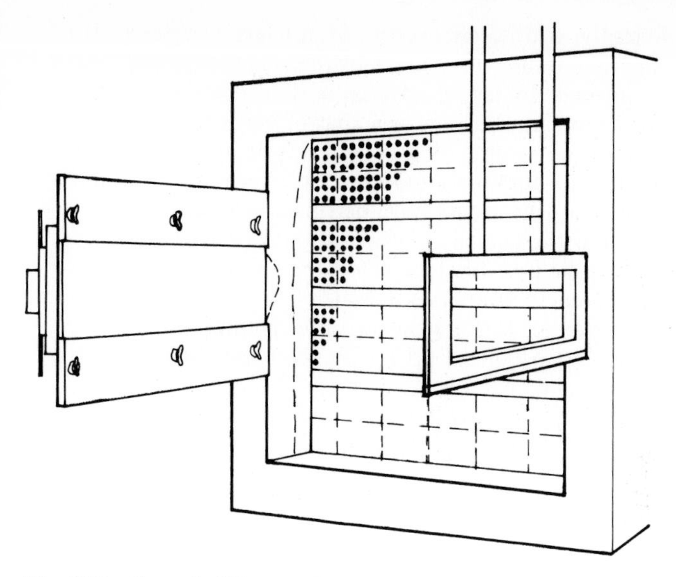

Fig. 6.23 Spread of flame apparatus

Table 6.5

Classification	Flame spread at $1\frac{1}{2}$ mins		Flame spread at 10 mins	
	Limit (mm)	Tolerance for one specimen sample (mm)	Limit (mm)	Tolerance (mm)
1	165	25	165	25
2	215	25	455	45
3	265	25	710	75
4	Exceeding Class 3 limits		Exceeding Class 3 limits	

Table 6.6

Specimen No.	Flame spread at $1\frac{1}{2}$ minutes (mm)	Final flame spread (mm)	Time	
			min	secs
1	NIL	20	10	00
2	NIL	40	10	00
3	NIL	80	10	00
4	NIL	80	10	00
5	NIL	20	10	00
6	NIL	120	10	00

which is inherently, with a few exceptions, a Class 3 material. It is possible to upgrade timber to Class 1 by various treatments. Similarly this untreated Class 3 material may be downgraded to Class 4 by the simple application of a coat of varnish or paint.

On the question of treatments it should not be overlooked that while it is possible to upgrade a material's surface spread of flame characteristics, some treatments have detrimental side effects on some materials and the choice of system should be thoroughly researched. It may be possible to apply a treatment under strictly controlled conditions of temperature and relative humidity in a laboratory, but whether this is possible on a building site is questionable. Similar problems arise with pretreated boards that are cut and jointed on site.

It should be noted with this test that:

1. Flame spread is measured laterally and does not indicate rate of vertical flame spread up the specimen. It is important that the vertical flame spread characteristics be known because once the flames impinge upon the ceiling, fire growth develops rapidly.
2. Flame spread is not measured in the most onerous situation, e.g. Fig. 6.24 shows a corner in a room with the lining material ignited on one side. Cross-radiation from the adjacent surfaces will promote rapid flame spread but the possibility of cross-radiation is ignored in this test.

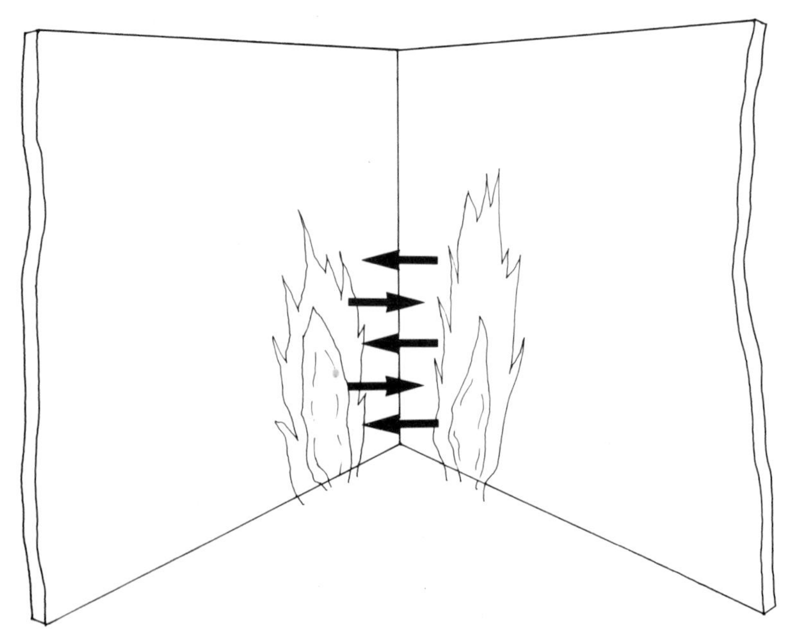

Fig. 6.24 Corner effect of flame spread

3. In the test the material is fixed in a hinged wing against an insulation board. However, in practice the material may be applied to a backing structure which acts as a heat sink and consequently the material may perform much better than it would in the test. This would be particularly so with thin materials used as linings to walls.

The value of this test is simply that it allows different wall lining materials to be ranked according to their performance under standard conditions. It does not give any absolute measure of performance in fire. Indeed, acceptance of the results of this test is in fact little more than relying on experience of how materials subjected to this test behave in full-scale fires. Given that this experience was gained at a time when most of the materials available were 'traditional' (wood, wood-based products, etc.), with the advent of synthetic building products, application of test results to material selection became somewhat dubious.

Some thermoplastics cannot be tested by this method. It was also found that even those materials which obtained Class 1 in this test could not all be accepted for use in particular applications simply because the test was not severe enough or sensitive enough to subdivide the Class 1 category further. For this reason the Fire Propagation Test BS 476: Part 6 was developed.

6.7 BS 476: PART 8: 1972 – TEST METHODS AND CRITERIA FOR THE FIRE RESISTANCE OF ELEMENTS OF BUILDING CONSTRUCTION

Having considered the possible effect of fire on materials and roofs regarding ignition, flame spread, fire propagation and development, elements of construction must also be considered in order to determine their capability to function in a fire situation. The fire-resistance test was one of the first carried out in the early days of fire-test development, consolidated as BS 476: Part 1: 1953, amended and reproduced as BS 476: Part 8: 1972.

Elements of construction are considered as: walls and partitions (loadbearing and non-loadbearing), floors, flat roofs, columns, beams, suspended ceilings protecting steel beams, door and shutter assemblies, and glazing, and this test specified the heating cycle, method of test and the criteria for the determination of fire resistance of elements of building construction.

All samples to be tested must be:

1. Representative of the element of construction

2. Conditioned to approximate strength and moisture content conditions expected in use
3. Fixed as in use
4. Restrained as in use.

It is necessary, therefore, to know how an element will behave in a fire situation for a period of time regarding its stability, integrity and insulation.

The heating cycle of the furnace is given by

$$T - T_0 = 345 \log_{10} (8t + 1)$$

where t = time of test in minutes
T = furnace temperature in °C at time t and
T_0 = initial furnace temperature in °C.

The time–temperature curve, Fig. 6.25, is determined by the above formula and is used as the furnace control parameter with predetermined tolerances.

The tolerances permissible during the test in respect of the area under the curve of the mean furnace temperature and that of the area under the standard curve are:

Time	*Tolerance*
First ten minutes	±15%
First thirty minutes	±10%
After first thirty minutes	5%

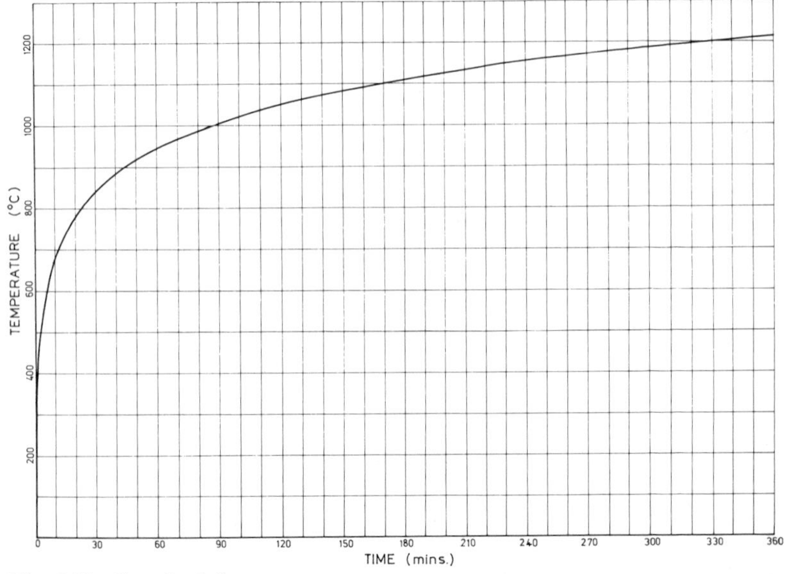

Fig. 6.25 Standard fire curve

and the mean furnace temperature must not differ from the standard curve by more than $\pm 100\,°C$.

Stability

It is essential that an element of structure performs its required function for a known predetermined time in a fire situation. Thus a column and a beam must carry the loads applied from floors above, for example, without failure for a given time and likewise floors or circulation spaces must be available for use as a means of escape for a known time.

Failure occurs for a non-loadbearing element when collapse of the test sample takes place, and for loadbearing elements if:

1. The element fails to support the test load during the heating periods, or
2. The element fails to support the test load when the sample is reloaded 24 hours after the heating period.

Should the element collapse during the heating period or during the reload test then the maximum period of stability for the element is deemed to be 80 per cent of the time to collapse or the duration of heating if collapse occurs in the reload test. The failure of certain elements, i.e. floors, beams and flat roofs, occurs when a sample being tested experiences a maximum deflection greater than $L/30$ where $L=$ clear span.

Stability, therefore, is a measure of a structural element's ability to function in a fire situation for a prescribed period of time.

Integrity

It must be understood that fire resistance is a different characteristic of a component than non-combustibility. A sheet of glass may be non-combustible in terms of BS 476: Part 4, but its fire-resistance properties are very, very limited. This is because the material will crack and allow flames and hot gases to pass through it in a very short time. Hence it would fail the second criterion of the fire-resistance test – its integrity is suspect. A fire-resisting component must not allow hot gases and flames to pass through its structure and cause ignition in another part of the building or compartment. The integrity of a sample is determined by holding a cotton pad close to, but not more than 30 mm from, the orifice at regular intervals for not more than ten seconds and observing if ignition of the pad occurs. The cotton pad should be 100 mm square, 20 mm thick, weigh 3–4 grams, be dired in an oven at 100 °C for not less than half-an-hour and must not be re-used. The integrity of the sample is determined, therefore, by observation and occurs when cracks or other openings exist through which flames or hot gases can pass causing flaming of the cotton-wool pad. Consequently the integrity criteria is also expressed in terms of time.

However, the type of element under consideration will influence the severity of the integrity criterion, e.g. a compartment wall may be required to have thirty minute stability and thirty minute integrity, whereas the integrity criteria may not be at all appropriate when applied to a beam or column.

Insulation

For some components such as doors, which by definition are expected to be opened and closed on a regular basis, it would be inconsistent to impose the insulation criterion of fire resistance. But for components with a separating function, such as a compartment or separating wall, it is essential that heat is not conducted through the wall sufficiently to cause ignition of combustible material in the adjoining compartment or building. This is achieved by ensuring that the mean temperature of the unexposed surface of the sample does not increase by more than 140 °C above the initial temperature, or the temperature of the unexposed surface does not increase by more than 180 °C above the initial temperature.

Fire resistance, therefore, has three components, stability, integrity and insulation, expressed in terms of time.

A compartment floor may be required by building regulations to have a fire resistance of a half-hour, meaning:

stability ... 30 mins
integrity ... 30 mins
insulation ... 30 mins.

But a floor in a two-storey dwelling may also be required to have a (modified) half-hour fire resistance, meaning:

stability ... 30 mins
integrity ... 15 mins
insulation ... 15 mins,

and a door will not be required to have any insulation characteristic whatsoever.

Therefore, while the term fire resistance may appear all embracing, in the end it is the appropriateness of the test criteria to a particular situation or component which determines and limits the extent of its application.

Furnace tests are specified in almost all building regulations and codes as the primary means of compliance with prescriptive requirements for fire resistance and consequently have tended to acquire a degree of infallibility. Test results are accepted without question, notwithstanding that test results can vary between different test furnaces. The geometry of the furnace and nature of its construction influence the convective heat transfer to the specimen because of induced turbulence. The thermal inertia of the furnace-boundary construction controls its responses to the gas temperatures and hence the heat transfer to the specimen by radiation.

The type of fuel also can affect results because gas flames are less radiative than oil flames. The heat transfer to the specimen is not measured directly. Thermocouples record gas temperatures at predetermined locations, but although thermocouples in two furnaces may record the same temperature the heat fluxes to the respective specimens under test may be different. Reference was made earlier to permissible tolerances in following the standard time–temperature curve. A closer scrutiny will show that the permissible deviations from the standard time–temperature curve are generous, e.g. the mean furnace temperature may differ from the standard curve within a band of 200 °C. Consequently the heating regime for any specimens being tested in the same or different furnaces may not be identical. Many factors must be considered in the design of furnace tests if variability in test results is to be improved. As indicated, the level of variability in furnace tests depends upon the design of the test and method of conducting it. Systematic variability may be considered as part of the variability built into the test by virtue of its design. The part of the variability due to differences in operation and sampling may be considered as random variability. Malhotra[6] considered the factors which could be used as a basis for improving the reproducibility of the furnace test. These factors are listed in Table 6.7.

He also considered the factors which contribute to systematic and random variability in furnace tests (Table 6.8).

Reference to the listing of methods of test under the BS 476 heading given earlier will show that a new breed of furnace tests will soon be available. Even though attempts will have been made to overcome many of the existing shortcomings, furnace tests in themselves will always retain a degree of artificiality.

6.7.1 Future developments

It is perhaps worthwhile at this juncture to comment on some future changes which will occur with the introduction of:

BS 476: *Fire Tests on Building Materials and Structures:*

- Part 20 – Method for the determination of the fire resistance of † elements of construction (General Principles). (Part revision BS 476: Part 8: 1972.)

† Draft Document.

Table 6.7 Main factors in tests and their consequences

Factor	Consequence
A. Construction of samples	
1. Size limitation due to equipment	1. Modelling problems
2. Single elements tests	2. Absence of interaction
3. Conditioning – strength	3a. Attainment of loadbearing capacity
– moisture content	b. Effect of excess moisture
– wear and tear	c. Representation of service damage
4. Loading – method	4a. Response of hydraulic systems
– constant loads	b. Unrepresentativeness of constant loads
5. Support and boundary conditions	5. Lack of simulation of actual conditions
6. Workmanship, high standard	6. Enhancement of performance
B. Heating	
1. Standard curve	1. Lacks realism, limited application
2. Uniform temperature	2. Effect of unequal heating ignored
3. Exposure conditions	3. Façades and special conditions omitted
4. Furnace design	4. Lack of specification leads to variability
5. Thermocouple type	5. Does not measure heat transfer
6. Thermocouple location	6. Can be affected by flames
C. Performance criteria	
1. Stability, loadbearing elements	1a. Precise collapse point may depend upon the response of loading system
Non-loadbearing elements	b. The precise point of instability difficult to define
2. Integrity – gas leakage through openings	2a. Present system does not have good reproducibility
– heat transfer	b. Importance of temperature hot spots not investigated

- Part 21 – Method for determination of the fire resistance of load-bearing elements of construction (Part revision BS 476: Part 8: 1972.)
- Part 22 – Method for determination of the fire resistance of non-loadbearing elements of construction (Part revision BS 476: Part 8: 1972).

Table 6.8 Variability in furnace tests

A. Systematic variability	
(a) Tolerance on standard temperature curve	
(b) Tolerance on furnace thermocouple response	Factors affecting repeatability and reproducibility
(c) Location of furnace thermocouples	
(d) Ageing of furnace thermocouples	
(e) Response of measuring systems	
(f) Tolerance on the loading system sensitivity	
(g) Variability in ambient conditions (draught, etc.) and their effect on surface cooling	
(h) Variability of the cotton pad test	
B. Random variability	
(a) Inconsistency of samples	
(b) Variability in moisture conditions	Factors affecting reproducibility
(c) Variability in material properties	
(d) Variability in erection standards	
(e) Variability due to furnace construction	
(f) Variability due to thermocouple design	
(g) Method of load application	
(h) Method of application of cotton pad test	

BS 476: Part 20 – Method for the determination of the fire resistance of elements of construction (general principles)

Part 20 contains some interesting definitions. For example, fire resistance is defined as the ability of an element of building construction to withstand exposure to a standard temperature – time and pressure regime without loss of its fire-separating function or loadbearing function, or both for a given time.

Integrity is defined as the ability of a specimen of a separating element to contain a fire to specified criteria for collapse, freedom from holes, cracks and fissures and sustained flaming on the unexposed face.

Sustained flaming is defined as flaming that is visible with the naked eye and that remains visible for an uninterrupted period of not less than ten seconds.

Loadbearing capacity is defined as the ability of a specimen of a loadbearing element to support its test load, without exceeding specified criteria with respect to either the extent, or rate, of deformation, or both.

Insulation is defined as the ability of a specimen of a separating element to restrict the temperature rise of the unexposed face of the specimen to below specified levels.

A completely new term – imperviousness – is introduced and defined as the ability of the specimen to prevent the egress of hot

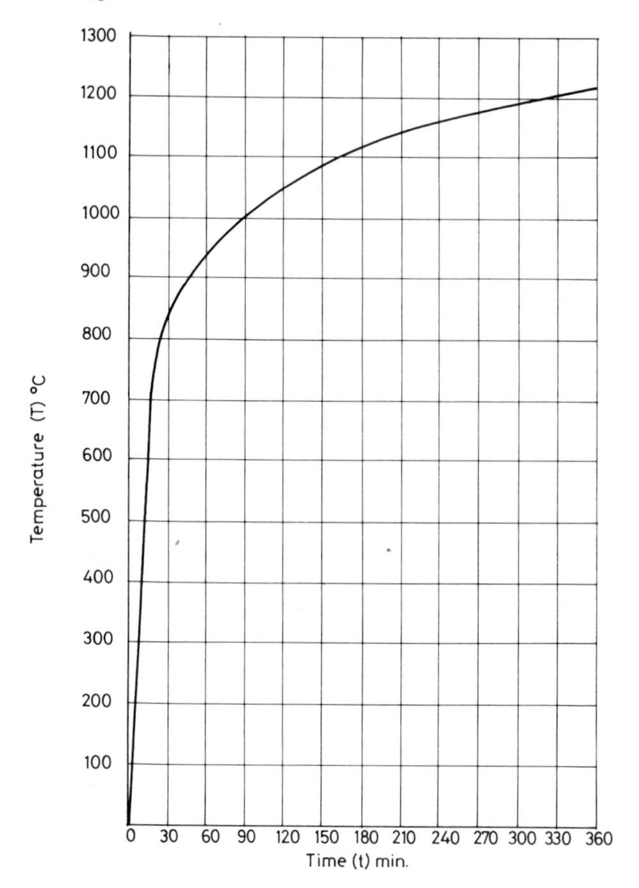

Fig. 6.26 Standard temperature time curve

gases from the unexposed face of the specimen exceeding specified limits.

The heating regime of the furnace is given by the expression:

$$T = 345 \log_{10}(8t + 1) + 20 \qquad \text{(Fig. 6.26)}$$

and is slightly different from that in Part 8 in that the furnace temperature is assumed to be 20 °C at the commencement of the test, irrespective of its original value, in order that all furnaces will operate to the same actual temperature – time programme. The actual initial furnace temperature is required to be within the limits of 5 °C to 35 °C. The reason for this change is illustrated in Fig. 6.27. Using the Part 8 temperature–time relationship:

$$T_f - T_0 = 345 \log_{10}(8t + 1)$$

differences in the initial furnace temperatures can result in a variety of temperature curves which, for example, with a critical specimen

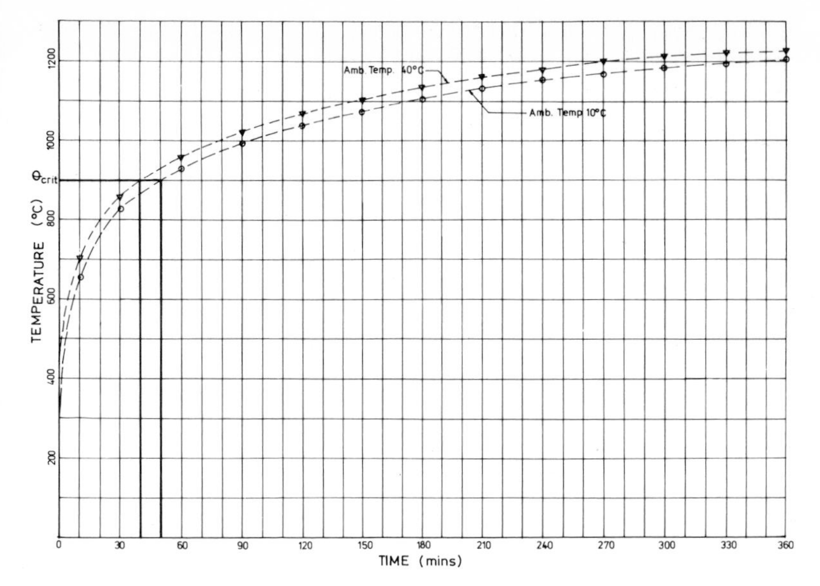

Fig. 6.27 Variation in standard fire temperature curve due to differences in ambient temperature

temperature of 900 °C would give different fire-resistance ratings for the specimen (Fig. 6.27).

Performance in the test is assessed in terms of:

1. *Loadbearing capacity* (*stability*). This criterion is further subdivided, i.e. loadbearing capacity of:
   (a) loadbearing vertical elements
   (b) loadbearing vertical non-separating elements
   (c) loadbearing horizontal elements.
2. *Integrity*. Failure of the test construction to maintain integrity (as previously defined) is deemed to have occurred when collapse, sustained flaming, or failure under the imperviousness requirements occurs.
3. *Imperviousness*. This is a rather curious term which by definition is a measure of the ability of the specimen under test to prevent the egress of hot gases from the unexposed face of the specimen exceeding specified levels.

Surprisingly the levels referred to are *not* defined, e.g. volume flow of hot gases. Failure under this criterion is deemed to have occurred when flames or hot gases cause flaming or glowing of a cotton-fibre pad; or in certain conditions if a 6 mm wide gap occurs in the specimen and extends for a distance of 1500 mm; or when a gap ⩾25 mm ∅ penetrates the complete thickness of the specimen. In effect this amounts to an Integrity failure and it is anticipated that the term imperviousness will be deleted when the standard is published in its final version.

BS 476: Part 21 – Method for determination of the fire resistance of load bearing elements of construction

As the title suggests, Part 21 sets out in detail a method of determining the fire resistance of loadbearing elements of construction. It does not provide a method for the evaluation of the fire resistance of composite construction of more than one element.

The fire resistance of loadbearing elements of construction is determined by the performance criteria shown in Table 6.9.

Table 6.9

Element	Performance criteria for determining fire resistance
Loadbearing horizontal element without any separating function	Loadbearing capacity
Loadbearing vertical element without any separating function	Loadbearing capacity
Loadbearing horizontal element of a floor or flat roof	Loadbearing capacity – integrity and insulation
Loadbearing wall which has a separating function	Loadbearing capacity – integrity and insulation

BS 476: Part 22 – Method for determination of the fire resistance of non-loadbearing elements of construction

Part 22 sets out in detail a method of determining the fire resistance of non-loadbearing elements of construction which include partitions, vertical doorsets and shutter assemblies, and ceiling membranes.

The definitions contained in Part 22 are very important and should be examined with care. For example, a ceiling membrane is defined as a non-loadbearing element of a building construction designed to provide horizontal fire separation as distinct from protection to any roof or floor above. Ceilings which form an integral part of a floor can be tested in accordance with BS 476: Part 23, which is new in concept and will contain procedures designed to evaluate the contribution made to the fire resistance of elements of construction by materials or components used in their assembly.

A doorset is defined as an assembly (including any frame or guide) for the closing of permanent openings in separating elements and includes shutter assemblies. A partition is defined as a non-loadbearing element of building construction designed to provide vertical fire separation.

Table 6.10

Elements	Performance criteria for determining fire resistance
Partitions and non-loadbearing walls providing vertical separation	Integrity Insulation
Doorsets and shutter assemblies	Integrity Imperviousness Insulation
Ceiling membranes	Integrity Insulation

The fire resistance of non-loadbearing elements is determined by the performance criteria shown in Table 6.10.

6.8 BS 476: PART 11: 1982 – METHOD FOR ASSESSING THE HEAT EMISSION OF BUILDING MATERIALS

This test is part of a new series of methods designed specifically to evaluate the various properties of building materials, components and structures under fire conditions. The apparatus used is essentially that which is used in the Part 4 non-combustibility test with some modifications (see Fig. 6.10). The method of testing is not suitable for composites, but the individual materials may be assessed separately.

In this test the specimen is in the form of a cylinder rather than a cube. Details regarding specimen construction and preparation, apparatus setting and calibration are given in BS 476: Part 11. The feature which distinguishes Part 11 from Part 4 is in the taking and interpretation of the results.

The test procedure is as follows:

1. The furnace is stabilised in accordance with the calibration requirements and the initial mean furnace temperature T_f is recorded.
2. The specimen is conditioned, weighed, mass and physical dimensions recorded and the specimen thermocouple is located centrally in the specimen.
3. When the specimen is inserted into the furnace the time is recorded.
4. The temperature T_f of the furnace thermocouple and T_c of the specimen thermocouple are continuously recorded.
5. The test is terminated when final equilibrium temperatures are established as shown in Figs 6.28(a) and (b).

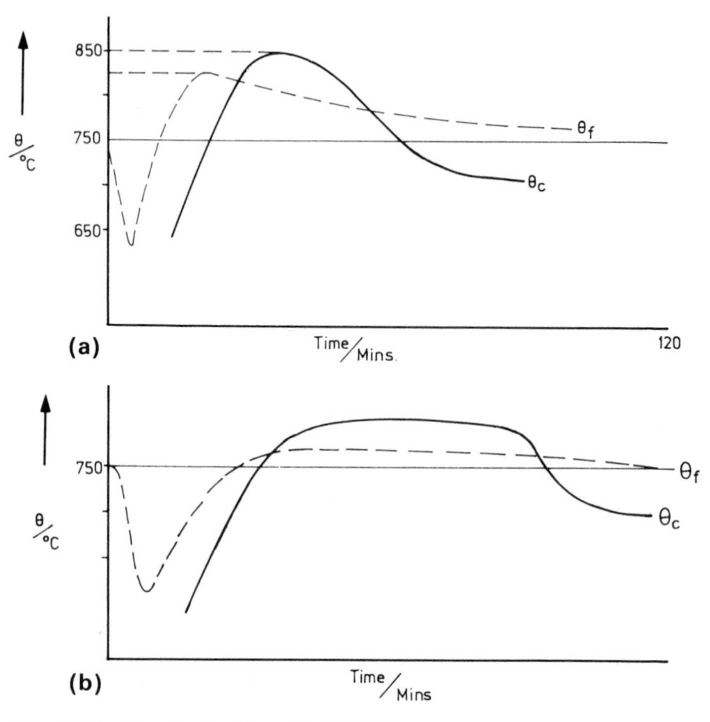

Fig. 6.28 Graphs for Part 11 BS 476 test

6. From the graphs produced, $T_{f\,max}$ (maximum furnace temperature), and $T_{f\,final}$ (furnace temperature at the end of the test) are recorded with $T_{c\,max}$ (maximum specimen temperature) and $T_{c\,final}$ (temperature of the specimen at the end of the test).
7. After cooling the remains of the specimen are carefully removed, weighed and the mass recorded.
8. During the test the occurrence of sustained flaming for a period of more than five seconds is noted.
9. The test procedure is carried out for five specimens. The results of the test are expressed as follows:
   (a) Temperature rises:

$$T_F = T_{f\,max} - T_{f\,final}$$

$$T_C = T_{c\,max} - T_{c\,final}$$

The arithmetic mean of the furnace and specimen temperatures are then obtained as follows:

$$\overline{T_F} = \sum_{i=1}^{5} \frac{T_{fi}}{5}$$

$$\overline{T_C} = \sum_{i=1}^{5} \frac{T_{ci}}{5}$$

(b) The arithmetic mean of the sum of the sustained flaming periods for each specimen is determined.
(c) The density for each specimen is obtained and the arithmetic mean of the densities of the fire specimen is evaluated.
(d) The mass loss of each specimen tested is expressed as a percentage of the initial mass and the arithmetic mean of the mass loss of the fire specimen is determined.

The problem with this test, like many others, lies with the interpretation of the results obtained and the derivation of a meaningful classification methodology for materials, which will be suitable for inclusion in building regulations or as mechanisms to assist designers in the correct choice and location of materials used in buildings.

Unfortunately, the title is misleading, i.e. Method for assessing the heat emission from building materials, because in reality it does not assess quantitatively the heat emission from building materials. In fact the test identifies four characteristics of materials associated with heat emission during a fire situation. These four characteristics are used for comparing the performance of the materials under test.

In the USA the fire tests are covered by the ASTM Standards.

American Fire Standards

The fire tests listed in the ASTM Standards for 1985[7] are:

Test Methods for:

E136-82	Behaviour of Materials in a Vertical Tube Furnace at 750°C
*E119-83	Building Construction and Materials, Fire Tests of
E970-83	Critical Radiant Flux of Exposed Attic Floor Insulation using a Radiant Heat Energy Source
E648-84	Critical Radiant Flux of Floor-covering systems using a Radiant Heat Energy Source
E152-81a	Door Assemblies, Fire Tests of
E160-80	Combustible Properties of Treated Wood by the Crib Test
E 69-80	Combustible Properties of Treated Wood by the Fire-tube Apparatus
E906-83	Heat and Visible Smoke Release Rates for Materials and Products
E108-83	Roof Coverings, Fire Tests of
E662-83	Specific Optical Density of Smoke Generated by Solid Materials
E 84-84a	Surface Burning Characteristics of Building Materials
E286-84	Surface Flammability of Building Materials Using an 8 ft (2.44 m) Tunnel Furnace

E162-83	Surface Flammability of Materials Using a Radiant Heat Energy Source
E814-83	Through Penetration Fire Stops, Fire Tests of
E163-84	Window Assemblies, Fire Tests of

Practice for:

E931-85	Assessment of Fire Risk by Occupancy Classification
E535-83	Preparation of Fire Test Standards

Terminology Relating to:

E176-85	Fire Standards

Guides for:

E800-81	Measurement of Gases Present or Generated During Fires
E603-77 (1983)	Room Fire Experiments

Proposed Test Method for:

E-5 Proposal P108	Fire Performance of Public Ground Transportation Seat Assemblies, Determining the

Some similarities exist between these tests and the British Standard fire tests. However, the ASTM fire tests have developed differently from the BS tests and significant differences exist in terms of application, test methodology and interpretation of results, Table 6.11

Table 6.11

ASTM	BS (476) UK	Comments on Variations
E108	Part 3	
E119	Part 8	The ASTM test induce a hose stream test when necessary
E136	Part 4	Differing criteria for passing BS < 50 °C rise in specimen; flaming not more than 10 seconds ASTM < 30 °C rise in specimen; flaming not more than 30 seconds
E152	Part 22	ASTM induce a hose stream test when necessary
E162	Part 7	ASTM measures downward flame spread over the specimen with a constant heat flux BS measures lateral flame spread with a variable heat flux
E906	Part 6	BS does not include smoke release rate

Table 6.11 highlights the differing philosophies in fire testing internationally. It also indicates that an understanding of fire test methodology is essential if technically sensible interpretations of test data is to be made.

This chapter is not intended to be an expose of UK fire tests *per se*. The tests discussed are used as examples to convey an understanding that the physical relationships in the testing environment may not represent completely what happens in reality.

6.9 CONCLUSION

In order for a method of testing to have any relevance to reality it must as a prerequisite have a clearly defined objective. This means, in effect, that the problem must first be identified, its measurable characteristics enumerated, and a suitable method of assessment designed in the knowledge of the following features:

1. *Environmental conditions*. The test should incorporate all of the physical conditions likely to be experienced in practice.
2. *Range of applicability*. The environmental conditions should be capable of variation.
3. *Flexibility*. The test should be capable of reproducing different orientations in which the material can be utilised.
4. *Reproducibility*. The test should be able to be repeated with a variance not greater than 10 per cent.
5. *Meaningful expression of results*. The results should be expressed in units which make comparison easy.

The present situation regarding the application of the results of fire tests is unsatisfactory. Fire tests have become an integral part of building control and as such have formed attitudes to the extent that the primary objectives of the attainment of fire safety in buildings have become obscured. The fact that test results are unquestionably accepted together with inflexibility in the enforcement of building legislation is a contributing factor.

Materials and building components should not be considered in isolation from the environment in which they may be located. This in itself demands as a prerequisite an analysis of the probable type and magnitude of the fire to which they could be exposed. A systems approach to fire safety in buildings is required and the new breed of tests currently being developed are more solidly based on the principles of fire behaviour, e.g. the ISO Ignitability Test (British Standard Draft for Development DD70 – 1981).[8] This test ascertains the time taken for a material to ignite when its surface is subjected to a defined level of radiated heat and when there is a small flame available to ignite of any volatile gases given off from the

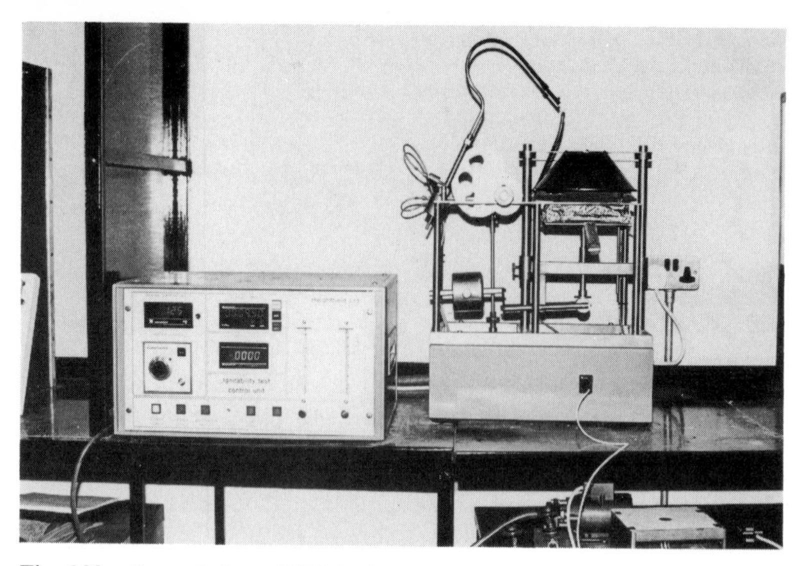

Fig. 6.29 General view of ISO ignitability test

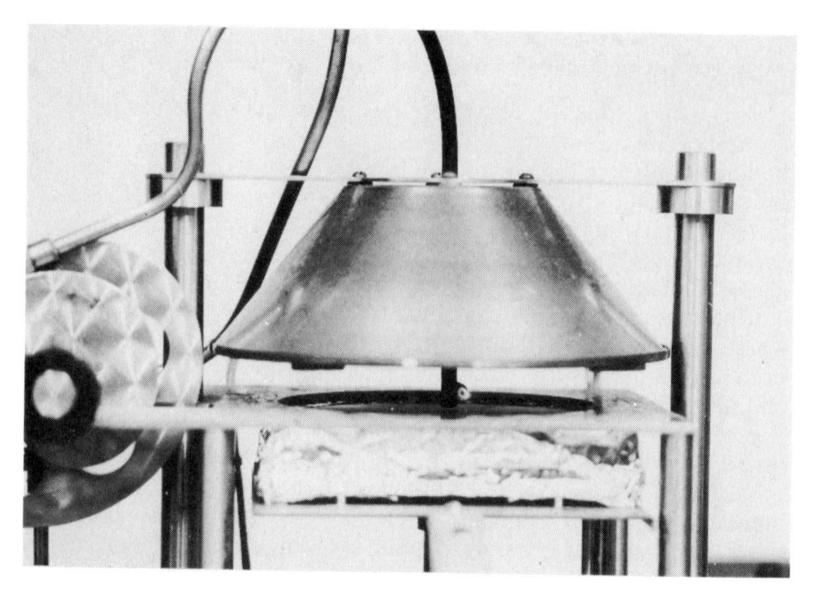

Fig. 6.30 Sample at beginning of test

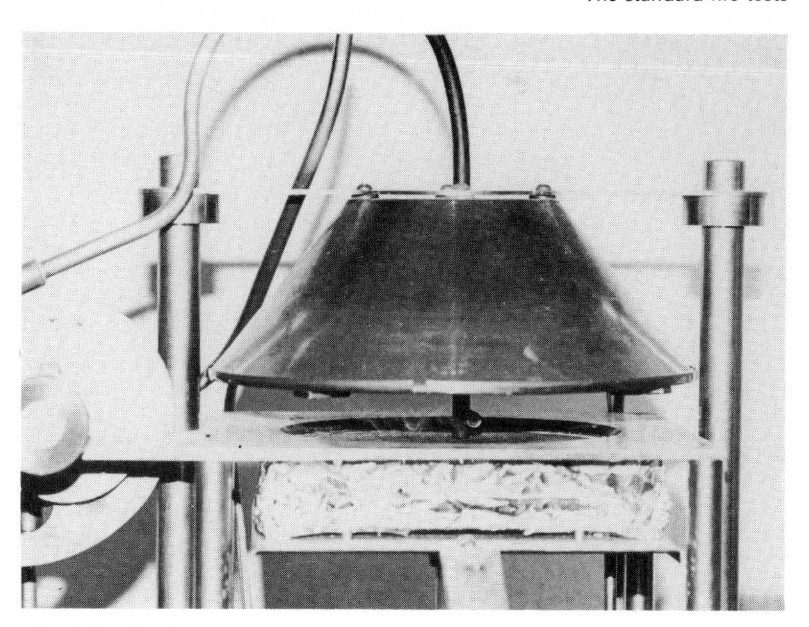

Fig. 6.31 Production of volatiles at surface during test

Fig. 6.32 Ignition of volatiles

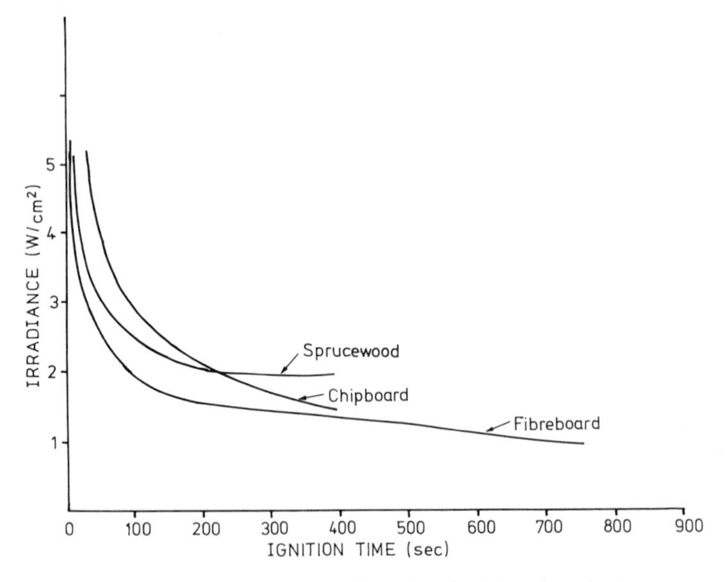

Fig. 6.33 Graph of irradiance (W-cm) vs ignition time (sec)

surface. The ignition performance of the material can be determined at different levels of irradiance ranging from 1 to 5 W cm^{-2}. The apparatus is shown in Fig. 6.29. Figures 6.29, 6.30, 6.31 and 6.32 show commencement of a test, through the production of volatiles until pilot ignition occurs. Using this apparatus it is possible to establish a linear relationship between irradiance and ignition time.

Figure 6.33 shows the relationship between irradiance and ignition time for different woods and wood products and Fig. 6.34 shows the graph of $\log_{10} I$ against $\log_{10} t$. Table 6.12 shows the data used for plotting the graphs $\log_{10} I$ versus $\log_{10} t$ for the three specimens.

Assume the relationship between irradiance and ignition time is in the form:

$$I = at^b$$

where a and b are constants

I = Irradiance (W/cm^{-2})

t = Ignition time (sec)

If this relationship is correct, then:

$$\log_{10} I = b \log_{10} t + \log_{10} a$$

i.e. in the form $y = mx + c$. *Compare:* $y = \log_{10} I$

$m = b$

$c = \log_{10} a$

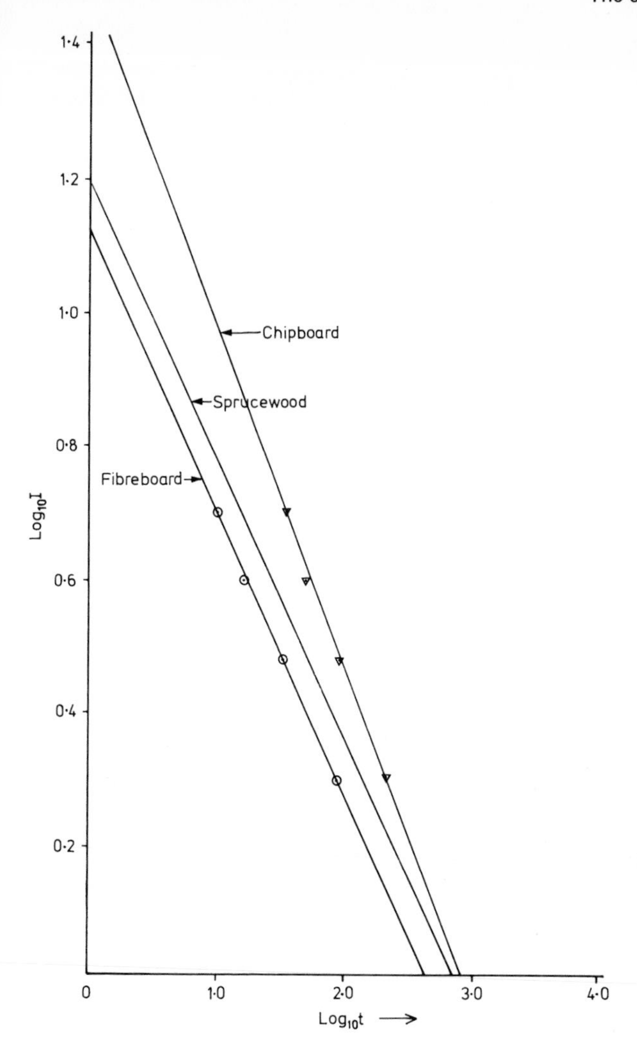

Fig. 6.34 Graph of $\log_{10} I$ against $\log_{10} t$

Using the relationship it is possible to predict the time for ignition to occur if the heat flux to the specimen is known.

Even with the current level of knowledge many unresolved problems exist with this test if meaningful results are to be obtained. For example the ignition time of thin materials may be a function of the thermal characteristics of the substrate. The emissivities of the surface of the material being tested may influence the test result. Also the spectral distribution of energy from the heater may vary with temperature. This factor may be important if the absorptivity of the sample surface is frequency selective.

Table 6.12

I = Irradiance (W/cm^{-2})	$\log_{10} I$	Spruce wood		Chipboard		Fibreboard	
		t = Ignition time (sec)	$\log_{10} t$	t	$\log_{10} t$	t	$\log_{10} t$
5	0.700	15.4	1.188	34.6	1.539	10.2	1.009
4	0.602	23.4	1.369	51.0	1.708	16.6	1.220
3	0.477	53.0	1.724	90.0	1.954	32.8	1.516
2	0.301	348.0	2.542	221	2.344	90.0	1.954
1	0.000					703.0	2.847

(1) Sprucewood

$$b = \frac{1.20}{2.85} = -0.421 \qquad \therefore \qquad I = 15.85t^{-0.421}$$

$$\log_{10} a = 1.20 \qquad \therefore \quad a = 15.85$$

(2) Chipboard

$$b = -\frac{1.48}{2.93} = -0.51 \qquad \therefore \qquad I = 30.20t^{-0.51}$$

$$\log_{10} a = 1.48 \qquad \therefore \quad a = 30.20$$

(3) Fibreboard

$$b = -\frac{1.125}{2.65} = -0.425 \qquad \therefore \qquad I = 13.34t^{-0.425}$$

$$\log_{10} a = 1.125 \qquad \therefore \quad a = 13.34$$

Another new test in course of development is a method of measuring the rate of heat release, RHR, of a material during exposure to fire, based on the oxygen consumption technique. In essence the RHR by oxygen consumption is based on the fact that the heat release per unit mass of oxygen consumed during complete combustion is approximately constant for most fuels encountered in fire.

The RHR of is calculated from:

$$\text{RHR} = \text{E} \cdot \varnothing \cdot \text{V}$$

where:

E = the heat of combustion per unit volume of oxygen consumed
$\varnothing$ = the fraction of oxygen depleted, and
V = the volume flow of gases a

Ostman *et al.*[9] compared three test methods for measuring the rate of heat release. These were: the Ohio State University box,[10] the Swedish Forest Products Research Laboratory,[11] and the National Bureau of Standards open cone calorimeter.[12] Ostman concluded that the three methods compared gave roughly the same results and that only a few experimental procedures required further study.

Future developments in fire resistance and reaction to fire testing in Europe will be influenced by the need within the European Economic Community to eventually obtain a general acceptance of harmonised classification. It is perhaps in the area of fire modelling that the way forward, toward harmonisation could be more productively pursued.

REFERENCES

1. Malhottra H L, *Fire Testing and fire hazard assessment, Fire Prevention No. 144 1985*. Fire Prevention Association.
2. *British Standard 476. Parts 3–31:* British Standards Institution.
3. *British Standard 476: Part 10, 1983*. British Standards Institution.
4. Malhotra H L, *The Rationale of Non-Combustibility Building Control Jan/Feb. 1985*. Institution of Building Control Officers.
5. Malhotra H L, *Fire Resistance versus fire behaviour, Fire Prevention No. 134*. Fire Prevention Association.
6. Malhotra H L, *Fire Resistance and Behaviour*, Fire Surveyor pp. 13–17, December 1984.
7. *Annual Book of A.S.T.M. Standards Section 4. Construction Vol. 04.07.* American Society for Testing and Materials 1985.
8. *British Standard Draft for Development D070-1981 The ISO Ignitability Test*, British Standards Institution.
9. Ostman B A L, Svensson R G and Blomquist J, *Comparison of Three Test Methods for Measuring Rate of Heat Release, Fire and Materials Vol. 9, No. 4, pp. 176–184*. (1985). Wiley Heden 1985.
10. *ASTM, Standard Test Method for heat and visible smoke release rates for materials and products*. Annual ASTMS Standards Part 18, 1980.
11. Svensson I G and Ostman B A L, *Rate of Heat Release by Oxygen Consumption in an open arrangement, Fire and Materials Vol. 8, No. 4, pp. 206–216*. (1984). Wiley Heden 1984.
12. Babranskas V, *Development of Cone Calorimeter. A bench-scale heat release apparatus based on oxygen consumption, Fire and Materials Vol. 8, No. 2, pp. 81–95*. (1984). Wiley Heden 1984.

CHAPTER 7

Building regulations

7.1 INTRODUCTION

The law in relation to fire safety, like the general body of law, has developed as society has changed. The controversy between those who believe that law should essentially follow, not lead and that it should do so slowly, in response to clearly formulated social sentiment – and those who believe that the law should be a determined agent in the creation of new norms is one of the recurrent themes of the history of legal thought. Suffice here to say that some fire safety legislation has developed in response to the perceived needs of society whilst some has developed in response to some tragic fire occurrence. The result is a complex web of legislation enforced by several different enforcing authorities. Current fire safety legislation tends to be fragmentary in nature in that no single piece of legislation considers all of the important aspects of fire safety. The main purposes of fire safety legislation may be stated as:

1. To impose a level of fire safety such that it is unlikely that people occupying a building would suffer hurt in the event of an unwanted fire, and
2. To protect the community at large from the consequences of fire in an individual building.

Thus the tactics of fire prevention and fire protection will be incorporated into fire safety legislation with differing degrees of emphasis. It may be simply stated that current fire safety legislation reflects the loss through fire of both life and property which society is prepared to accept or tolerate.

The applicability of fire safety legislation in the United Kingdom varies according to region, i.e. England and Wales, Inner London, Scotland and Northern Ireland.[1,2] The general legislation applying is listed in Appendix 7.1.

7.2 BUILDING REGULATIONS

The current set of building regulations[3] have their origin in *Post-War Building Studies*[4] and the objectives, although never expressly stated, have experienced a subtle change of emphasis in order to remain relevant to a modern social and industrial infrastructure. In this chapter it is intended to discuss the philosophy of the current prescriptive building regulations.

Building regulations assume that if certain components of fire safety can be identified and suitable standards applied to particular building types, a satisfactory level of fire safety will be achieved. There is no evidence to support this assumption. Indeed, some would argue that the available evidence points in the opposite direction.

Generally the structure of building regulations follow the pattern developed below:

1. Classify buildings by type
2. Compartment buildings
3. Prescribe fire resistance requirements for elements of structure
4. Limit unprotected areas of external walls
5. Prescribe constructional requirement for separating walls, compartment walls and floors
6. Prescribe constructional requirements for protected shafts
7. Specify the type and constructional requirements for fire-resisting doors
8. Control the penetration of fire barriers by services
9. Specify non-combustibility requirements for stairways in prescribed situations
10. Describe requirements for cavity barriers and fire stops
11. Control spread of flame over walls and ceilings
12. Control the use of plastics on ceilings
13. Relate the siting of buildings to roof constructional requirements.

7.3 BUILDING CLASSIFICATION

It is assumed that if the purpose for which buildings will be generally used can be determined then buildings used for a similar purpose can be classified as a particular building type. This method assumes that each building of a particular type will:

1. Have the same fire loading

2. Be of similar geometry
3. Experience a similar fire scenario
4. Be exposed to a fire of similar severity.

Given the foregoing another assumption follows, i.e. that standards for components of fire safety can be prescribed for building types.

Thus building classification becomes a factor in risk determination and the inaccuracies inherent in this assumption are compounded throughout building regulations. Because the method of classifying buildings is so crude it is possible that some buildings may be underprotected and other buildings may be overprotected.

7.4 COMPARTMENTATION

The factors which generally determine the requirement for mandatory compartmentation are:

1. Building type
2. Height of building
3. Floor area of the storey or compartment
4. Cubic capacity of building or compartment.

It can be seen that within the component compartmentation several subcomponents exist which apparently influence fire safety, e.g. the height of the building. When a sub-component such as the height of the building has been identified as a contributory factor towards the achievement of fire safety, it is implicit that measures have been taken to ensure that occupants on the topmost floor of a building are at no greater risk than those on the ground floor.

The concept of containment is therefore adopted to restrict the lateral and vertical fire spread through the building. In order to achieve the required degree of containment, compartmentation was invented, i.e. the sub-division of a building into fire-tight cells or compartments enclosed by boundary structural and constructural components with the necessary fire endurance characteristics. The factors influencing the degree of containment required are the:

1. Potential fire severity
2. Potential hazard to the occupants
3. Degree of difficulty experienced in fire-fighting.

The potential fire severity factor also has its origins in the *Post-War Building Studies*[4] where buildings are classified in terms of fire loading, Table 7.1.

Thus the quantity of combustibles in the building is related to

Table 7.1

Building hazard classification	Fire load (kg/m^{-2})
Low-hazard group	0–49
Medium-hazard group	49–100
High-hazard group	> 100

Table 7.2

Hazard	Combustible content		Equivalent severity of fire in hours of standard test
	($kg\ m^{-2}$)	Fire load ($kJ\ m^{-2}$)	
Low	49	186,080	1
Medium	49–100	372,160	2
High	> 100	> 372,160	4

Calorific value of materials 16,282–18,608 $kJ\ kg^{-1}$

hazard or risk and the fire loading is related to fire severity, Table 7.2.

It will be noted that a constancy has been assumed for the calorific value of the combustibles in buildings. Also fire severity is determined by reference to a standard test procedure which does not adequately reflect the behaviour of real fires. Suffice to say here that over the intervening years the test procedure has changed and the test itself is most useful for comparing the performance of components when exposed to standard conditions.

Given the potential fire severity as expressed in Table 7.2 related to each hazard classification it follows that the concept of containment is positive if the enclosing construction has a fire-resisting capability not less than the potential severity of a fire within the protected space. This means in effect that all the combustibles within a compartment could be consumed by fire, and the fire contained within the compartment, without the boundary structural or constructional compartments being unduly damaged. The compartment could withstand a total burnout of the combustibles contained within it.

Throughout the above process, uniform distribution of the combustibles in the building has been assumed. The building regulations then somehow relate the foregoing to spatial limitations for compartments in terms of area or cubic capacity. It is difficult to find the logic which supports these parameters other than the fact that larger spaces will accommodate larger fires and the extent of damage, the risk to occupants and the difficulty in fire-fighting will be greater. The structural fire precautions guidance note[5] published in 1975 perhaps gives the best insight when it states that it has now become traditional to accept 7,000 m^3 as the basic maximum size for a compartment. Few concessions are given in building regulations for the inclusion of active fire control systems in the design of buildings, passive fire protection remains the preferred method of containment.

Notwithstanding the difficulties in establishing correlations which would validate the factors used to determine mandatory compartmentation, there are positive consequences of the inclusion of compartmentation in the design concept. These may be considered as:

1. Limitation on the size of a potential fire, and
2. Retardation (prevention even) of fire spread to adjacent spaces.

It follows from the foregoing that life safety is enhanced, fire control is less difficult, and the amount of fire damage is limited. It has been suggested[6] that any implied link between fire resistance and hence fire severity and compartmentation requirements should be dispensed with and each should be specified on the particular needs for the building. The factors to be taken into account in applying the concept of compartmentation to a particular building design have been suggested as the:

1. Development of fully-protected zones to facilitate safe escape
2. Separation of fire risks and different occupancies from each other
3. Separation of communicating spaces
4. Protection of temporary refuges
5. Containment of fire to the floor of origin in a high rise building
6. Containment of fire originating in basements to the floor of origin
7. Provision of access zones for fire fighters
8. Possibility of trade-off of passive against active protection.

In essence the way forward requires a systems approach to determine the optimal utilisation of compartmentation to the attainment of stated objectives.

7.5 FIRE RESISTANCE REQUIREMENTS

Fire resistance is a term which is generally used to denote the extent to which a structural or constructional component will resist the impact of fire. For the purposes of BS 476: Part 8, fire resistance

Table 7.3

BS 476: Part 8: 1972	BS 476: Part 1: 1953
Stability	Collapse
Integrity	Passage of flame
Insulation	Insulation

means the ability of an element of building construction to withstand the effects of fire for a specified period of time without loss of its fire separating and loadbearing functions.

Fire resistance is determined on the basis of criteria specified in BS 476: Part 8: 1972[7] and BS 476: Part 1: 1953[8]. These criteria are given in Table 7.3.

Thus the criteria attempt to relate the ability of a structural component to:

1. Endure a fire without collapse
2. Prevent the penetration of flame due to loss of integrity
3. Resist the spread of fire by conduction through the component or by radiation from the face of the component not exposed to the fire.

The BS 476: Part 1 standard is still quoted in building regulations although it has been superseded by BS 476: Part 8. Thus building regulations apparently contain and encourage double standards.

Consider for a moment self-closing fire-resisting doors. In theory, fire-resisting doors are required to stop a fire from spreading through a building and the fire-resistance capability of doors like structural components is expressed in terms of time. In order to improve the performance of doors exposed to a fully-developed fire, the fire conditions must be standardised and the time–temperature relationship is expressed by:

† $T_f = T_0 + 345 \log_{10}(8t + 1)$

This time–temperature relationship was developed in the 1930s[9] and given the type, quantity and fire characteristics of modern furnishings there is some doubt as to the validity of this relationship. As well as the foregoing, the BS 476: Part 1 test operates with a negative pressure to allow the passage of smoke through the chimney outlets, but the BS 476: Part 8 test operates with a positive pressure differential across the door. Thus the two tests are significantly different. In effect, doors that had previously been subjected to the Part 1 test and obtained a satisfactory performance could not achieve the same level of performance when tested in accordance with the Part 8 procedure.

† See Chapter 6.

Thus it would appear that building regulations which are often quoted as containing minimum requirements in some cases obscure the real level of fire safety achieved. This kind of legislation leaves those occupying buildings whose components meet the requirements of BS 476: Part 1 with a false sense of security.

The factors which determine the level of fire resistance of structural components are:

1. Purpose grouping
2. Height of the building
3. Floor area of the storey or compartment
4. Cubic capacity of the building or compartment
5. Location of the component, i.e.
   (a) ground or upper storey
   (b) basement storey.

Since as shown earlier, the fire-resistance capability required of components is derived from a knowledge of fire severity likely to be experienced in use, it follows that the factors listed above somehow determine fire severity and consequently fire resistance. The question yet to be answered quantiatively is how?

Reference has been made to fire-resistance testing and the standardised time–temperature relationship.[10] It must also be recognised that only components are tested, i.e. the facilities currently available cannot test assemblies. Therefore while a testing facility exists for comparability purposes between components, the test procedure does not measure or indicate the likely performance of a component when exposed to real fire conditions. Neither does the test procedure on components measure the performance of complete buildings or structures since the connections between, e.g. column or beams, will not have been subjected to the test procedure.

Although many shortcomings exist with regard to determining the fire resistance of components (see Chapter 6), it is an important factor which contributes significantly to the attainment of fire safety in buildings. Fire resistance of components is an important performance criterion, e.g. of compartment walls and floor, and therefore contributes to the attainment of life safety and property protection.

7.6 UNPROTECTED AREAS IN EXTERNAL WALLS

Unprotected areas in external walls may be simply defined as those areas which have not the required degree of fire resistance.

Building regulations usually extend this definition to include combustible cladding of a specified thickness attached to the external

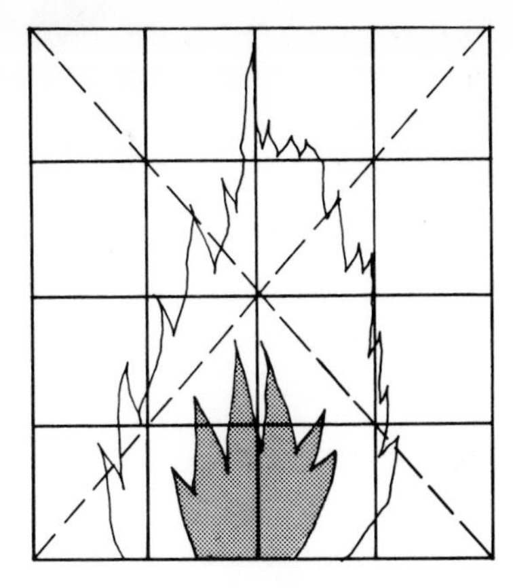

Fig. 7.1 Elevation – no compartmentation. Whole façade acts as a radiator

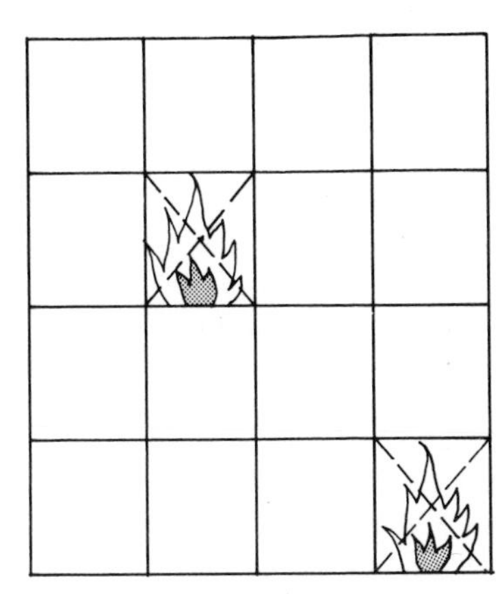

Fig. 7.2 Elevation – internal divisions are compartment walls and floors, therefore potential radiating surface area is considerably reduced

face of the walling. In order to minimise the possibility of fire spread between buildings, building regulations relate potential fire severity and unprotected areas to distance.

Figure 7.1 shows the potential effects of a fire in an uncompartmented building and Fig. 7.2 shows the potential effect of a fire in a compartmented building. Each unprotected area, in this case each window, acts as a radiator. If it is assumed that, e.g. the fire reaches a temperature of 1,000 °C and the windows act as black bodies, it is possible to calculate the heat flux at distance 'X' from the building on fire. Conversely it is possible to calculate the distance at which a given heat flux will occur.

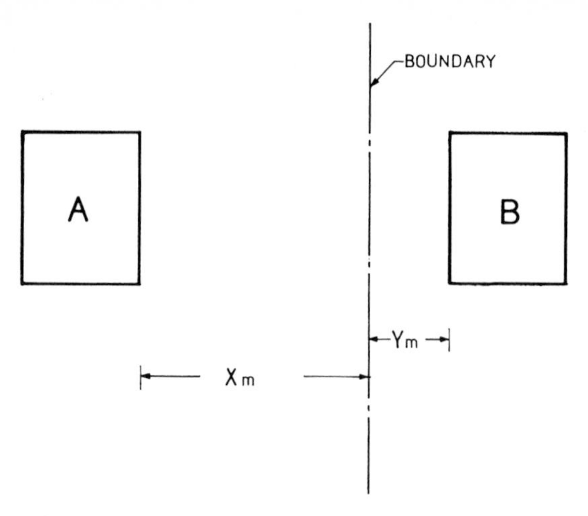

Fig. 7.3 Space separation between buildings

Taking 2.4 W cm^{-2} as the radiation incident on a combustible material for pilot ignition to occur (after a period of exposure), the distance at which the material can be placed from the heat source to avoid ignition can be calculated. Similarly with combustible claddings, etc., on the external façade of buildings. Building regulations generally introduce a factor of safety by siting buildings relative to boundaries. Thus as in Fig. 7.3, buildings A and B are sited so that the heat flux at the boundary is less than that required for pilot ignition.

It is of course assumed using this method that every building within given purpose groups will experience the same fully-developed fire and hence the same fire severity and that external climatic conditions do not vary. Thus building regulations appear to assume the worst case.

7.7 CONSTRUCTIONAL REQUIREMENTS FOR SEPARATING WALLS, COMPARTMENT WALLS AND FLOORS

The purpose of prescribing constructional requirements for separating walls, compartment walls, and floors is quite simply to ensure that the fire-resistance requirements specified are not negated by faulty construction. Since these components are utilised in the first instance to control the spread of fire within and between buildings, it is obvious that other factors must also be considered as well as their fire endurance capability, e.g.

1. Generally the component should be imperforate. There is no allowance generally in building regulations for these components to be other than imperforate, i.e. the regulations assume each component as constructed is 'perfect', in a fire endurance sense, thus the percentage of completeness tolerated is *nil*. In practice this assumption may not be correct.
2. Interspatial communications must be provided for. It is necessary to move people and goods about a building and occasionally between buildings. Where a separating wall, compartment wall or floor is so penetrated as a necessity, such penetration must be executed in a manner so that the overall fire endurance capability of the component is not diminished. This may be achieved in a variety of ways as shown in Fig. 7.4(a), (b), (c), (d), (e).
3. Component interactions must be taken into account. Having prescribed components and additional constructional requirements for these components it would be foolish not to consider the negative interactions between components which could reduce the level of fire safety initially achieved. Figure 7.4(a), (b), (c), (d), (e) gives examples of potential negative component interactions which could result in fire bypassing what was thought to be an effective barrier without appropriate protective measures being taken. Figure 7.5 shows a compartment or separating wall carried up through the roof of the building. This requirement is related to the fire-endurance capacity of the roof component in an effort to control fire spread from one space to another via the roof.

Given the obvious and continuing problem of weathering at the wall and roof junctions it is questionable whether this kind of flexibility should be permitted.

7.8 PROTECTED SHAFTS

Where the protection afforded to a compartment is penetrated to allow the movement of people, goods or anything else between compartments by means of, e.g. a staircase, lift or duct, the protection to the compartment must not be diminished. This is achieved by enclosing the means of transportation from one compartment to another in a shaft which is suitably protected so that the penetration of the construction protecting the compartment poses no direct or indirect threat to the linked spaces. Hence the terms 'protected shaft' and 'protected structure' for the structure enclosing such a shaft.

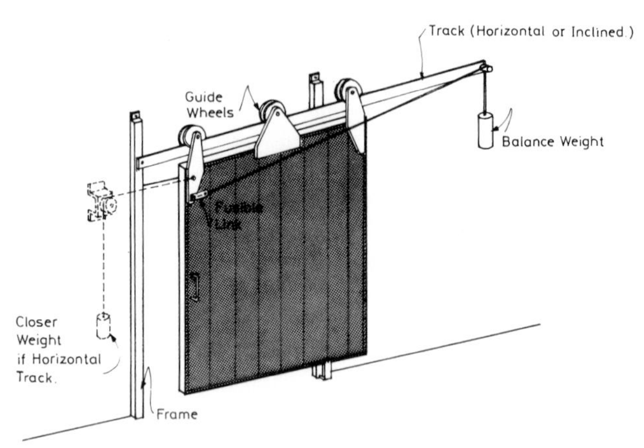
Track (Horizontal or Inclined.)
Guide Wheels
Balance Weight
Fusible Link
Closer Weight if Horizontal Track.
Frame
SLIDING SELF CLOSING (FUSIBLE LINK) FIRE RESISTANT DOOR

(a)

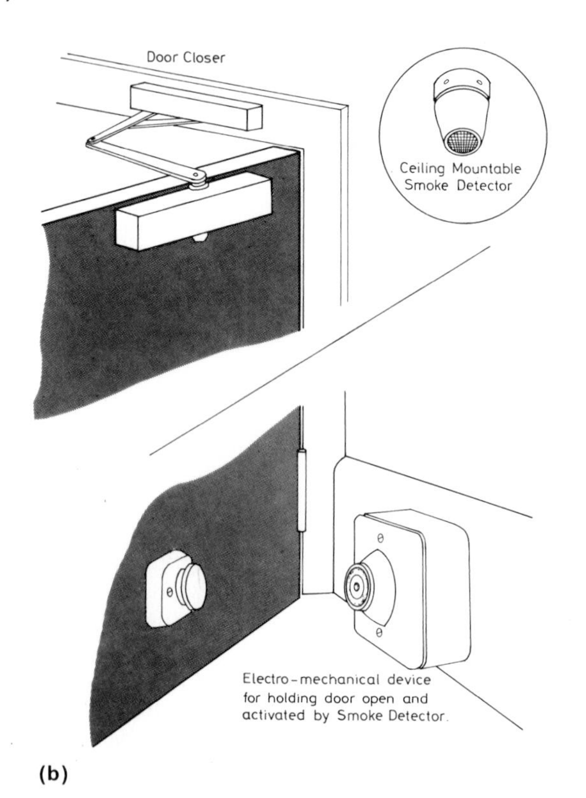
Door Closer
Ceiling Mountable Smoke Detector
Electro-mechanical device for holding door open and activated by Smoke Detector.

(b)

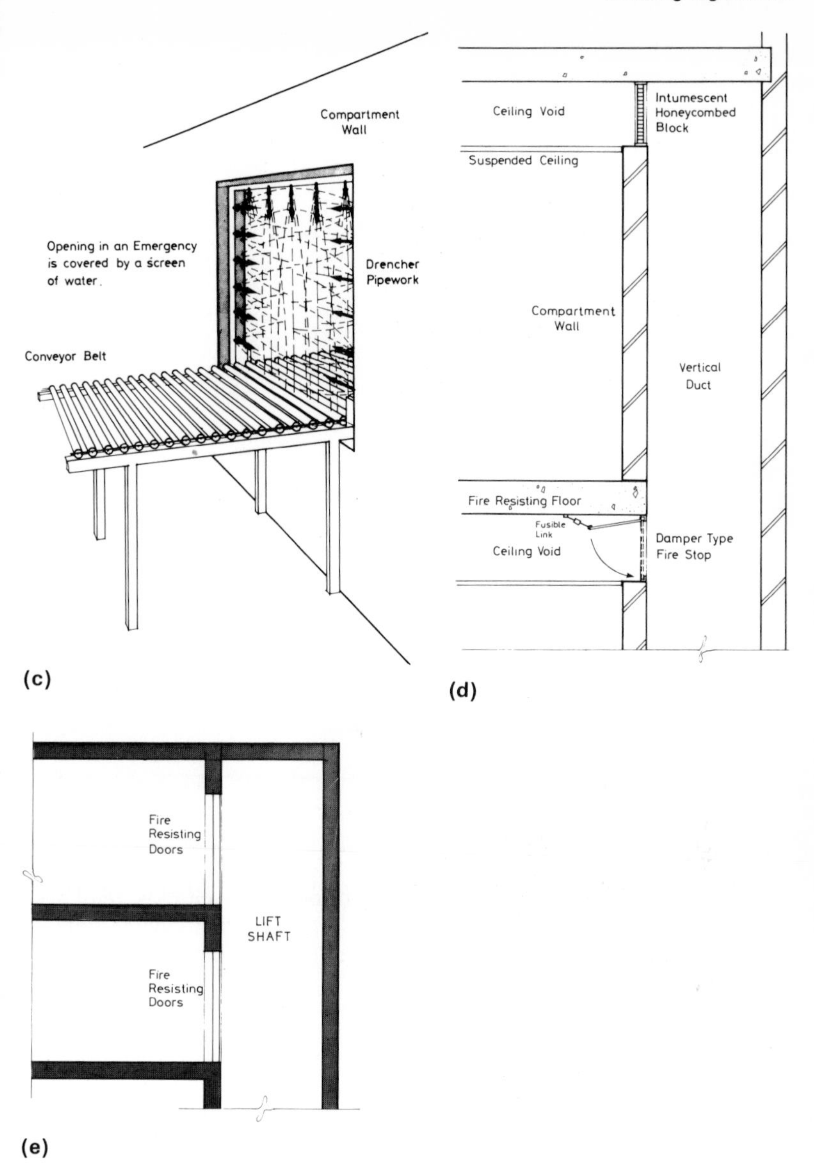

Fig. 7.4 (a)–(e) Example of component interaction and appropriate measures

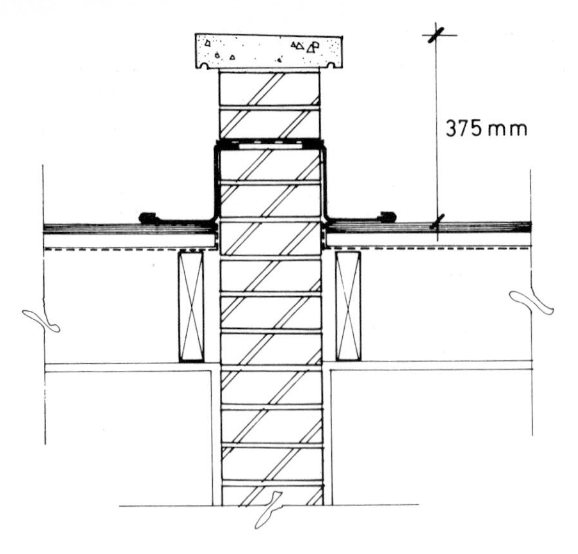

Fig. 7.5 Separating wall construction extending through roof

7.9 FIRE-RESISTING DOORS AND DOOR ASSEMBLIES

We have already discussed the anomaly in the method of testing for doors and other components in the section dealing with fire resistance and there can be no doubt that a uniform performance standard should be applied.

Fire-resisting doors are generally required to be self-closing. This means that they must be fitted with an automatic self-closing device. Rising-butt hinges may be acceptable as an automatic self-closing device when fitted to:

1. Exit doors from flats
2. Doors in houses over two storeys giving direct access from an internal stairway to a kitchen or habitable room
3. A door communicating between a dwelling and a garage.

It must be borne in mind that 'self-closing' permits the door to be held open and to close when a threat to the protected space is imminent. This is achieved by the closing mechanism being activated by any or one of a number of occurrences, e.g.:

1. Operation of a smoke detector dedicated to the purpose
2. Manual operation
3. Failure in the power supply
4. Operation of the alarm system.

When choosing the self-closing mechanism, consideration should be given to the force required to be exerted to open a door against the self-closing device, particularly in buildings such as old people's dwellings, schools, etc. It is worth mentioning here that a number of injuries have occurred to people as a result of having limbs trapped by doors fitted with self-closing mechanisms.

Fire-resisting doors must be considered as a complete assembly, i.e. the nature and quality of the door frame are important factors which influence the fire performance of the door assembly. Consequently the door must be tested as it would be fitted in practice. This latter point is extremely important. The door is provided for communication purposes and as such represents a potential weakness in the fire protection offered to a particular space. It is possible to purchase a door blank from builders' suppliers which may be described as a fire-resisting door. The purchaser does not know the framing in which the door has been tested, nor does he know if he has purchased a door tested to BS 476: Part 1: 1953 or a door tested to BS 476: Part 8: 1972.

Other factors which affect the fire performance of doors are:

1. Hinges.
   Hinges with a higher thermal diffusivity will, after a period of heating, conduct heat rapidly throughout the duration of a fire. Thus hinges which have a low thermal diffusivity are much better, and the hinge itself should not connect the face of the door exposed to a fire to the face of the door not so exposed.
2. Door furniture generally.
   In essence only door furniture of the type and quality fitted when the door assembly was tested should be fitted in practice. The foregoing applies to such items as letter plates. A door, for example, may have been submitted for testing and because it is an interior flush door, letter plates may not have been fitted. The same door in practice may then be fitted as exit doors from flats. Thus the fire performance of the door may be diminished by the inclusion of a letter plate.
3. Intumescents.
   If intumescent strips are fitted during the test procedure, then in practice intumescent strips should be fitted to the door edge or frame as shown in Fig. 7.6. The purpose of the intumescent is to prevent the passage of flame and hot gases past the door edges and possibly cause ignition on the side of the door not exposed to fire. When heated the intumenscent material expands providing a complete seal between the door and the frame.
4. Workmanship.
   Perhaps all that need be said here is that it is extremely doubtful if the same care and precision is taken when fitting the door furniture and fixing the frame as is executed in preparation for

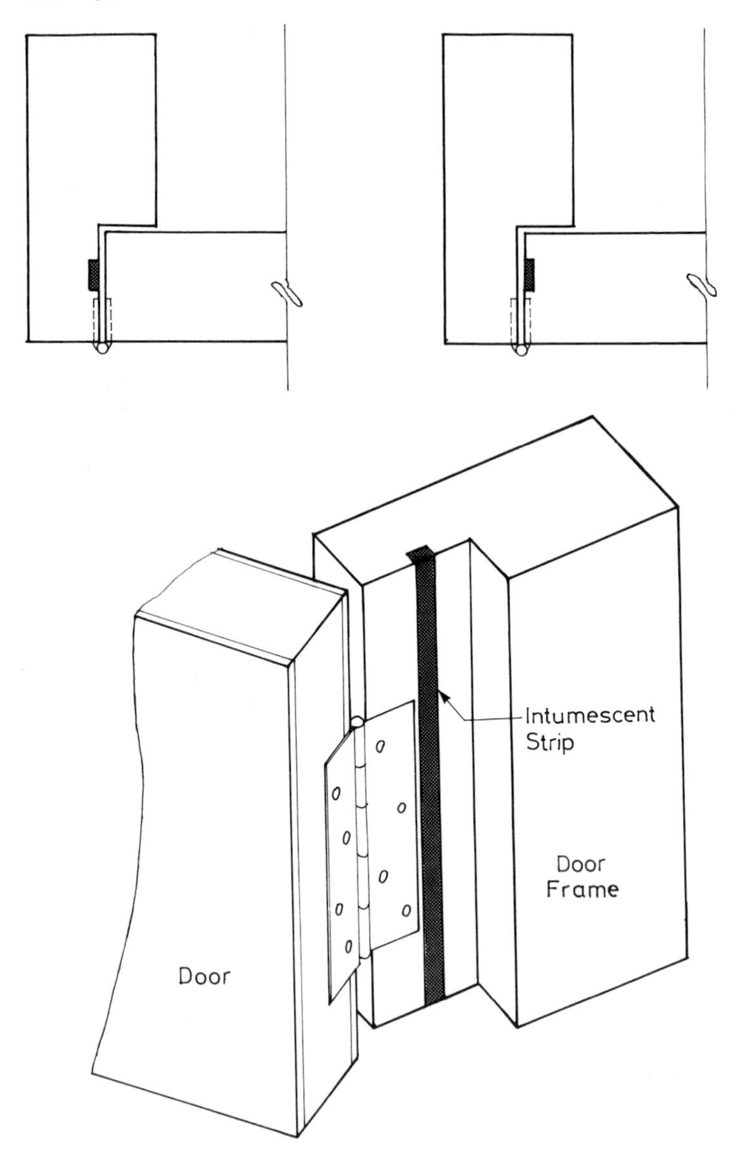

Fig. 7.6 Location of intumescent strips in fire door construction

testing. Gaps between the door and frames may be excessive, the wrong gauge of screw or nails may be used and mistakes of over-mortising may be 'rectified' by inserting a variety of infill materials. All of these factors and many others will inevitably reduce the fire performance of the door.

It is inevitable that doors in reality may not perform as well in a fire situation as the results of a carefully prepared specimen in a test situation might indicate.

7.10 PENETRATION OF FIRE BARRIERS

As stated earlier, barriers are provided to control the spread of fire and these may enclose, e.g. compartments. When the protecting barrier enclosing a compartment is breached it must be done in such a way that the level of fire safety originally provided is not reduced. Hence the earlier comments on protected shafts and protected structure. However, it would be unreasonable to expect a protected shaft, etc., to be provided to enclose a single pipe which penetrates the fire barrier. Building regulations attempt to reduce the risk of fire spread caused by pipes passing through fire barriers by controlling the:

1. Diameter of the pipes
2. Specification of the pipes
3. Installation of pipes.

Figure 7.7 gives examples of the method of control adopted. The permissible diameter of a pipe passing through a fire barrier is also related to the pipe specification.

7.11 NON-COMBUSTIBILITY REQUIREMENTS FOR STAIRWAYS IN PRESCRIBED SITUATIONS

Stairways are not generally required to be fire resisting, simply because they are not considered as elements of structure. As discussed earlier, the construction enclosing stairways may be considered as a protected structure and consequently be fire resisting but the stairway thus enclosed is not required to be fire resisting.

Stairways are, however, with a few exceptions, required to be non-combustible. It must be understood that the non-combustibility of a component is no indication of the fire-resisting capability of that component. For example a light metalwork stairway would be non-combustible, but in conventional terms would have little fire resistance. Conversely a component may be combustible and yet be fire resisting, e.g. timber beams or columns. Timber being a sacrificial material in a fire situation can be consumed and yet the component can be fire resisting in terms of stability, integrity and insulation for considerable periods of time.

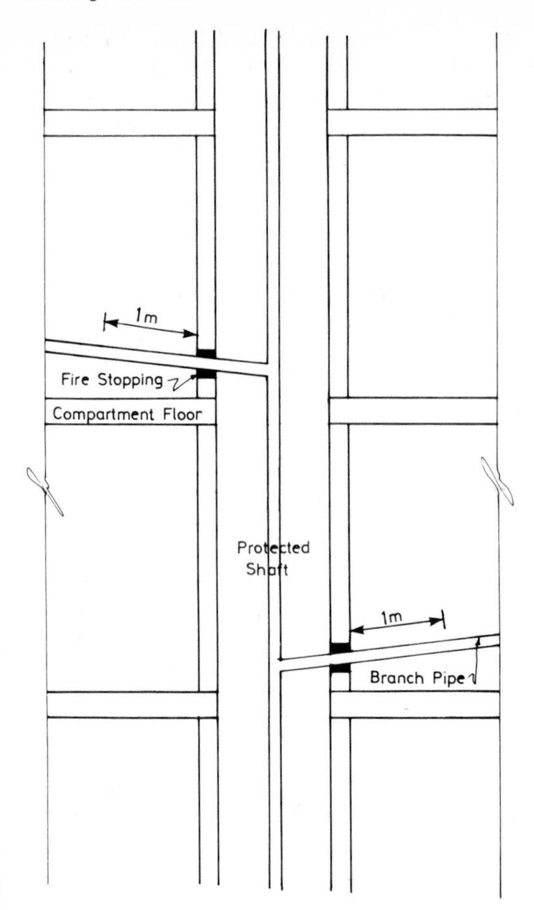

Fig. 7.7 Penetration of protected shaft related to pipe specification and diameter

It is perhaps surprising that stairways, which in themselves form essential components of escape routes, should not be required to be fire resisting. Stairways, however, must be considered within the totality of building regulations, which in effect require a high degree of protection to be afforded to stairways.

There is some doubt, however, as to what precisely the non-combustibility requirement can hope to achieve. If it is supposed to control the spread of fire up or down the stairway, it fails, because the addition of combustible material to the upper surface of the stairway and associated landings is usually permissible. If, however, it is supposed to guarantee sufficient inherent fire endurance in the stairway to maintain an effective escape route for a short time, then as described earlier, forms of component construction may be devised and used which render such a guarantee useless.

7.12 PROVISION OF CAVITY BARRIERS AND FIRE STOPPING

Cavities in this context include roof spaces, underfloor spaces, ceiling voids and cavities contained within structural and constructional components.

Obviously a fire starting or perhaps finding its way into such a void could travel undetected for quite some time, thus increasing the risk to the building's occupants. Building regulations attempt to mitigate this risk by requiring the introduction of cavity barriers when area or linear dimensions of cavities exceed specified limits. Cavity barriers may be flexible or rigid and may be required to be fire resisting. Figure 7.8(a), (b), (c), (d) gives examples of the application of cavity barriers.

Fire stopping is concerned with ensuring that the fire-resisting capability of a component is not diminished when penetrated, e.g. by a pipe or when two components abut. This imperfection of fit must be considered, as must the making good around pipework penetrating, for example, compartment walls. Figure 7.9 gives examples of fire stopping.

7.13 CONTROL OF FLAME SPREAD ON WALLS AND CEILINGS

Building regulations are of course concerned with the design and construction of buildings and attempt to control fire spread within buildings by prescribing requirements for those surfaces of constructional components most likely to be involved in a fire at a very early stage, e.g. surfaces of walls and ceilings. These exposed surfaces, if combustible, provide large continuous areas over which flame can spread rapidly. The factors which influence flame spread over combustible solids were listed in Chapter 5.

Two of these factors only will be considered here, namely thickness of the fuel and directions of propagation.

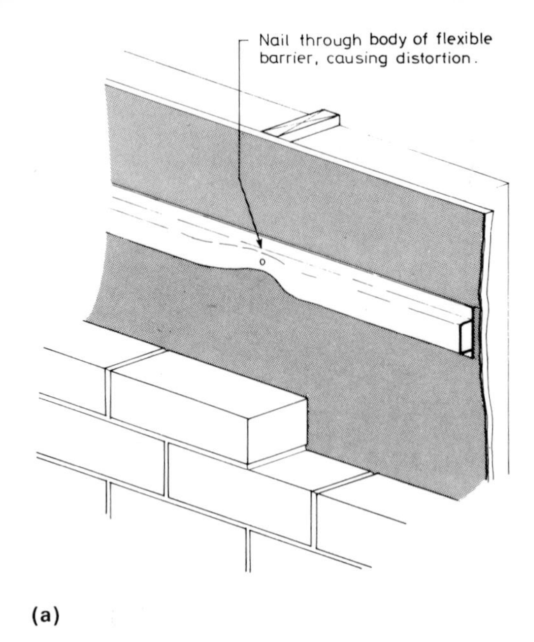

(a)

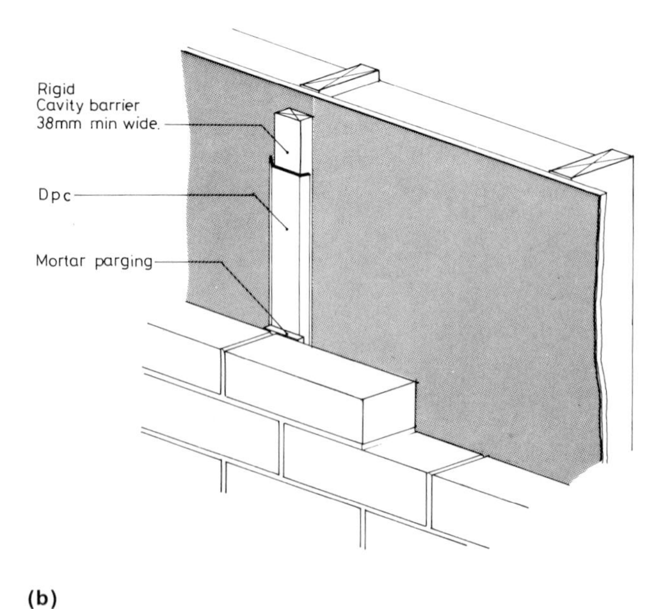

(b)

Fig. 7.8 (a), Bad practice – improper nailing; (b), good practice – gaps fire-stopped with mortar parging; (c) bad practice – incorrect lapping; (d), good practice – continuity at joints

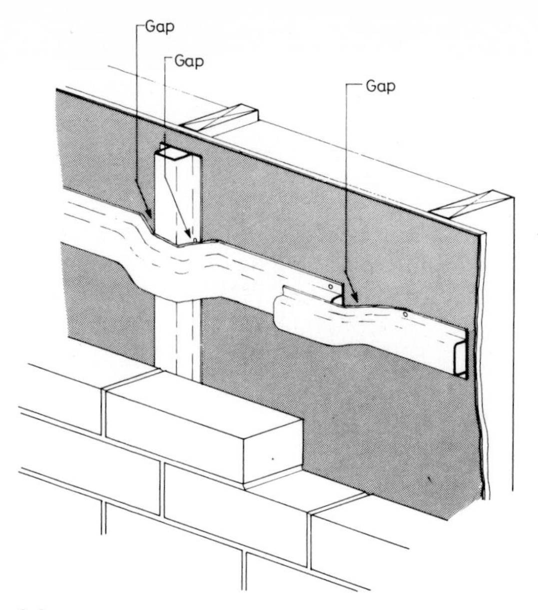

(c)

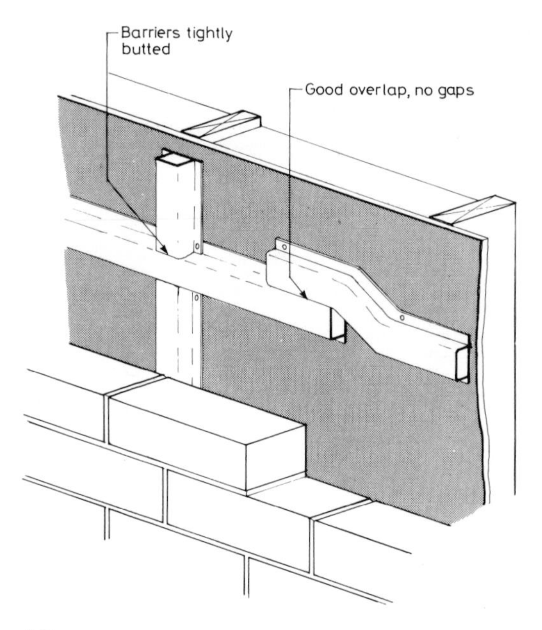

(d)

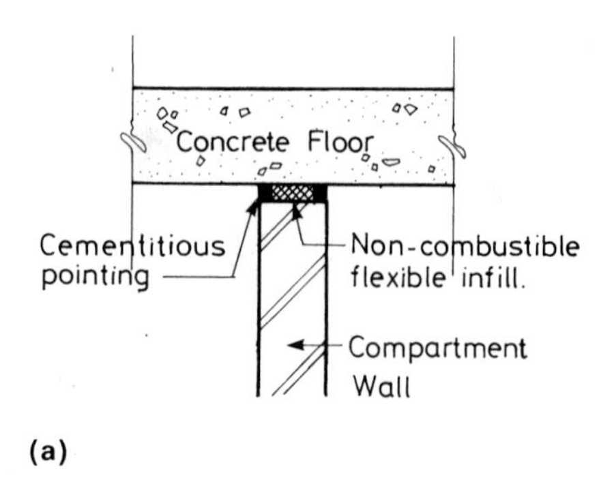

(a)

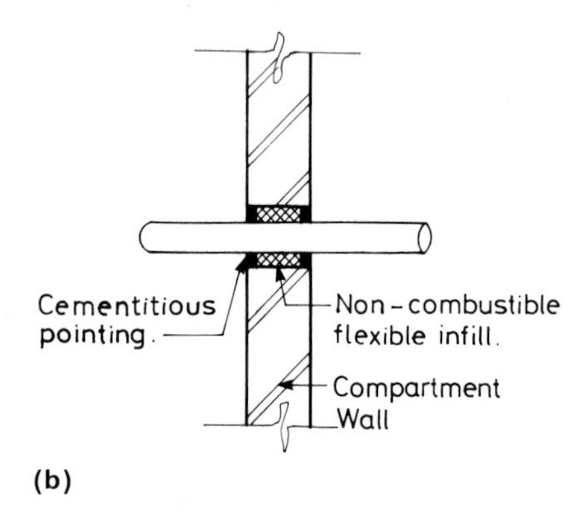

(b)

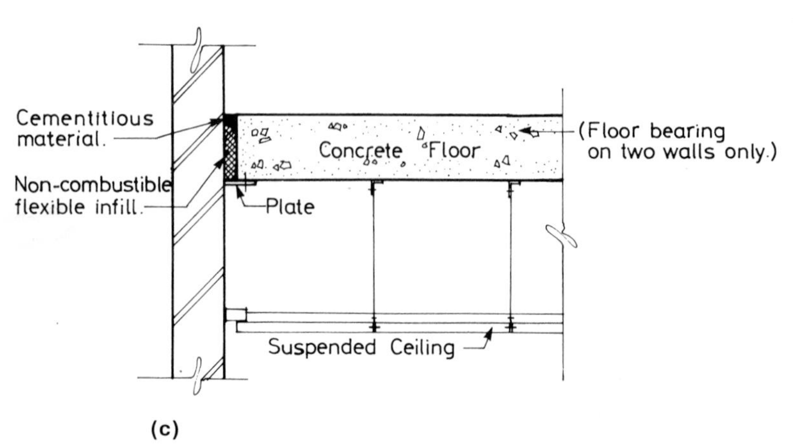

(c)

Fig. 7.9 (a)–(d) Examples of fire stopping

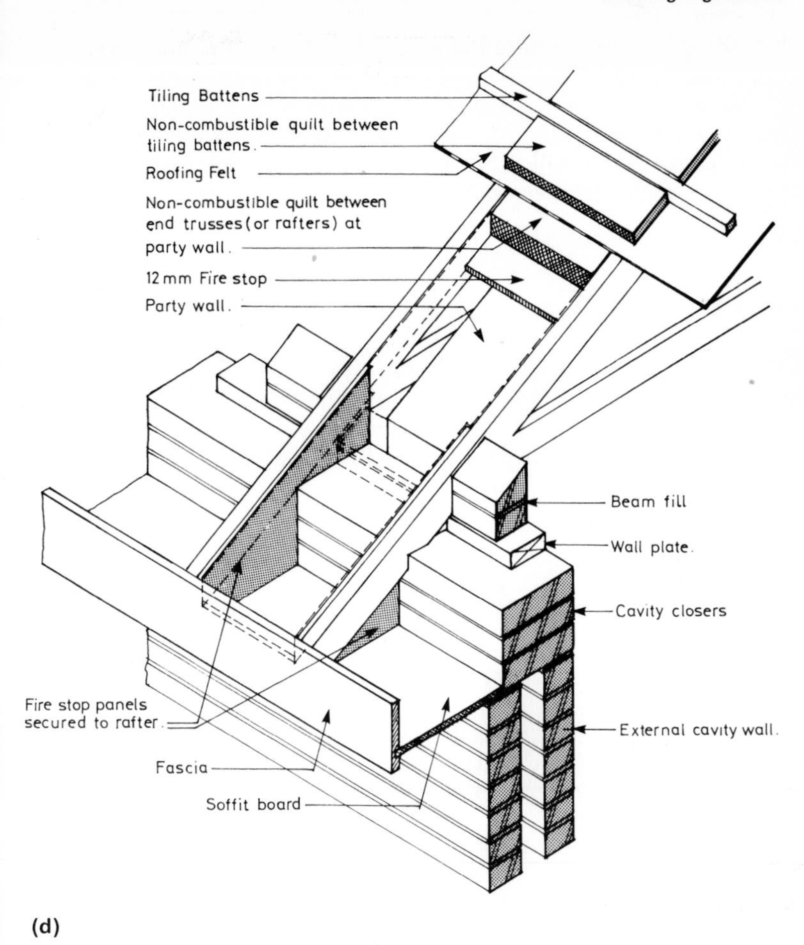

(d)

7.13.1 Thickness of fuel

Wall and ceiling linings may be considered as thermally 'thick' or thermally 'thin' materials or fuels. This means in effect that with thermally thin materials there is no significant temperature gradient through the material, i.e. between the front and rear surfaces. The material has a low thermal inertia ($\lambda \rho c$) and the surface exposed to fire will be quickly raised to fire point and flame spread over the surface of the material will be rapid.

With thermally thick materials there is a temperature gradient through the fuel and it is the depth of material which is required to be heated in order to raise the exposed surface to fire point which is important.

The effect of the foregoing on composites is that a thermally thin material may be backed by a material of high thermal inertia, i.e. a high heat capacity, which in effect acts as a heat sink and draws the heat across the facing material reducing the temperature at the surface.

7.13.2 Direction of propagation

The direction of propagation is related to the orientation of the material under consideration. For walls the direction will be vertical.

Figure 7.10 shows a material burning in a horizontal configuration. The principal component of heat transfer in this instance is conduction, hence the relatively slow rate of lateral flame spread.

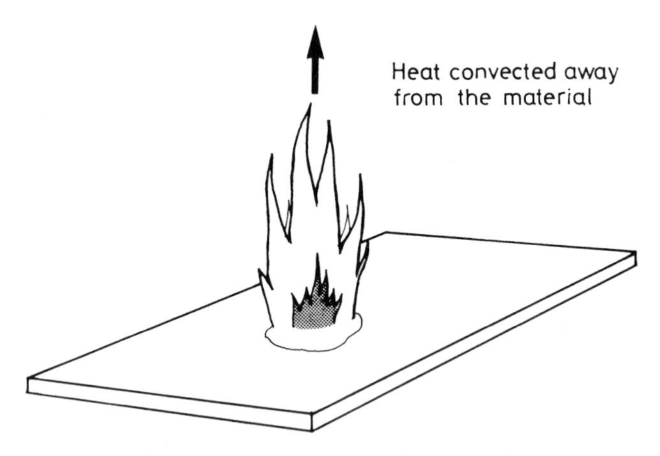

Fig. 7.10 Flame propagation on horizontal surface

Figure 7.11 shows the same material in a vertical orientation. Here the principal component of heat transfer is convection and the surface of the material is preheated, hence the rate of flame spread vertically will increase exponentially.

The ceiling component also can affect the time of flashover. This is because of the radiation from the burning surfaces or from the hot gases at the ceiling surface back to the fuel bed, which in turn increases the burning rate.

Building regulations attempt to control the flame spread over wall and ceiling surfaces by prescribing acceptable classifications for wall and ceiling surfaces. The acceptable classification for walls and

Fig. 7.11 Flame propagation on a vertical surface

ceilings may be limited as to their permissible extent and are determined by reference to:

1. Purpose group of the building or compartment
2. Location of the exposed surfaces
   (a) rooms
   (b) circulation spaces and protected shafts.

Materials are classified in accordance with BS 476: Part 7: 1971 and BS 476: Part 6: 1968. These tests were discussed in Chapter 6.

7.14 CONSTRAINTS ON THE USE OF PLASTICS ON CEILINGS

The dynamic nature of building design ensures the continuous introduction of new materials and methods of construction. Plastics materials are used for an increasingly wide variety of applications and in many cases act as substitutes for traditional materials such as timber.

The methods of evaluating the fire properties of traditional materials may not be suitable for evaluating the behaviour of plastics materials in a fire situation. Building regulations permit the use of plastics materials used as ceilings provided that prescribed conditions are met, e.g.:

1. Specified melting point is not exceeded
2. Burning does not exceed that specified
3. The degree of flammability specified is not exceeded.

Plastics materials are assessed for the purpose of building regulations in accordance with BS 2782: 1970.[11]

The potential spread of fire over the surface of plastics is controlled by:

1. Prescribing a sufficiently low melting-point for the plastics material forming the ceilings
2. Limiting the thickness of the material
3. Ensuring that any surface within the void exposed above the material meets the requirements prescribed for surface spread of flame characteristics for walls and ceilings (Fig. 7.12)
4. Limiting the permissible area of plastics panels (Fig. 7.13).

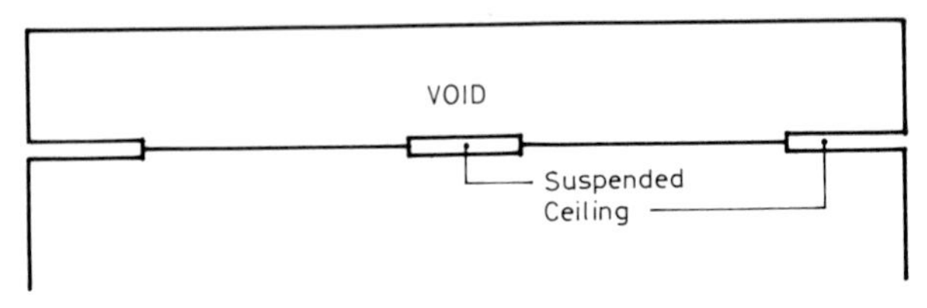

Section

Internal surfaces to comply with requirements re surface spread of flame characteristics of linings.

Fig. 7.12 Requirements for concealed ceiling voids

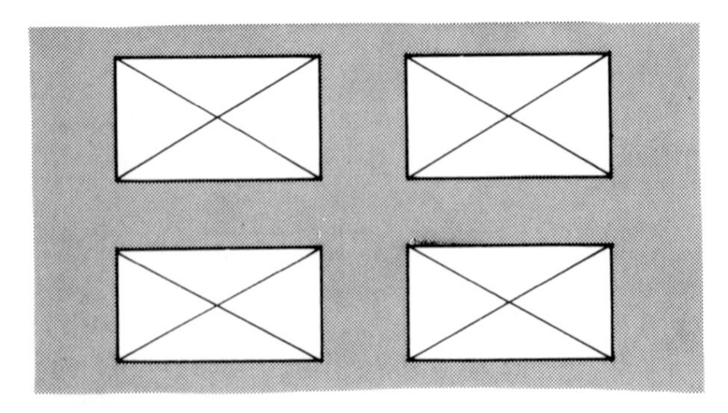

Plan

Minimum distances X prescribed between panels and the area of individual panels is limited.

Fig. 7.13 Limitations on permissible areas of plastic panels

7.15 THE SITING CONSTRAINTS ON BUILDINGS AND ROOF CONSTRUCTION

The general theory of fire spread between buildings was outlined in Chapter 3, particularly with regard to unprotected areas and the transmission of heat by radiation. The risk of fire spread by burning debris being carried on convection currents from the building on fire to another building must also be considered. Thus the closer buildings are sited to one another, the greater is the required fire-endurance capability of the roof construction.

Here again it can be noted that risk is somehow related to distance only. One may be tempted to think that fire statistics will indicate the probability of fire spread between buildings as outlined above. Unfortunately the UK Fire Statistics[12] do not provide such information and therefore it could be concluded that the risk of fire spread via radiation and burning debris on a roof is negligible. It is time that a fresh approach to this aspect of fire safety engineering was adopted.

7.16 CONCLUSIONS

Building Regulations are statutory instruments which set down minimum requirements for the design and construction of buildings. Minimum requirements are established to safeguard the health and safety of society and generally represent a compromise between optimum safety and economic feasibility. Building developers can of course establish their own requirements which may exceed the minimum requirements of building legislation. Very often it would appear that fire safety requirements are applied to a building design after the design process has been virtually completed with all the attendant problems this creates. Compliance with minimum code requirements should follow naturally as a consequence of good design and not be seen as an appendage thereto. Current fire safety legislation is by nature prescriptive, i.e. the type of materials which can be used are specified, areas and volume of spaces are laid down. The building regulations in their current form have been under review and the proposed changes for England and Wales indicate a move towards, eventually, a regulatory system based on performance specifications rather than prescription. Performance-type regulations should specify the objectives to be attained and criteria for determining if the established objectives have, in fact, been met. The building designer thus has freedom to develop a scheme using materials and components of his choice provided the performance criteria are satisfied. It is possible to incorporate prescriptive requirements into a performance based system for example as approved methods of construction, and hence a design could succeed because it has been engineered correctly or because it meets the requirements of an approved method of construction.

Another way of achieving an acceptable level of fire safety in the design of buildings is to consider fire safety as a sub-system in the design process, like the building services sub-system or the structural, mechanical and electrical sub-systems.

The systems approach to the attainment of fire safety provides the opportunity for the application of engineering methodology to achieve the required objectives and will be discussed in Chapter 8.

Suffice to point out here that potential positive and negative interactions between components of fire safety must be taken into account. Also similar interactions between components of building regulations, i.e. fire and thermal regulations must also be considered at the developmental stage of the legislation.

As well as satisfying statutory requirements the designer and building owner may have to satisfy the requirements of other external agencies, e.g. insurance requirements.[1,2]

Future developments

The current Building Regulations in England and Wales[13] are based on functional requirements. Regulation B2 for internal fire spread (surfaces) is as follows:

In order to inhibit the spread of fire within the building, surfaces of materials used on walls and ceilings:

(a) shall offer adequate resistance to the spread of flame over their surfaces; and
(b) shall have, if ignited, a rate of heat release which is reasonable in the circumstances.

As a statement of the necessary functional requirements the regulation serves its purpose. Unfortunately there is no definition of what is adequate and reasonable in the prevailing circumstances. Further reference to the previous chapter will show that some considerable work has yet to be completed before an acceptable rate of heat relase test is available.

Functional requirements in regulations are by themselves of limited use, serving only to add to confusion and creating unnecessary tensions between the design team and the enforcing bodies. What is required to complement the functional regulations, assist design, and enable the enforcing authorities to assess applications for approval under functional requirements is the development of a methodology which would evaluate the level of fire safety provided in individual proposals, against what is considered acceptable in the circumstances.

The thrust of much of the work in the future regarding regulations will be the development of quantitative methods of assessing the level of fire safety provision in buildings in order to demonstrate compliance with functional requirements.

REFERENCES

1. Read R E H and Morris W A, *Aspects of fire precautions in Buildings*', British Research Establishment, HMSO, London 1983.
2. *Fire and Buildings; A Guide for the design team.* Aqua Group Granada Publishing Limited, London 1984.
3. *The Building Regulations (England and Wales) 1976*, S.R.O. No. 1676 HMSO, London 1976.
   The Building Standards (Scotland) Regulations 1981, S.R.O. No. 1596 (S.169). HMSO, London 1981.
   The Building Regulations (N.I.) 1977, S.R.O. No. 149. HMSO, Belfast 1977.
4. Ministry of Works, *Post-War Building Studies, Special Report No. 20*, HMSO, London 1946.
5. Department of The Environment, Guidance Note, *Structural Fire Precautions*, HMSO, London 1975.
6. Malhotra H L, 'Fire compartment requirements', *Fire Surveyor*, pp. 37–41, 1983.
7. BS 476: Part 8: 1972. *Fire Tests on Building Materials and Structures, Part 8: Test methods and criteria for the fire resistance of elements of building construction*, British Standards Institution, London 1972.
8. BS 476: Part 1 1953, *Fire Tests on Building Material and Structures*, British Standards Institution, London 1953.
9. Hamilton S B, '*A Short History of the Structural Fire Protection of Buildings*'. National Building Studies Special Report No. 27, HMSO, London 1958.
10. Hinkley P, 'Standard Time/Temperature Curve – Its behaviour', *Fire Surveyor*, June 1984.
11. BS 2782: 1970. *Methods of Testing Plastics*, British Standards Institution, London 1970.
12. *Fire Statistics – United Kingdom*, Home Office, London.
13. The Building Regulations 1985. *FIRE B2/3/4 Fire Spread.* D.O.E., HMSO, London 1985.

7.17 APPENDIX 7.1 FIRE-RELATED LEGISLATION

Legislation for England and Wales

Building Regulations 1976
The Health and Safety at Work (etc.) Act 1974
The Fire Precautions Act 1971
The Public Health Act 1937 amended by the 1961 Act
The Offices, Shops and Railway Premises Act 1963
The Factories Act 1961 (amended by Statutory Instruments (1976))
The Housing Acts 1961 and 1969
Education Act 1944
Children and Young Persons Act 1969
Children's Act 1948
Nurseries and Child Minders Regulations Act 1948
Health Service and Public Health Act 1968
The National Assistance Act 1948
The Mental Health Act 1959
Nursing Homes Act 1975
The Cinematographic Acts 1909 and 1952 and Regulations Made Thereunder
The Private Places of Entertainment (Licensing) Act 1967
The Theatres Act 1968
The Sunday Theatres Act 1972
The Chronically Sick and Disabled Persons Act 1970
The Licensing Act 1964
The Gaming Act 1968
The Petroleum (Consolidation) Act 1928
The Explosives Acts 1975 and 1923

Legislation for Inner London

London Building Acts (Amendment) Act 1935
London Building Acts (Amendment) Act 1939
London Building (Constructional) Byelaws 1972
London Building (Constructional) Amending Byelaws 1979
The London Government Act 1963
The Greater London Council (General Powers) Act 1966
GLC (General Powers) Act 1968
GLC (General Powers) Act 1975
GLC (General Powers) Act 1976
GLC (General Powers) Act 1978
London Gas Undertaking (Regulations) Acts 1939 and 1954

Legislation for Scotland

Building Standards (Scotland Regulations) 1971–1974
Fire Precautions Act 1971
The Health and Safety at Work (etc.) Act 1974
Gas Safety Regulations 1972
Fire Precautions (Special Premises) Regulations 1976
The Civic Government (Scotland) Bill (in preparation)
The Housing (Scotland) Act 1966
Education Scotland Acts 1962 and 1969
Social Work (Scotland) Act 1968
Nursing Homes Registration (Scotland) Act 1938 and Amendments
The Cinematograph (Safety) (Scotland) Regulations 1955
The Theatres Act 1968 and Borough Police (Scotland) Act 1892 and Amendment
Safety of Sports Ground Act 1975 and Safety of Local Authority Sports Grounds (Scotland) Regulations 1976
Gaming Act 1968
Licensing (Scotland) Act 1976
Petroleum (Consolidation) Act 1928

Legislation for Northern Ireland

The Housing (NI) Order 1981 S109
Education Act 1944
Office and Shop Premises Act (NI) 1966 Sections 28–40
The Local Government Act 1934 – (Provisions for Public Entertainment Licensing)
Fire Services Act (NI) 1969 Sections 10–14A and SI 1975 No. 601
The Factories Act (NI) 1965 and SRO NI 1967 Nos 47 and 48
The Cinematograph Act 1909–59 and Regulations
The Petroleum Act 1929
The Highly Inflammable Liquids and LPG Regulations (NI) 1974. SRO256
The Health and Safety at Work Order (NI) 1978
Fire Services Order (NI) 1985

CHAPTER 8

Calculation of fire resistance

8.1 INTRODUCTION

In order successfully to design fire safety into buildings, the fire system as an unwanted sequence of events should be considered as a set of interrelated discrete components. Figure 8.1 illustrates the components of a fire safety system which, if considered in the design and construction phases, should result in the attainment of fire safety objectives. Essentially the objectives are achieved by preventing ignition and by controlling the effect of fire. The shaded portion of Fig. 8.1 indicates the components of the traditional approach to fire safety embodied in building legislation.

Figure 8.2 illustrates more fully the traditional conceptual view of the fire problem.

8.2 TEMPERATURE CRITERIA ANALYSIS

Figure 8.2 shows three methods incorporated in the traditional approach to the attainment of fire endurance of structural and constructural components. It is intended to concentrate on the shaded portion of this diagram, dealing with the theoretical estimation of fire resistance based on the standard temperature–time curve.

In using any mathematical technique to solve a physical problem it is always necessary to model the physical interactions and energy exchanges occurring within the system. In the furnace test BS 476: Part 8, Fig. 8.3, it is assumed that all the heat reaches the component in the form of radiation. The net amount of this radiation actually absorbed by the component depends upon the configuration factor Φ for this radiation process, which in turn is

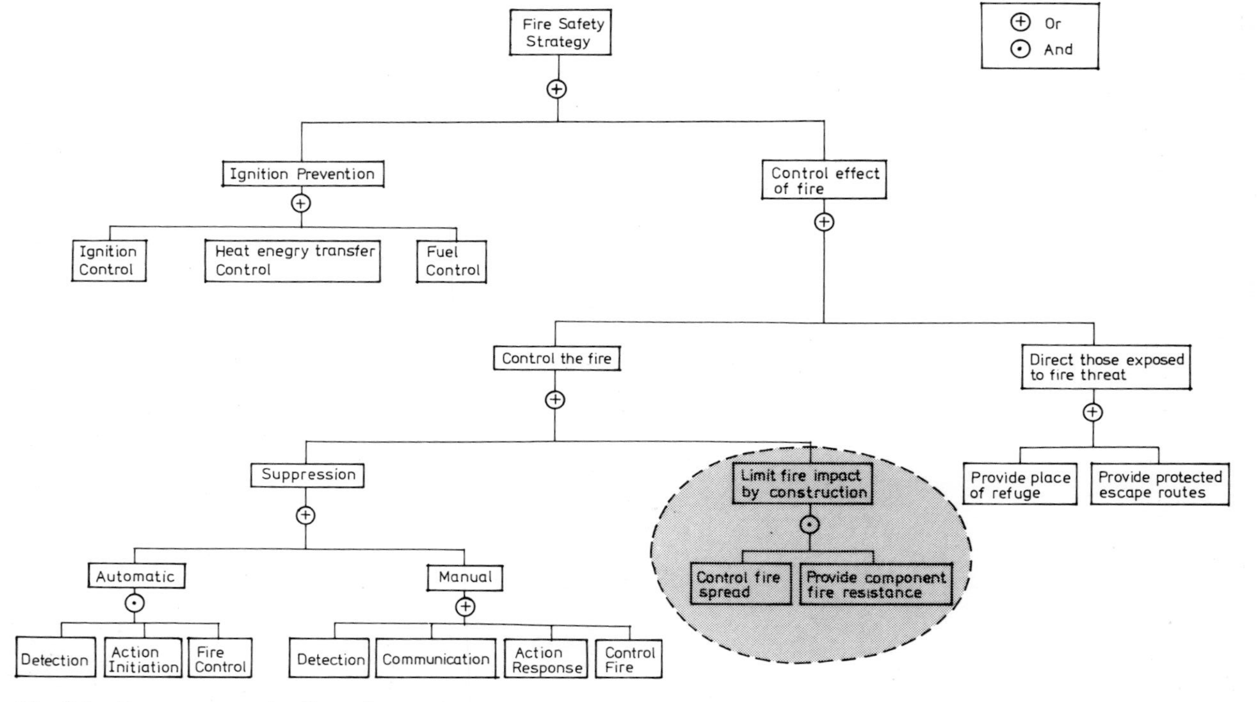

Fig. 8.1 Components of a fire safety system

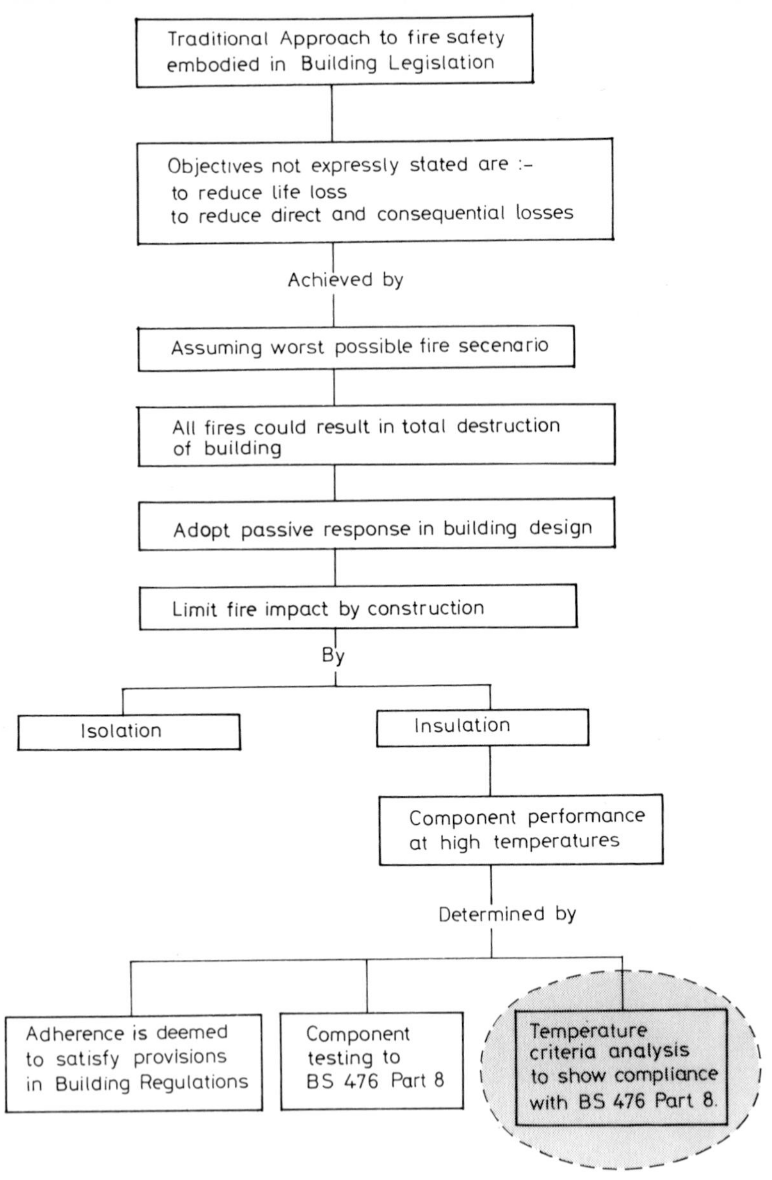

Fig. 8.2 Conceptual approach to traditional fire safety

controlled by the emissivity and geometry of the system. A geometry has been used enabling a simple model of the system to be developed. The net radiation absorbed by the component goes to increasing its thermal storage capacity thus causing a temperature rise.

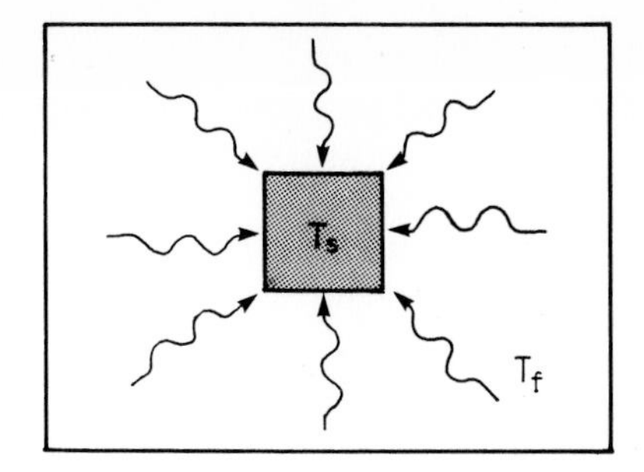

Fig. 8.3 Radiation heat exchange within furnace

In this simple case the component is considered to behave as a lump of material having the same temperature throughout its volume. In formal terms:

Rate of radiation absorption = Rate of change of thermal energy stored by the body

$$\Phi A_s(T_f^4 - T_s^4) = M_s C_s \cdot \frac{dT_s}{dt} \qquad \ldots [1]$$

where:

A_s = surface area of the component exposed to radiation (m^2)
Φ = configuration factor of the system
M_s = Mass of the component (kg)
C_s = specific heat capacity of the component's material ($J\,kg^{-1}\,°C^{-1}$)
T_f = absolute temperature of the gas in the furnace (K)
T_s = absolute temperature of component (K)

Equation [1] can be solved analytically but it is rather time consuming. Consequently it is convenient to use a time-step iteration process the outline of which is given below:

Since $\dfrac{\Delta\theta}{\Delta t} = \dfrac{d\theta}{dt}$

limit as $\Delta t \rightarrow 0$

If small enough time intervals Δt are used it is possible to write equation

$$\frac{\Delta T_s}{\Delta t} = \frac{\Phi A_s \sigma (T_f^4 - T_s^4)}{M_s C_s}$$

$$\therefore \quad \Delta T_s = \frac{\Phi A_s \sigma (T_f^4 - T_s^4)}{M_s C_s} \cdot \Delta t \qquad \ldots [2]$$

Given:

$$T_f = T_0 + 345\log_{10}(0.133t + 1) \qquad \ldots [3]$$

and the following assumptions are made:

C_s = a constant or an average value over the range of temperature of the component
Φ = a simple algebraic format (dimensionless)

Consider the following worked example.

WORKED EXAMPLE 1

Estimate the fire resistance of an unprotected hollow steel column as shown in Fig. 8.4 given that the critical temperature for steel is 550 °C. Assuming the following:

ε_s (emissivity of steel) = 0.9
ε_f (emissivity of furnace surface) = 0.7
Volumetric heat capacity for steel = 4.1×10^6 J m^{-3} K^{-1}
Ambient temperature = 27 °C
Stefans constant = 5.67×10^{-8} W m^{-2} K^{-4}

* Let $\rho_s C_s$ = volumetric heat capacity for steel

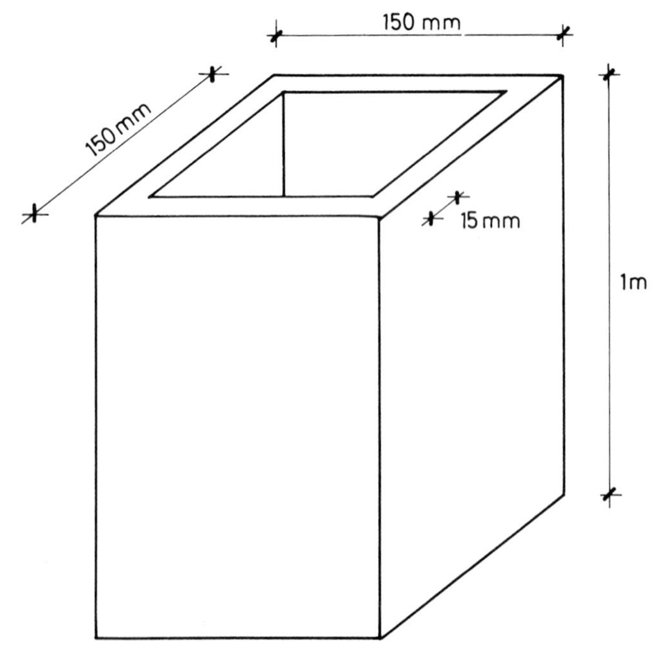

Fig. 8.4 Unprotected hollow steel column

Solution

$$\Delta T_s = \frac{\Phi A_s \sigma}{\rho_s V_s C_s}(T_f^4 - T_s^4) \cdot \Delta t$$

or

$$\Delta T_s = \frac{\sigma A_s}{\rho_s C_s V_s}(\varepsilon_f T_f^4 - \varepsilon_s T_s^4) \cdot \Delta t$$

It is necessary to determine A_s and V_s, where V_s = volume of steel.
NB. In all examples used from No. 1 on, a one metre length of element will be used for the sake of calculations.

$$A_s = 4 \times 1.5 \times 10^{-1}\,\text{m}^2$$

$$V_s = \{(1.5)^2 \times 10^{-2} - (1.2)^2 \times 10^{-2}\}\text{m}^3 = 8.1 \times 10^{-3}\,\text{m}^3$$

A modified form for Φ gives

$$\Delta T_s = \frac{A_s}{\rho_s C_s V_s}\sigma \times (\varepsilon_f T_f^4 - \varepsilon_s T_s^4) \cdot \Delta t \qquad \ldots [4]$$

$$\text{or} \quad \Delta T_s = \frac{0.6 \times 5.67 \times 10^{-8}}{4.1 \times 10^6 \times 8.1 \times 10^{-3}}(0.7T_f^4 - 0.9T_s^4) \cdot \Delta t$$

$$\Delta T_s = 0.102 \times 10^{-11}(0.7T_f^4 - 0.9T_s^4) \cdot \Delta t$$

$$\Delta T_s = 10^{-12}(0.7T_f^4 - 0.9T_s^4) \cdot \Delta t$$

In order to give the reader a better understanding of the iteration process a simple hand-calculated estimation is as follows:

Assume $\Delta t = 300$ seconds

Using $\Delta T_f = 300 + 345 \times \log_{10}(0.133t + 1)$

$\therefore$ After 400 seconds, $T_f = 856$ K

Substituting into eqn [4]:

$$\Delta T_s = 10^{-12}[0.7 \times (856)^4 - 0.9(300)^4] \times 300$$

$$= 113\ \text{K}$$

$$\therefore \quad T_s(300) = 300 + 113 = 413\ \text{K}$$

After 600 seconds,

$$T_f = 958\ \text{K}$$

$$\Delta T_s = 173\ \text{K}$$

$$T_s(600) = 513 + 173 = 586\ \text{K}$$

likewise for the remainder of time steps.

Results are given in tabular form in Table 8.1.

Assuming a critical temperature for steel = 550 + 273 = 823 K, it is shown that failure occurs in the material within 900–1200 seconds, i.e. between 15 and 20 minutes.

Table 8.1

Time	T_f (K)	ΔT_s	T_s (K)
0	300	113	300
300	856	173	413
600	958	198	586
900	1,018	167	704
1,200	1,060		951

8.3 COMPUTER AIDED ANALYSIS OF EXAMPLE 1

Using a computer this procedure can be extended to consider iterations for small time steps which will enable a more realistic estimate to be obtained.

A listing of the program used and the resulting output is shown in Figs. 8.5 and 8.6, respectively.

```
READY.

 1 REM UNPROTECTED HOLLOW  STEEL
 3 PRINTCHR$(147)
 5 INPUT"TIME STEP";Q
 10 INPUT"AMBIENT TEMPERATURE";TS:S1=TS
 12 PRINT
 13 IF D=5THENOPEN3,4:CMD3:TS=S1
 14 PRINT:PRINT"UNPROTECTED HOLLOW STEEL":PRINT:PRINT:PRINT"TIME STEP="Q"SEC."
 17 PRINT"AMBIENT TEMPERATURE="TS"K":PRINT
 18 PRINT" AMB.TEMP.,K","FURN. TEMP.,K","TIME,SEC."
 20 FOR T=0TO3000STEPQ
 30 TF=345*LOG(.133*(T+Q)+1)/LOG(10)+300
 40 DTS=10↑-12*(.7*TF↑4-.9*TS↑4)*Q
 50 PRINTTS,TF,T
 60 TS=TS+DTS
 65 REM IFTS>823THEN75
 70 NEXTT
 75 IFD=5THENPRINT#3:CLOSE3,4
 100 PRINT:INPUT"DO YOU WANT A HARD COPY";A$
 110 IF A$="N"THEN200
 120 IFA$="Y"THEND=5:GOTO5
 200 END
```

Fig. 8.5 Program listing for unprotected hollow steel column

WORKED EXAMPLE 2

Estimate the theoretical fire resistance of an unprotected solid steel column as given in Example 1.

Solution

Using $$\Delta T_s = \frac{A_s \sigma(\varepsilon_f T_f^4 - \varepsilon_s T_s^4)}{\rho_s C_s V_s} \cdot \Delta t$$

UNPROTECTED HOLLOW STEEL

TIME STEP= 300 SEC.
AMBIENT TEMPERATURE= 300 K

AMB.TEMP.,K	FURN. TEMP.,K	TIME,SEC.
300	856.044542	0
410.585899	958.05692	300
579.835741	1018.18901	600
775.016503	1060.98221	900
943.709787	1094.22946	1200
1030.61992	1121.42239	1500
1058.12117	1144.42997	1800
1079.88964	1164.37036	2100
1098.70468	1181.96598	2400
1115.11985	1197.7107	2700
1129.76866	1211.95712	3000

READY.

UNPROTECTED HOLLOW STEEL

TIME STEP= 100 SEC.
AMBIENT TEMPERATURE= 300 K

AMB.TEMP.,K	FURN. TEMP.,K	TIME,SEC.
300	698.590933	0
315.943081	797.113634	100
343.306763	856.044542	200
379.647551	898.229754	300
423.344603	931.109802	400
473.067837	958.05692	500
527.534699	980.88842	600
585.364695	1000.69648	700
644.992988	1018.18901	800
704.650322	1033.85116	900
762.431908	1048.02995	1000
816.468453	1060.98221	1100
865.17536	1072.90336	1200
907.504498	1083.94552	1300
943.094779	1094.22946	1400
972.250721	1103.85266	1500
995.762658	1112.89491	1600
1014.65627	1121.42239	1700
1029.96984	1129.49056	1800
1042.61339	1137.14639	1900
1053.3117	1144.42997	2000
1062.60511	1151.37584	2100
1070.87846	1158.01393	2200
1078.39789	1164.37036	2300
1085.34447	1170.46806	2400
1091.84085	1176.32729	2500
1097.97079	1181.96598	2600
1103.79257	1187.40014	2700
1109.34793	1192.64409	2800
1114.66786	1197.7107	2900
1119.77629	1202.61156	3000

Fig. 8.6 Data output for unprotected hollow steel column

here V_s = volume of solid steel

$$= (1.5)^2 \times 10^{-2}\,\text{m}^3 = 2.25 \times 10^{-2}\,\text{m}^3 \quad \text{(assuming unit length)}$$

$$A_s = 0.6\,\text{m}^2$$

$$\therefore \quad \Delta T_s = \frac{0.6 \times 5.67 \times 10^{-8}}{4.1 \times 10^6 \times 2.25 \times 10^{-2}}(0.7T_f^4 - 0.9T_s^4) \cdot \Delta t$$

$$\Delta T_s = 0.369 \times 10^{-12}(0.7T_f^4 - 0.9T_s^4) \cdot \Delta t$$

Using the computer to run the iteration process (see Appendix 8.1) it can be clearly seen that this column has taken some 2200 seconds or 36 minutes to reach the critical value of 550 °C or 873 K.

8.4 METHODS USED TO INCREASE FIRE RESISTANCE OF CONSTRUCTIONAL COMPONENTS

There are several methods of increasing the fire-resisting characteristics of such a column, which include the following:

1. Filling the void in the hollow column with materials having a high thermal capacity (i.e. providing an adequate heat sink)
   (a) water-filled column
      (i) non-replenishment system

      Non-replenishment means in effect that water in the column is non-flowing and acts as a stationary heat sink whereas a replenishment system, as illustrated in Fig. 8.7, is one where the water is flowing and acts as a heat exchanger.
      For a non-replenishment system the following factors must be considered:
      - water boils at 100 °C
      - pressure release valves must be fitted
      - the conversion of water to steam results in a reduction in water

the gradual loss of water from the column as steam may result in the formation of hot spots on the column resulting in localised failure

      - given the foregoing, the water-filled column will nevertheless perform better than the unfilled column in a fire situation. This is illustrated in Fig. 8.8.

      (ii) replenishment system
      Figure 6.7 illustrates such a system.
      In this case the steel temperature should never exceed 100 °C if the water is pumped through the system at an adequate rate.
   (b) Concrete
2. The external application of suitable thermal insulants
   Figure 8.9 illustrates the same hollow column with a 25 mm

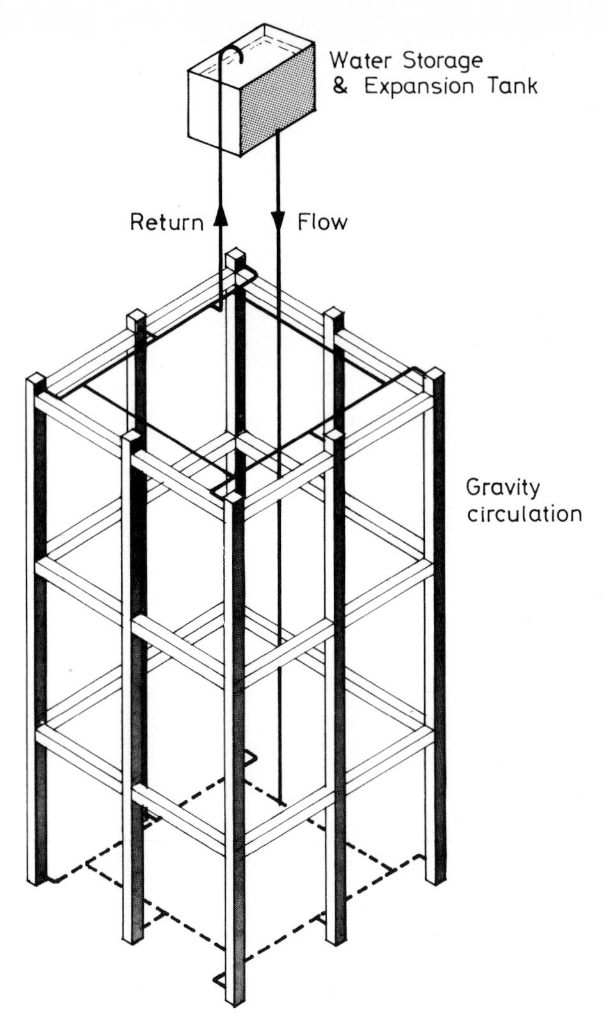

Fig. 8.7 Replenishment system for water-cooled column

insulating board fixed to its outer surface. In this case the outside temperature of the insulation is assumed to be the gas temperature of the furnace. The heat flux then conducts through the insulation to the steel behind it. This heat flux then enters the steel to increase its heat capacity and to raise its temperature. This is illustrated in Fig. 8.10.

Again it is assumed that the steel behaves as an isothermal lump.

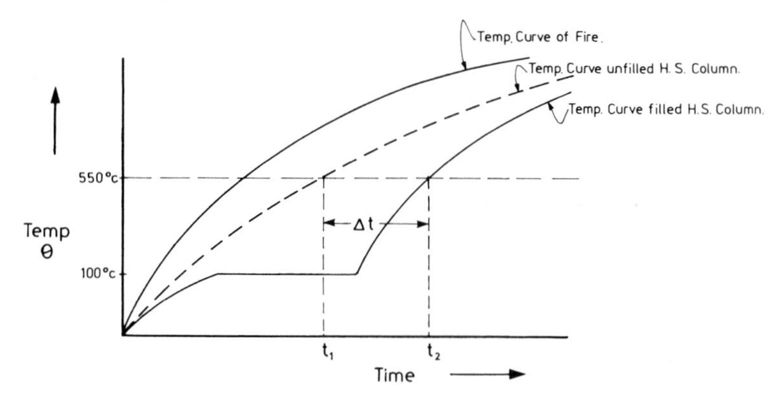

Fig. 8.8 Comparison of temperature gradients in water-filled and un-filled hollow columns

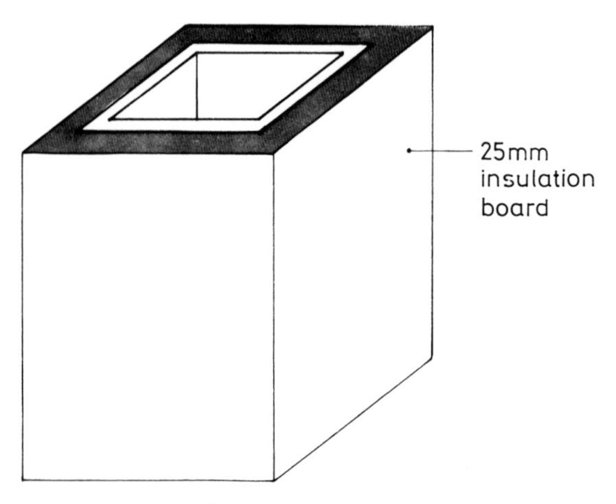

Fig. 8.9 Externally insulated hollow steel column

Thus:

Rate of heat conducting through insulation

= *Rate of increase of thermal storage of steel*

i.e. $$\frac{\lambda_i A_I (T_f - T_s)}{d_i} = M_s C_s \cdot \frac{dT_s}{dt}$$

Again in these calculations the specific heat capacity of steel and the thermal conductivity of insulation are taken to reflect the average value over the temperature ranges experienced by the materials during the test:

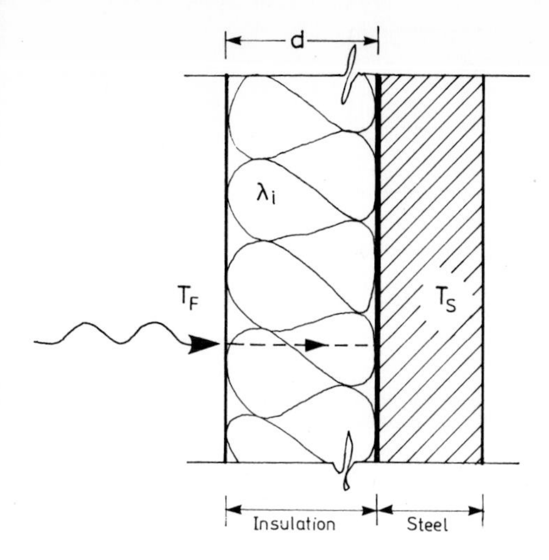

Fig. 8.10 Heat flux to insulated steel column

$$\frac{dT_s}{dt} = \frac{\lambda A_i(T_f - T_s)}{d_i M_s C_s}$$

Let $M_s = \rho_s V_s$

$$\therefore \quad \Delta T_s = \frac{\lambda_i A_i}{d_i \rho_s V_s C_s}(T_f - T_s)\,\Delta t$$

Using a time-step iteration process as before, the temperature rise of the protected steel can be estimated with reasonable precision. An example of the estimation of the fire resistance of a hollow steel column insulated with 25 mm layer of fire insulation is dealt with in Appendix 8.2 which gives a fire resistance in excess of two hours for the example chosen.

N.B. This estimation is based on temperature criteria analysis only, structural analysis is not considered here.

8.5 THE ALTERNATIVE APPROACH

The alternative to the traditional attitude of fire safety is to adopt a fire safety engineering approach, the basic components of which are illustrated in Fig. 8.11.

The basis of this approach is that if enough details are provided about the total fire system, namely:

1. Enclosure shape
2. Enclosure ventilation

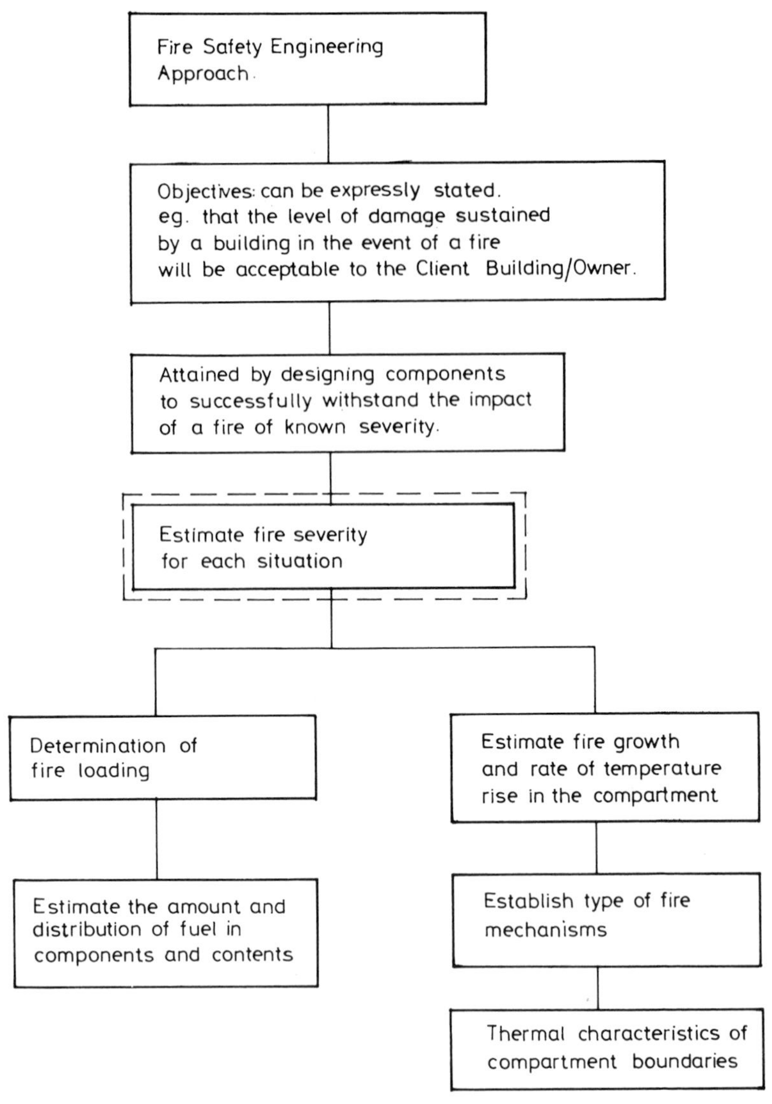

Fig. 8.11 Fire safety engineering approach

3. Nature and distribution of fuel
4. Thermal properties of the fuel
5. Thermal characteristics of the enclosure perimeter construction,

it is possible using an energy balance concept to determine the temperature–time variation of the fire gases within the enclosure.

Petterson *et al.*[1] have derived fire-gas-temperature profiles for different situations taking into account the variables listed above.

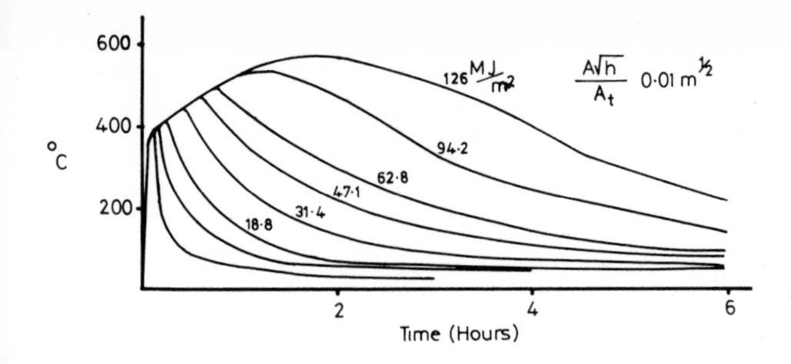

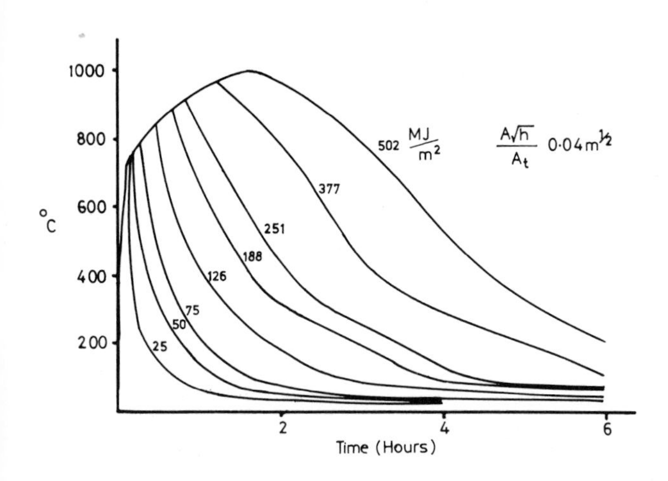

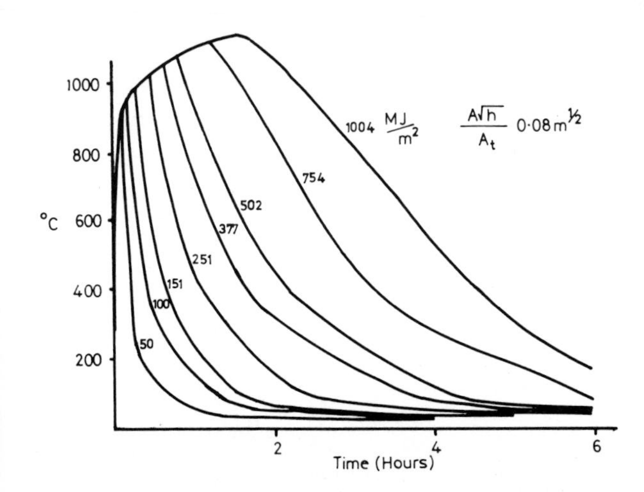

Fig. 8.12 Generated gas temperature curves related to fire loading and ventilation factors

Table 8.2(a)

	(i) $(A\sqrt{h}/A_t = 0.04\ m^{1/2})$							
	q (Mcal/m^{-2})							
	6.0	12.0	18.0	30.0	45.0	60.0	90.0	120.0
	q (MJ/m^{-2})							
t (h)	25	50	75	126	188	251	377	502
0.05	504	504	504	504	504	504	504	504
0.10	745	745	745	745	621	621	621	621
0.15	422	747	747	747	681	681	681	681
0.20	360	696	767	767	777	7777	777	777
0.25	268	587	784	784	776	776	776	776
0.30	164	472	734	799	793	793	793	793
0.35	162	437	665	814	808	808	808	808
0.41	155	389	593	828	822	822	822	822
0.45	148	337	513	841	836	836	836	836
0.51	142	281	481	779	848	848	848	848
0.60	128	259	397	682	874	874	874	874
0.65	120	246	352	626	882	882	882	882
0.70	114	232	307	565	839	894	894	894
0.80	100	204	285	527	785	912	912	912
0.90	86	178	260	483	720	862	928	928
1.00	71	149	235	437	645	827	942	942
1.10	54	118	208	388	589	787	955	955
1.20	51	85	183	337	555	740	967	967
1.30	49	82	156	316	518	688	942	977
1.40	46	77	128	296	480	632	931	987
1.50	45	74	98	276	441	602	919	996
1.60	43	70	94	255	400	571	895	1,004
1.70	41	68	89	235	358	540	870	981
1.80	40	65	85	214	343	507	843	973
1.90	39	62	82	194	328	575	813	963
2.00	38	60	79	174	313	440	781	953
2.20	36	56	73	131	288	369	718	923

Table 8.2(a) *continued*

	(ii) $(A\sqrt{h}/A_t = 0.06\ m^{1/2})$							
	q (Mcal m^{-2})							
	9.0	18.0	27.0	45.0	67.5	90.0	135.0	180.0
	q (MJ m^{-2})							
t (h)	38	75	113	188	283	377	565	753
0.05	575	575	575	575	575	575	575	575
0.10	858	858	858	858	704	704	704	704
0.15	493	861	861	861	784	784	784	784
0.20	404	802	879	879	882	882	882	882
0.25	296	679	898	898	889	889	889	890
0.30	175	538	838	914	908	908	908	908
0.35	174	490	761	928	923	923	923	923
0.40	166	430	669	942	936	936	937	937
0.45	159	369	572	954	949	949	949	949
0.50	151	303	532	877	961	961	961	961
0.60	136	277	433	762	982	982	982	982
0.65	128	262	402	694	992	992	992	992
0.70	120	247	326	620	939	1,001	1,001	1,001
0.80	104	215	300	574	872	1,018	1,018	1,018
0.90	89	185	272	520	795	954	1,032	1,032
1.00	71	152	243	466	705	909	1,044	1,044
1.10	51	116	213	409	657	858	1,054	1,054
1.20	48	80	184	343	593	803	1,064	1,064
1.30	45	76	155	327	550	742	1,029	1,072
1.40	43	72	123	303	505	675	1,013	1,080
1.50	41	68	89	281	460	640	996	1,087
1.60	40	65	86	259	413	603	966	1,093
1.70	38	62	81	236	364	567	935	1,062
1.80	37	59	78	213	348	529	902	1,049
1.90	36	56	74	191	332	491	866	1,036
2.00	35	54	71	169	317	452	830	1,022
2.20	33	50	66	121	289	371	756	984

Table 8.2(b)

	(iii) $(A\sqrt{h}/A_t = 0.08\ m^{1/2})$							
	q (Mcal m^{-2})							
	12.0	24.0	36.0	60.0	90.0	120.0	180.0	240.0
	q (MJ m^{-2})							
t (h)	50	100	151	251	377	502	754	1,004
0.5	622	622	622	622	622	622	622	622
0.10	935	935	935	935	766	766	766	767
0.15	532	937	937	937	853	853	853	853
0.20	432	869	955	955	959	959	959	959
0.25	314	734	973	973	965	965	965	965
0.30	181	575	903	987	981	981	981	982
0.35	180	521	818	1,001	995	995	995	995
0.40	171	454	720	1,013	1,008	1,008	1,008	1,008
0.45	163	387	611	1,024	1,020	1,020	1,020	1,020
0.50	155	314	561	937	1,031	1,031	1,031	1,031
0.60	139	285	454	807	1,050	1,050	1,050	1,050
0.65	131	269	396	732	1,058	1,058	1,058	1,058
0.70	122	253	336	651	996	1,066	1,066	1,066
0.80	106	219	306	598	920	1,081	1,081	1,081
0.90	89	186	275	539	833	1,005	1,092	1,092
1.00	70	151	245	479	735	953	1,102	1,102
1.10	47	113	214	417	659	897	1,111	1,111
1.20	44	73	185	352	612	836	1,119	1,119
1.30	42	70	151	328	564	769	1,077	1,126
1.40	40	66	117	304	516	695	1,058	1,132
1.50	39	62	81	281	466	657	1,038	1,138
1.60	37	59	78	257	415	618	1,004	1,143
1.70	36	56	74	234	363	577	969	1,105
1.80	35	53	71	210	347	537	932	1,090
1.90	34	51	68	187	331	496	893	1,074
2.00	33	49	65	163	316	454	853	1,058
2.20	31	46	59	112	287	368	774	1,016

Table 8.2(b) *continued*

(iv) $(A\sqrt{h}/A_t = 0.12\ m^{1/2})$

	q (Mcal m^{-2})							
	18.0	34.0	54.0	90.0	135.0	180.0	270.0	360.0
	q (MJ m^{-2})							
t (h)	75	151	226	377	565	754	1,130	1,507
0.05	670	670	670	670	670	670	670	670
0.10	1,027	1,027	1,027	1,027	847	847	847	847
0.15	581	1,033	1,033	1,033	1,033	933	933	933
0.20	465	951	1,049	1,049	1,051	1,051	1,051	1,051
0.25	333	799	1,063	1,063	1,057	1,057	1,057	1,057
0.30	186	620	981	1,076	1,071	1,071	1,071	1,071
0.35	186	536	882	1,088	1,083	1,083	1,083	1,083
0.40	176	480	774	1,098	1,094	1,094	1,094	1,094
0.45	168	404	650	1,107	1,103	1,103	1,103	1,103
0.59	159	324	593	1,004	1,112	1,112	1,112	1,112
0.60	142	292	472	856	1,127	1,127	1,127	1,127
0.65	133	275	407	774	1,133	1,133	1,133	1,133
0.70	124	257	341	681	1,060	1,139	1,139	1,139
0.80	106	221	309	622	971	1,150	1,150	1,150
0.90	88	186	276	556	873	1,062	1,159	1,159
1.00	67	149	244	490	765	1,001	1,166	1,166
1.10	41	107	211	422	680	937	1,173	1,173
1.20	39	63	178	351	628	868	1,178	1,178
1.30	37	60	145	327	575	794	1,128	1,183
1.40	36	56	108	301	523	713	1,106	1,188
1.50	34	53	69	277	469	672	1,082	1,192
1.60	33	50	67	253	414	629	1,043	1,195
1.70	32	47	63	228	358	585	1,003	1,151
1.80	31	45	60	203	343	542	962	1,133
1.90	30	43	57	180	327	497	919	1,114
2.00	29	42	54	155	311	452	874	1,096
2.20		39	49	100	283	361	789	1,048

Figure 8.12 shows gas-temperature profiles derived by Petterson *et al.* for a standard fire compartment.

A standard fire compartment is defined as one having its perimeter construction composed of materials having thermal properties corresponding to the average values for concrete brickwork and lightweight concrete. Tables 8.2a and 8.2b show a selection of the profiles of the same compartment in tabular form. The graphs and tables are so produced that the opening factor $A\sqrt{h}/A_t$ is constant and the fire loading q (MJ m^{-2}) varies.

Figure 8.13 illustrates how the opening factor $A\sqrt{h}/A_t$ may be calculated for real situations for a multiple ventilation opening.

To develop an understanding of this approach to the solution of fire engineering problems consider the following examples.

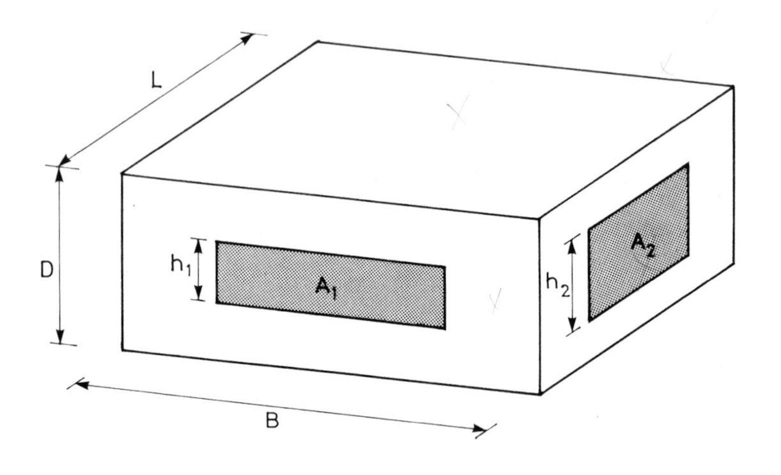

A = Total opening area $= A_1 + A_2$

h = Height of the opening $= \left(\dfrac{h_1 A_1 + h_2 A_2}{A_1 + A_2}\right)$

A_t = Total perimeter area of compartment

when $A_t = 2(B \times D + L \times B + D \times L)$

$$\text{Opening Factor} = \frac{A\sqrt{h}}{A_t} \quad (m^{1/2} \text{ UNITS})$$

$$= \frac{(A_1 + A_2)\sqrt{\dfrac{h_1 A_1 + h_2 A_2}{A_1 + A_2}}}{2(B \times D + L \times B + D \times L)}$$

Fig. 8.13 Calculation of opening factor

WORKED EXAMPLE 3

Estimate the fire endurance of the steel column considered in Example 2 if the column is subjected to a fire in a room 10.0 × 5.0 × 2.5 m high with a ventilation opening 2.5 m wide and 1.25 m high on one long wall if:

1: The fire load is 75 MJ m^{-2} in a standard fire compartment.
2: The fire load is 75 MJ m^{-2} in a Type C fire compartment.

Solution 1

$$A\sqrt{h}/A_t = 0.04\ \text{m}^{\frac{1}{2}}$$

Given that: $\Delta T_s = \dfrac{\lambda_i A_i}{d_i \rho_s V_s C_s}(T_f - T_s)\cdot\Delta t$

Taking $\Delta t = 180$ seconds

$$\Delta T_s = 0.178(T_f - T_s)\cdot\Delta t$$

With reference to Table 8.2a for $A\sqrt{h/A_t}$ (Opening factor) = 0.04 m$^{\frac{1}{2}}$ and the column under $q = 75$ MJ m^{-2} (*Fire loading*) gives the calculated gas temperature–time variation for the compartment under consideration.

Table 8.3 shows the calculated steel temperature based on the temperature above.

Table 8.3

Time	Secs/mins	T_f (°C)	ΔT_s (°C)	T_s (°C)
	$1\frac{1}{2}$			27
180	3	504	85	
	$4\frac{1}{2}$			112
360	6	745	113	
	$7\frac{1}{2}$			225
540	9	747	93	
	$10\frac{1}{2}$			318
720	12	767	78	
	$13\frac{1}{2}$			376
900	15	784	60	
	$16\frac{1}{2}$			456
1,080	18	734	49	
	$19\frac{1}{2}$			505
1,260	21	665	28	
	$22\frac{1}{2}$			533
1,440	24			
	$25\frac{1}{2}$			544

Solution 2

For practical reasons and to make the manipulations more manageable the Tables have been produced for standard fire compartments. They can, however, be used to correlate fires in other compartment types using conversion factors (K_f). These conversion factors take into account how the materials forming the perimeter construction of the compartment affect the fire-temperature profile. Thus if a non-standard fire enclosure is to be analysed then the fire load and opening factor have to be manipulated by the conversion factor to simulate a fire in a corresponding standard compartment. In this example referring to Table 8.4 the conversion factor K_f for a Type C is 3.0.

Thus the new standard fire compartment will have an opening factor $= 3 \times 0.04\,\text{m}^{\frac{1}{2}}$

$A\sqrt{h/A_t} = 0.12\,\text{m}^{\frac{1}{2}}$ and a fire load $= 3 \times 75 = 225\,\text{MJ m}^{-2}$

Table 8.2(b) gives the gas temperature–time curve for $A\sqrt{h/A_t} = 0.12\,\text{m}^{\frac{1}{2}}$ and $q = 225\,\text{MJ m}^{-2}$.

Table 8.5 shows the calculated steel temperature based on the gas temperature curve indicated above. From Fig. 8.14 it can be seen that in Case 2 the critical steel temperature is usually taken as 550 °C, and that this temperature is exceeded after a period of some 14 to 15 minutes.

However, this critical temperature is never reached for Case 1 which is for the standard fire compartment.

CONCLUSIONS FOR WORKED EXAMPLES 3

These examples show clearly that, all things being equal, the material properties of the boundary walls have a dramatic effect on the overall severity of the fire, which in turn affects the fire resistance of the structures located within such an enclosure.

8.6 ENGINEERING RELATIONSHIPS FOR FIRE RESISTANCE – PART 1

Law[2] used the data obtained from a collaborative research programme, executed under the auspices of a commission of the CIB (International Council for Building Research Studies and Documentation) to develop an equivalent fire-resistance concept by relating the temperature rise of internal steel sections in experimental fires and in furnace tests. From the experimental work it was shown

Table [illegible] Conversion to equivalent fire load and opening factor. Note: Equivalent fire load [illegible]
opening factor = K_f·actual opening factor

Type	Description of enclosing construction (Fire compartment)	Factor K_f: Actual opening factor ($m^{1/2}$) 0.02	0.04	0.06	0.08	0.10	0.12
A	Thermal properties corresponding to average values for concrete, brick and lightweight concrete (standard fire compartment)	1.0	1.0	1.0	1.0	1.0	1.0
B	Concrete	0.85	0.85	0.85	0.85	0.85	0.85
C	Lightweight concrete	3.0	3.0	3.0	3.0	3.0	2.5
D	50 % concrete 50 % lightweight concrete	1.35	1.35	1.35	1.50	1.55	1,65
E	50 % lightweight concrete 33 % concrete 17 % { from the inside outwards 13 mm gypsum plasterboard 100 mm mineral wool brickwork }	1.65	1.50	1.35	1.50	1.75	2.00
F[b]	80 % uninsulated steel sheeting 20 % concrete	1.0–0.5	1.0–0.5	0.8–0.5	0.7–0.5	0.7–0.5	0.7–0.5
G	20 % concrete 20 % concrete 80 % { 2 × 13 mm gypsum plasterboard 100 mm air gap 2 × 13 mm gypsum plasterboard }	1.5	1.45	1.35	1.25	1.15	1.05
H	100 % { steel sheeting 100 mm mineral wool steel sheeting }	3.0	3.0	3.0	3.0	3.0	2.5

Table 8.5

Time	Secs/mins	T_f (°C)	ΔT_s (°C)	T_s (°C)
780	$1\frac{1}{2}$			27
	3	670	115	
360	$4\frac{1}{2}$			142
	6	1,027	158	
540	$7\frac{1}{2}$			300
	9	1,033	130	
720	$10\frac{1}{2}$			430
	12	1,049	110	
900	$13\frac{1}{2}$			540
	15	1,063	93	
1,080	$16\frac{1}{2}$			633
	18	981		

that:

$$t_f = k\frac{L}{(A_w A_t)^{\frac{1}{2}}} = \text{effective fire resistance (mins)}$$

where $k = 1.3 \text{ min M}^{-2} \text{kg}^{-1}$

L = total fire-load (wood cribs) (kg)
A_w = area of ventilation opening (m^2)
A_t = area of internal surface to which heat is lost (excluding A_w) (m^2).

Further large scale experiments in brick and concrete compartments and different fire loads gave a value for k approximating to unity. Hence:

$$t_f = \frac{L}{(A_w A_t)^{\frac{1}{2}}} = \frac{L}{A_f} \times \frac{A_f}{(A_w A_t)^{\frac{1}{2}}}$$

where A_f = the floor area (m^2) and L/A_f = conventional fire-load density (kg/m^{-2}).

Figure 8.15 shows that increasing the window from 25 to 100 per cent halves the effective fire resistance. If the window area is 50 per cent an increase in floor area from 200 to 2,000 m^2 doubles the effective fire resistance (Fig. 8.16).

It can be seen that the first term in the above equation is the conventional fire load and the second term is a geometric one determined by building design. Consequently by making some assumptions about:

1. The potential ventilation area, i.e. the failure of window glazing

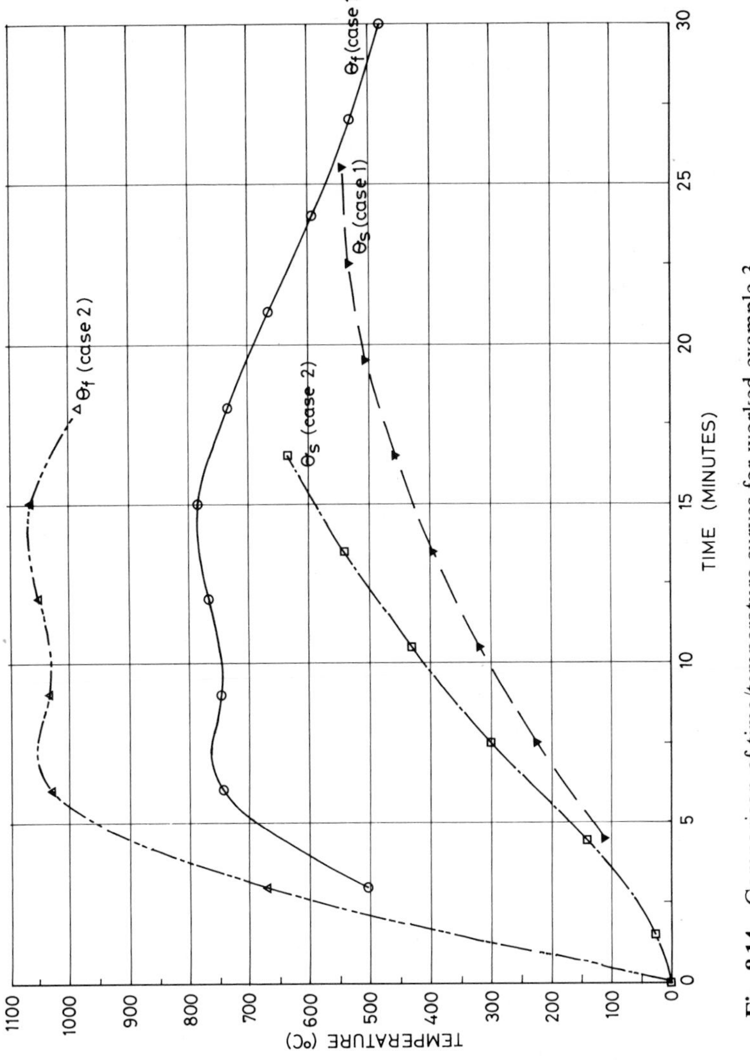

Fig. 8.14 Comparison of time/temperature curves for worked example 3

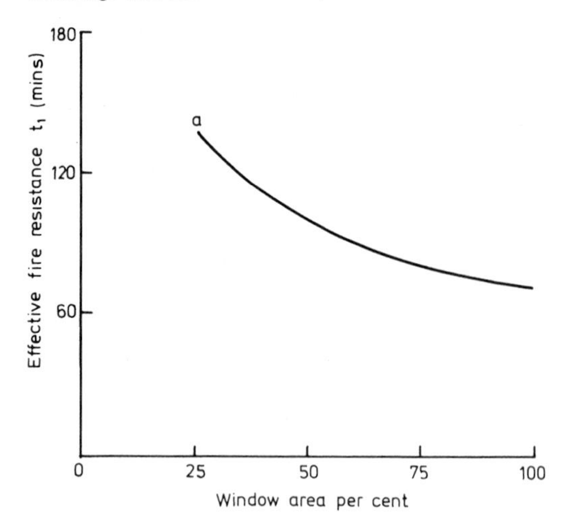

Fig. 8.15 Relationship between window area and fire resistance. (Note: a = square plan. Window area is expressed as percentage of wall which contains it)

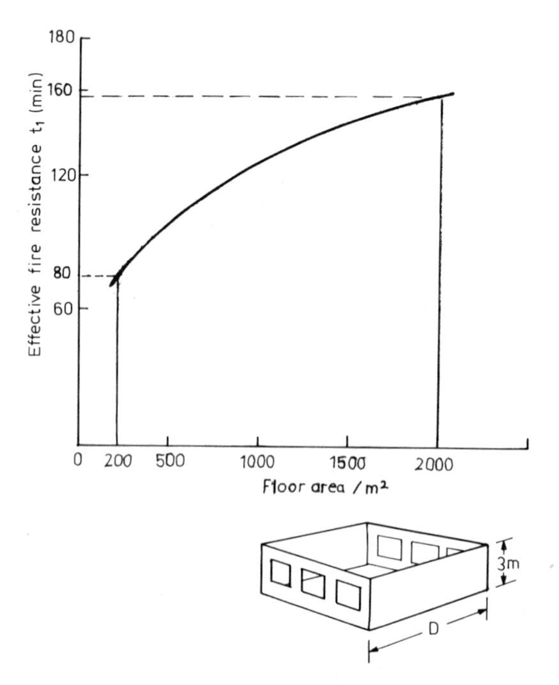

Fig. 8.16 Relationship between floor area and fire resistance. (Note: Fire load-density = 60 kg^{-2} m. Percentage window area constant = 50 per cent on each of two walls)

and other unprotected areas (non-fire-resisting) of external walling

2. The anticipated fire load accepting the variation in fire loading within and between occupancies
3. The extent of the 'fire compartment', this equation may be used:

$$t_f = \frac{L}{A_f} \times \frac{A_f}{(A_w A_t)^{\frac{1}{2}}}$$

Malhotra[3] gave two further equations based on work by Petterson[4] for determining the equivalent fire resistance necessary for constructional components located in fire compartments. These equations are:

$$t_e = 0.067 q_f \left(\frac{A_t}{A_v \sqrt{H_v}} \right)^{\frac{1}{2}} \quad \text{(min)} \qquad \text{and}$$

$$t_e = \frac{0.057 q_f \cdot A_t}{\sqrt{A_v (A_t - A_v)}} \quad \text{(min)}$$

where:

t_e = equivalent fire resistance (min)
q_f = fire load-density $(L/A_t) \cdot \rho$ (MJ m^{-2})
ρ = calorific value of fuel (MJ kg^{-1})
L = total fire load (kg)
A_t = the internal surface area of the compartment (m^2)
A_v = area of the ventilation opening (m^2)
H_v = height of ventilation openings (m).

WORKED EXAMPLE 4

Determine the equivalent fire resistance (t_e) using the three equations as given for a compartment of specified dimensions, construction and fire loading.

Compartment size:	10 × 6 × 3 m high
Windows in one wall:	2, each 4 m wide × 1.5 m high
Construction:	brickwork and concrete
Fire load:	50 kg m^{-2} of floor area (cellulosic materials/equipment)
Window height:	1.5 m
Window area:	12 m^2

Total internal surface area $= (10 \times 6 \times 2) + (6 \times 3 \times 2) + (10 \times 3 \times 2)$

$= 216 \text{ m}^2$

Calorific value of cellulosic contents = 18 MJ kg^{-1}

Fire load-density/unit surface area, $q_f = \frac{50 \times 60 \times 18}{216} = 250 \text{ MJ m}^{-2}$.

(a) using

$$t_f = \frac{L}{A_f} \times \frac{A_f}{(A_w A_t)^{\frac{1}{2}}} \quad \text{(mins)}$$

$$= \frac{3{,}000}{60} \times \frac{60}{(12 \times 204)^{\frac{1}{2}}}$$

$$= \underline{60.6} \text{ mins.}$$

(b) using

$$t_e = 0.067 q_f \left(\frac{A_t}{A_v \sqrt{H_v}} \right)^{\frac{1}{2}} \quad \text{(mins)}$$

$$= 0.067 \times 250 \left(\frac{216}{12\sqrt{1.5}} \right)^{\frac{1}{2}}$$

$$\equiv \underline{64.6} \text{ mins.}$$

(c) using

$$t_e = 0.057 q_f \left(\frac{A_t}{A_v \sqrt{(A_t - A_v)}} \right) \quad \text{(mins)}$$

$$= 0.057 \times 250 \left(\frac{216}{\sqrt{12(216 - 12)}} \right)$$

$$= \underline{62.18} \text{ mins.}$$

8.7 ENGINEERING RELATIONSHIPS FOR FIRE RESISTANCE – PART 2

Witteven[5] showed that the temperature which an unprotected steel section reaches when exposed to a standard heating regime for some time depends on the ambient temperature (standard time–temperature curve is used) and the shape of the cross-section. The first factor was discussed previously and is not considered further here.

The increase in temperature of the steel is controlled by the quantity of heat supplied to the section and by the quantity of steel in which the heat accumulates. The larger the surface area, F, through which the heat is supplied, the quicker the temperature rise in the steel. Conversely the greater the quantity of steel, A, to which heat is supplied, the slower the temperature rise. The increase in temperature is greater as the ratio F/A increases. Thus in a steel section with a large external area and small volume, the temperature will increase more quickly than in a section with a relatively small external area and a large volume.

According to Witteven, for a steel section of volume A (m^3) and external surface area F (m^2), the quantity of heat, ΔQ, added in a time interval Δt is given by the expression:

$$\Delta Q = (T_f - T_s)\,\Delta t\,\alpha\,F \quad \text{(kJ)}$$

and assuming that the steel is uniformly heated, the temperature rise in the steel is given by:

$$\Delta T_s = \frac{\Delta Q}{C_s \rho_s A} \quad (^\circ\text{C})$$

Substituting gives

$$\Delta T_s = (T_f - T_s)\,\Delta t\,\frac{\alpha}{C_s \rho_s}\,\frac{\text{F}}{\text{A}} \quad (^\circ\text{C})$$

where:

T_f = ambient temperature (standard time–temperature curve) (°C)
T_s = temperature of the steel (°C)
C_s = specific heat of steel (kJ kg $^\circ C^{-1}$)
ρ_s = specific mass of steel (kg m^{-3})
α = surface coefficient of heat transfer (kJ m^{-2} h^{-1} $^\circ C^{-1}$).

A number of furnace tests were made to test the derived relationships previously expressed for steel sections with different values of F/A. The measured temperatures for two test-pieces with values of F/A of 94 and 151 which were heated in a gas-fired furnace according to the standard time–temperature curve are shown in Fig. 8.17.

From the equation above, α can be calculated. One value of α was found for each test-piece and α was also found to be practically independent of the furnace temperature.

For the test furnace used, α was found to approximate to 125 kJ m^{-2} h^{-1} °C. Using this value of α, the steel temperature of the test-pieces were recalculated and compared with the measured steel temperature from Fig. 8.17. It can be seen that there is good agreement between the calculated and measured temperature. Thus the equation which includes the ratio F/A as well as α gives a good description of processes which control the temperature rise in the steel section.

Figure 8.18 shows the temperature curve for unprotected steel sections calculated for a number of values of F/A when α = 125 kJ m^{-2} h^{-1} °C. Using the single criterion for failure as the critical steel temperature, the fire resistance of a section with

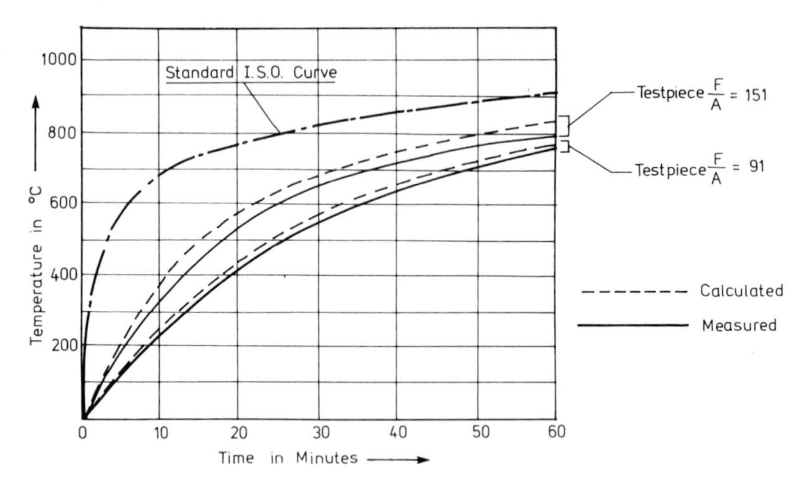

Fig. 8.17 Relationship between time and temperature of two unprotected steel sections HE 200 M and HE 200 B, respectively, $F/A = 91$, $F/A = 151$

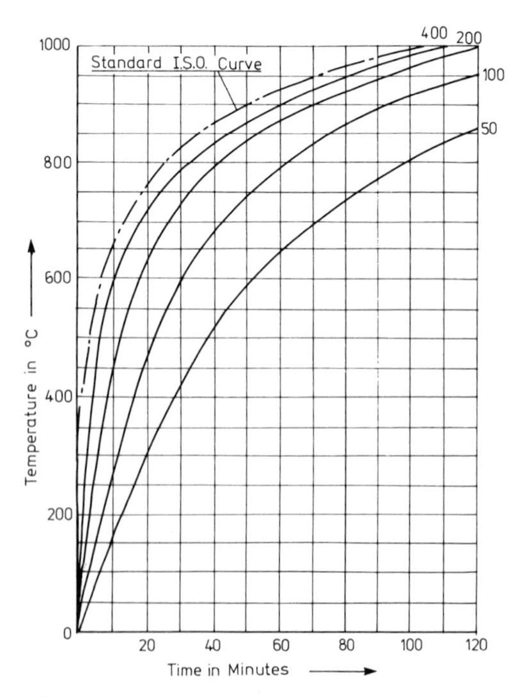

Fig. 8.18 Temperature in unprotected steel sections for different values of F/A ($\alpha = 125\,\text{kJ m}^{-2}\,\text{h}^{-1}\,°\text{C}$)

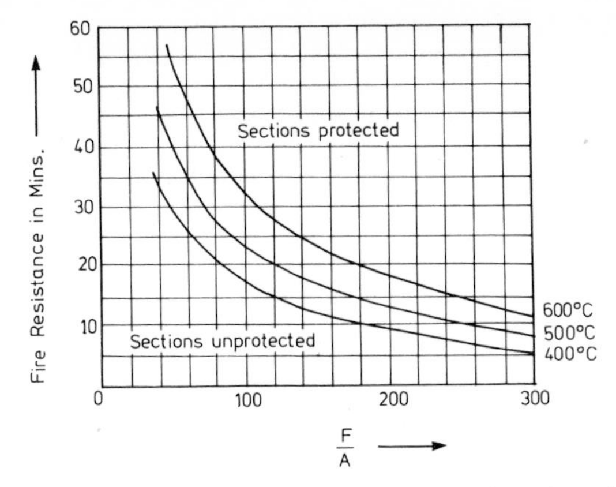

Fig. 8.19 Fire resistance of unprotected steel sections as a function of F/A at different critical steel temperatures ($\alpha = 125\,\text{kJ}\,\text{m}^{-2}\,\text{h}^{-1}\,°\text{C}$)

$F/A = 50$ is 38 minutes at a critical temperature of 500 °C. Figure 8.19 shows the relationship between fire resistance at critical temperatures of 400, 500 and 600 °C and the factor F/A. The section under consideration has to be protected if the required fire resistance is greater than that indicated by the curve which indicates the critical steel temperature selected.

The fire resistance of a protected steel section can be calculated by several methods as seen earlier. Using scaling laws, the fire resistance of a small-section steel column can be derived from the results of a fire-resistance test on a similar column of heavier section using the same type of protection. From the following equation:[6]

$$\left(\frac{T_2}{T_1}\right)^{1.25} = \frac{X_2 A_2 P_1}{X_1 A_1 P_2}$$

the fire resistance of the smaller-section column can be found where:

T = fire-resistance time (mins)
X = thickness of encasement (m)
A = cross-sectional area of the steel section (m^2)
P = average perimeter of encasement (m)
Subscript 1 denotes the test condition
Subscript 2 denotes the predicted condition.

WORKED EXAMPLE 5

A building designer wishes to use a small universal column (152 mm × 152 mm × 30 kg m^{-1} having a cross-sectional area of 38 cm^2) protected with hollow encasement to give a fire resistance of $1\frac{1}{2}$ hours.

Reference to the 'deemed to satisfy' requirements of the Building Regulations would indicate that 25 mm thick vermiculite–cement slabs of 4:1 mix reinforced with wire mesh and finished with plaster skim are needed on a 254 mm × 254 mm × 73 kg m^{-1} column having a cross-sectional area of 93 cm^2 to achieve 2 hours fire resistance:

$$\left(\frac{T_2}{T_1}\right)^{1.25} = \frac{X_2 A_2 P_1}{X_1 A_1 P_2}$$

Rearranging and solving for X_2:

$$X_2 = \left(\frac{T_2}{T_1}\right)^{1.25} \times \frac{X_1 A_1 P_2}{A_2 P_1}$$

$$= \left(\frac{90}{120}\right)^{1.25} \times \frac{25 \times 93 \times 0.608}{38 \times 1.016}$$

$$= 25.6\ mm$$

The thickness of insulation slab required is 12 mm.

This equation may also be expressed[7] as:

$$\left(\frac{T_2}{T_1}\right)^{1.25} = \frac{t_2}{t_1} \cdot \frac{W_2}{W_1} \cdot \frac{P_1}{P_2}$$

where T = fire resistance time (mins)
t = thickness of protective material (mm)
W = weight (unit length of steel and insulation)
P = mean perimeter of the insulation.

Using this expression $t_2 = 25.4$ mm.

With modern light protection systems it is sufficiently accurate to take the perimeter and weights of the steel section only and ignore those of the protection.

Then the ratio $W_2/W_1 = A_2/A_1$ and the ratio:

$$\frac{\text{perimeter}}{\text{cross-sectional area}} = \frac{\text{surface area}}{\text{volume}} = \frac{F}{V}$$

which is the factor previously discussed.

In paragraph 8.12 F/V was referred to as F/A.

Scaling laws should only be used with considerable caution. They should not be used where:

1. The predicted fire resistance is considerably greater than that indicated by the test result, or
2. Where the steel sections are vastly different in size and shape.

8.8 CRITICAL STEEL TEMPERATURE AND CRITICAL FIRE LOAD

Reference to Chapter 2 will indicate that the strength and rigidity of steel decreases when it is heated to elevated temperatures. Figure

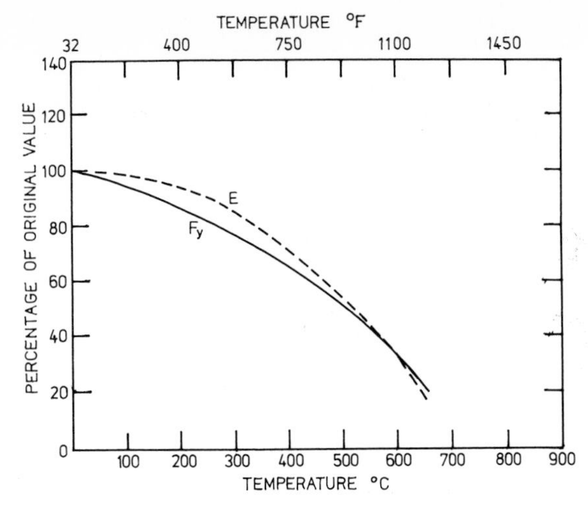

Fig. 8.20 Yield strength (F_y) and modulus of elasticity (E) of carbon steels as a function of temperature

8.20 illustrates how the yield strength and modulus of elasticity of carbon steels vary as a function of temperature. If the duration of the fire is sufficiently long, a point will be reached at which the steel member can no longer perform its loadbearing function. Lie[7] defined the 'critical temperature' as the average temperature of the cross-section of a steel member at which the steel member can no longer perform its loadbearing function. He also defined the 'critical fire-load' as the fire load that is just sufficient to raise the steel temperature to the critical temperature. This definition recognises the dependence of the temperature rise in the steel member on the fire load, i.e. the higher the fire load and the longer the duration of the fire, the higher the maximum temperature of the steel. If the fire load is sufficient the critical temperature of the steel will be reached.

Several factors influence the critical temperature of steel members in a fire situation. For steel beams the most important are:

1. Loading on the member
2. Properties of the steel employed
3. End conditions of the member
4. Degree of exposure of the member.

In general the critical temperature of beams fixed at the ends is higher than that of simply supported beams.

For steel columns the most important factors are:

1. Loading on the member
2. Properties of the steel employed
3. End conditions of the member

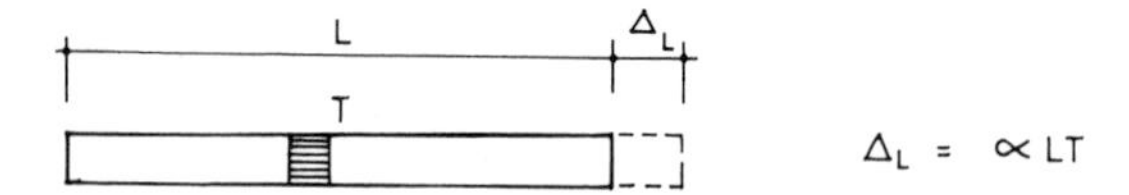

Fig. 8.21 Expansion due to uniform temperature distribution

4. Degree of exposure of the member (e.g. may be built into a wall)
5. Slenderness of the column.

The longitudinal expansion of a component of length L (Fig. 8.21) which has been raised uniformly through temperature T when unrestrained is given by:

$$\Delta L = \alpha_s LT$$

where:

α_s = coefficient of linear expansion
L = original length of the component
T = temperature.

Figure 8.22 shows an unheated flexible bar which just fits between its end supports which are maintained a constant distance H apart.

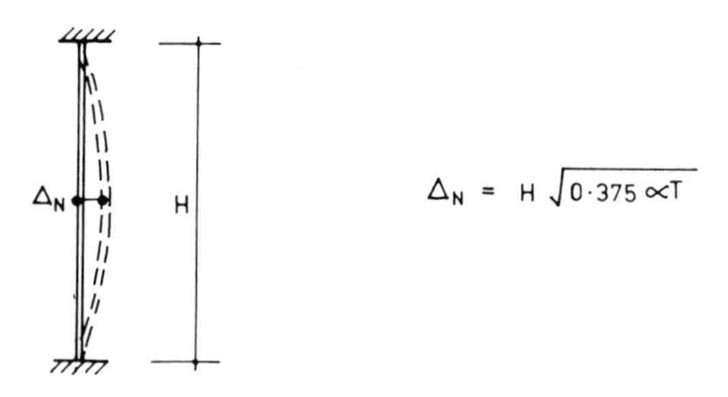

Fig. 8.22 Lateral deflection due to end restraint

If the bar is heated uniformly across the section and along its length, the end restraint restricts the expansion of the bar causing a lateral deflection. According to Cooke,[8] the deflection ΔN normal to the bar at mid-height is given by:

$$\Delta N = H\sqrt{0.375\alpha T}$$

Taking $\alpha = 0.000014/°\mathrm{C}$ for structural steel

$$\Delta N = 0.0023H\sqrt{T}$$

Figure 8.23 indicates the values of ΔN per metre height of the member for different temperature rises. From the gradient of the

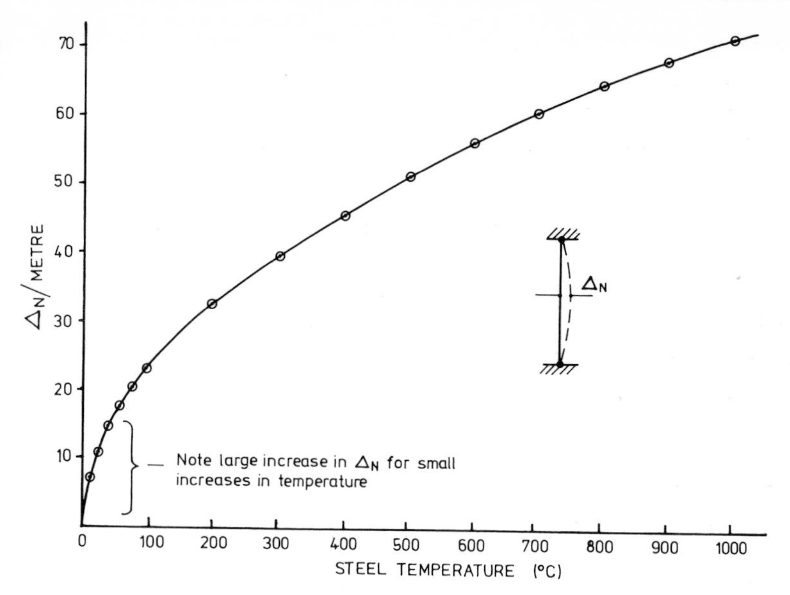

Fig. 8.23 Variation in bowing deflection with temperature

graph it can be seen that large lateral deflections result from small initial temperature increases in the steel, e.g. a member 3 m long raised through 150 °C bows by roughly 80 mm. If sufficient longitudinal clearance can be provided then bowing of the member can be prevented.

For example, suppose a maximum temperature rise in the steel member which can be otherwise tolerated is 550 °C, the longitudinal expansion of the member 3 m long is given by:

$$\Delta L = LT$$

$$= 0.000014 \times 3 \times 550$$

$$= 0.023 \text{ m}$$

Thus a longitudinal clearance of 8 mm per metre length of member if provided would be sufficient to ensure that bowing will not occur.

Bowing of this kind will only occur where the member is flexible. Consequently the axial forces are minimal and easily restrained by the end supports.

In reality structural and constructural components are not exposed to heat on all sides. This exposure of an element of construction in a fire compartment results in non-uniform temperature across the thickness of the element. Where the temperature gradient across an element is large and assuming that α

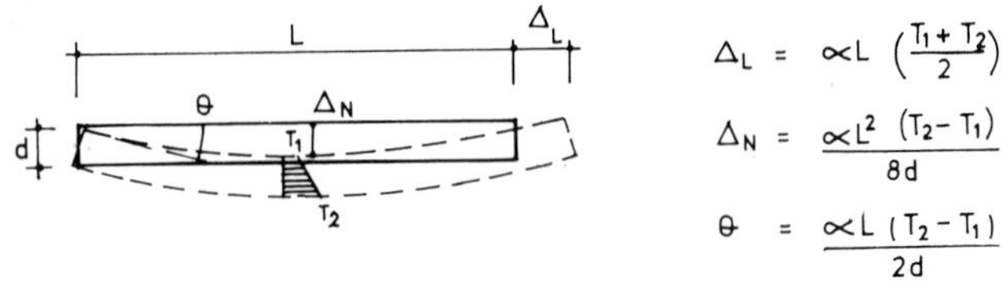

$$\Delta_L = \alpha L \left(\frac{T_1 + T_2}{2}\right)$$

$$\Delta_N = \frac{\alpha L^2 (T_2 - T_1)}{8d}$$

$$\theta = \frac{\alpha L (T_2 - T_1)}{2d}$$

Fig. 8.24 Thermal bowing of structural steel

does not vary greatly with temperature, large thermal bowing deflections will occur unless opposing forces are present.

It has been shown[8] that the thermal bowing of structural steel is described by the equation:

$$\Delta N = \frac{\alpha T L^2}{8d}$$

and is valid for linear and curve-linear temperature profiles (Figs. 8.21 and 8.24).

Thus:

$$\Delta L = \alpha L \frac{(T_1 + T_2)}{2}$$

$$\Delta N = \alpha L^2 \frac{(T_2 - T_1)}{8d}$$

where T_2 and T_1 = surface temperature of the steel
d = depth of the section.

Since T_m represents the mean temperature rise in the section,

$$T_m = (T_2 - T_1) \qquad \text{and}$$

$$\Delta N = \alpha \frac{T_m L^2}{8d}$$

8.9 CONCRETE CONSTRUCTIONAL COMPONENTS

The fire resistance of concrete components incorporating reinforcing or prestressing steel depends upon the ability of the concrete to prevent the steelwork from reaching a critical temperature and its resistance to spalling. Structural steelwork protected by encasement in concrete is similarly dependent on the thermal characteristics of the concrete casing. Assuming uniform heating of a component exposed to fire on all sides, a critical temperature for the steel can be taken as 550 °C. The thermal inertia ($\lambda \rho c$) of the concrete controls

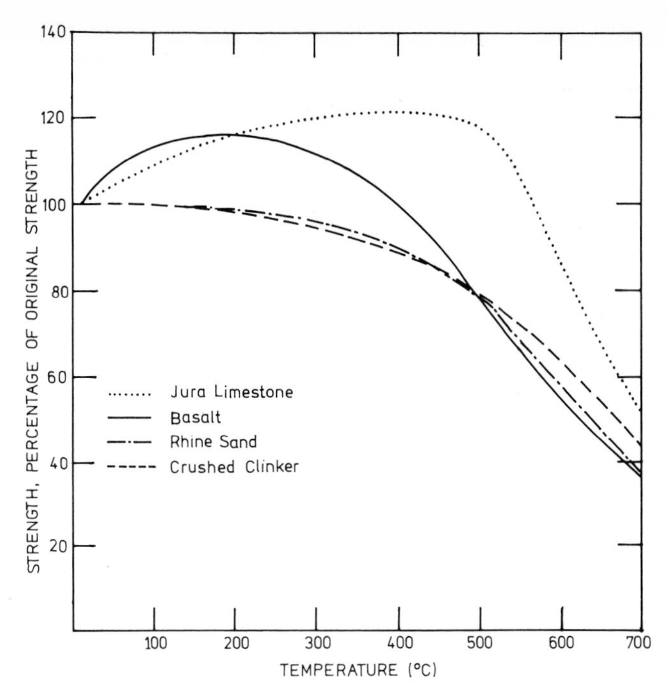

Fig. 8.25 Influence of the aggregate on the compressive strength of concrete at elevated temperatures

the heat-transfer process and the thermal conductivity (λ) is the dominant factor in this case. As discussed in Chapter 2, ordinary Portland cement concrete begins to lose its strength at temperatures around 100 °C and at temperatures around 300 °C the concrete is permanently damaged. At temperatures around 600 °C the concrete has little residual strength (Fig. 8.25). It follows therefore that it is necessary in the design of structural and constructional components to determine a critical thickness of concrete cover which will ensure that the protected steel remains below the appropriate critical temperature in a fire for a prescribed period of time.

It is not usually necessary to calculate the fire resistance of concrete components as the design of such components will usually follow the procedures set down in CP 110[9] which provides guidance as to the fire resistance of concrete components without the necessity for further calculation. The guidance provided is in the form of tables which indicate the minimum concrete cover necessary for particular components to achieve a prescribed fire resistance. Malhotra[3] has adequately dealt with the fire-related design of concrete elements and it is not intended to repeat this work here.

Generally there are a number of factors which affect the fire performance of concrete components. These factors are:

1. Type of concrete
   Reference has been made already to the affect of temperature on the strength of ordinary Portland cement concrete. The choice of aggregate will also influence fire performance, e.g. a suitable lightweight aggregate may reduce the thickness of cover necessary to the steelwork than if a heavy aggregate were used. The thermal properties of concrete are greatly influenced by the choice of aggregate Figs 8.25 and 8.26. The resistance of concrete

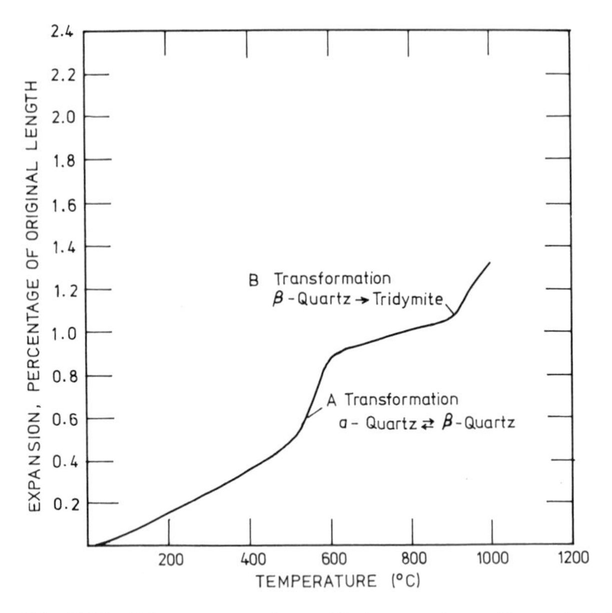

Fig. 8.26 Linear expansion with temperature of a material of mainly quartz (heating rate 5 °C per minute)

to spalling also depends largely on the type of aggregate used, e.g. aggregates which have been exposed to high temperature during the manufacturing process (e.g. foamed-slag, pulverised fuel-ash) can be expected to be more resistant to spalling damage than, for example, flint gravel. Spalling of concrete may be considered as the explosive breaking off of pieces of the material and in general it reduces the fire resistance of the component. The spalling of reinforced or prestressed concrete components may lead to the direct exposure of the steel to fire, which, in turn, reduces the time for the steel to reach its critical temperature and hence causes a reduction in the fire resistance of the component. Spalling of concrete is a complex phenomenon and a number of factors which promote spalling need to be considered.[10] These are:

(a) high content of free water
(b) restraint (beam restrained at the ends)
(c) low porosity of the material
(d) low permeability of the material
(e) rapid temperature rise at the exposed surfaces of the component
(f) closely-spaced reinforcement.

2. Type and quality of reinforcement or prestressing steel
   It is known that all steels lose strength when heated. However, the rate at which loss of strength occurs varies with the type and quality of steel used, e.g. mild steel loses strength less rapidly than prestressing steel and work-hardened high-yield reinforcement. It is necessary to know for each steel the temperature at which the yield or ultimate stress equates to the actual stress in the steel, i.e. the critical temperature of the steel.
3. Size and shape of the component
   As with structural steel components, the rate of temperature rise in concrete components can be related to the ratio of fire-exposed surface area to the cross-sectional area. It follows that the smaller the surface area exposed to fire and the larger the cross-sectional area the slower the rate of temperature rise in the section than in a section where the converse is true.
4. Concrete cover
   The purpose of the concrete cover is really to act as an insulant so that the critical temperature of the steel is not achieved in a fire situation. Consequently the thermal conductivity of the concrete is an important factor which in effect determines the thickness of cover to be provided. In situations where the thickness of the cover required exceeds 50 mm, additional reinforcing steel-mesh may have to be provided and located at mid-thickness of the cover concrete so as to ensure that the cover concrete is not displaced during a fire. It will be appreciated that difficulties will be encountered in placing the concrete with something like 50 mm clearance between the mesh reinforcement and the shuttering.

With the changes in the system of building control in England and Wales which utilises approved documents there was some uncertainty in the application of appropriate covers to reinforcement tendons and in corresponding minimum widths of flexural members. This uncertainty arose from the existence of four publications.[9,11,12,13] Consequently in an effort to bridge the time gap between the publication of approved documents and to remove ambiguities, the '*1984 Guidance for the application of tabular data for fire resistance of concrete elements*' was published.[14] This 'Guidance' which is a very useful practical work made three changes to existing tabular data.[11] These were:

1. Minimum practical width of a beam was determined as 200 mm and covers to beams were adjusted in accordance with existing recommendations.[15]
2. Minimum practical width of a rib in a 'trough' or 'waffle' floor was determined as 125 mm and covers to ribs were also adjusted as above, and
3. Measures recommended for additional protection against spalling were extended and the use of supplementary reinforcement was recommended to be removed from Code recommendations. These changes are briefly explained hereafter.

Practical width of a beam

Existing tabular data will indicate a range of beam widths capable of providing a prescribed degree of fire resistance. A common fault in published fire data is that many widths, thicknesses, etc., of components and materials are specified, but rarely used in practice. The minimum width of beams was determined simply on the basis that beam widths of less than 200 mm rarely occurred in practice. Increasing the minimum width of the beam can result in a decrease in the cover provided and Table 8.6 indicates the trade-off of beam width against decrease in cover.

For beam widths of less than 200 mm the table can be used in reverse, i.e. decrease in beam width corresponds to an increase in cover.

Table 8.6

Minimum increase in width	Decrease in cover	
	Dense concrete (mm)	Lightweight concrete (mm)
25	5	5
50	10	10
100	15	15
150	15	20
200	15	25
250	20	25
350	20	30
450	20	35

Care must be exercised in the use of this Table, however, so as to ensure that the cover to the beam is not less than that which would be provided for an equivalent plain soffit floor, i.e. as the beam becomes shallower and broader it becomes in effect a slab. Table 8.7 shows the cover required to achieve various periods of fire resistance for beams of differing concrete mixes and beam widths. Reference to Note 8 of Table 8.7 will indicate cover dimensions which have been underlined twice, thus indicating that the cover indicated cannot be reduced. Table 8.8 indicates the covers for plain soffit concrete floors.

In situations where members are heavily reinforced, e.g. two or more layers of reinforcement, the 'effective cover' C_e is given by:

$$C_e = \frac{A_1C_1 + A_2C_2 + \cdots A_nC_n}{A_1 + A_2 \ldots A_n}$$

where $A_1 - A_n$ = individual areas of the reinforcing bars
$C_1 - C_n$ = individual cover corresponding to each bar.

The practical width of a rib

The general considerations above also applied to the determination of the minimal practical width of ribs in 'trough' or 'waffle' floors. This width was determined as 125 mm and Table 8.6 indicates the required cover necessary for achieving the required period of fire resistance.

Additional protection against spalling

The 'Guidance' rejects the current Code recommendations *re* the use of mesh reinforcement as a means of maintaining the concrete cover to main reinforcement because:

1. Use of mesh might be contrary to recommendations on cover for durability
2. As mentioned previously, practical difficulties exist regarding maintaining the mesh in place during the placement of the concrete
3. Other methods could be used to reduce the occurrence of spalling.

In the latter case the other methods referred to include the use of additional reinforcement, false ceilings and applied coatings such as lightweight plaster.

Table 8.7 Concrete beams

Nature of construction and materials		Minimum dimensions (mm) excluding any finish for a fire resistance of					
		½ h	1 h	1½ h	2 h	3 h	4 h
1. Reinforced concrete (simply supported)							
(a) dense concrete	width	200	200	200	200	250	300
	cover	15	20	30	50	70	80
(b) lightweight concrete	width	200	200	200	200	200	250
	cover	15	15	25	40	55	65
2. Reinforced concrete (continuous)							
(a) dense concrete	width	200	200	200	200	250	300
	cover	15	20	25	40	50	60
(b) lightweight concrete	width	200	200	200	200	200	250
	cover	15	15	20	25	35	45
3. Prestressed concrete (simply supported)							
(a) dense concrete	width	200	200	200	200	250	300
	cover	20	30	45			
(b) lightweight concrete	width	200	200	200	200	200	250
	cover	20	20	35	50		
4. Prestressed concrete (continuous)							
(a) dense concrete	width	200	200	200	200	200	250
	cover	20	20	30	45	70	80
(b) lightweight concrete	width	200	200	200	200	200	200
	cover	20	20	25	35	45	60

Notes:

1. Cover in this table is the distance from the main tensile reinforcement or tendon to the nearest surface exposed to fire.
2. In multilayered reinforcements or tendons the effective cover is the average of the values for cover for the individual bars or tendons, each value weighted according to the area.
3. When the cover to outermost steel (including links, lacers, etc.) exceeds 400 mm for dense concrete or 50 mm for lightweight concrete, then measures to protect against spalling should be adopted.
4. Density range is dense concrete $>2{,}000>$ lightweight concrete (kg m^{-3}).
5. Minimum widths have been raised to 200 mm and then in units of 50 mm fire period.
6. Figures without underlines in this table accord exactly with draft CP 110 revised 83/12036 and BRE *Guidelines* by Reads Adams and Cooke (1980) Table 9.
7. A number underlined once in this table is a change to width/cover either to accord with the 50 mm step in width or by use of the M Law Chart.

8.10 TIMBER CONSTRUCTIONAL COMPONENTS

Timber is a combustible non-homogeneous and variable material. On initial heating it undergoes expansion until a temperature of around 80 °C is reached, when the process of expansion is reversed and shrinkage occurs. To determine the fire resistance of timber components it is necessary to know:

1. Rate of charring
2. Temperature gradient in the uncharred part of the section
3. Strength and deformation properties of the material as a function of temperature.

The charring rate of wood depends on various factors which include the density of the wood, its moisture content and permeability. Figure 8.27 shows the relationship between charring rate, density and moisture content. Notional charring rates for various timbers are given in BS 5268: Part 4: Section 4.1. Method of calculating the fire resistance of timber members[16] and are given in Table 8.9.

Reference is made in Table 8.9 to Table 1 of CP 112,[17] and charring rates are given for the species listed in the latter Table. This Table is reproduced here as Table 8.10. However CP 112 will eventually be superceded by BS 5268: Part 2: 1984[18] and BS 5268: Part 4: 1978 will have to be amended accordingly. From the literature,[19,20] an average charring rate of 0.6 mm min^{-1} can be derived. It has been suggested[21] that when the wood is light and dry a value of 0.8 mm min^{-1} is a better average value for the charring rate of wood, and that when the wood is dense and moist the value should be 0.4 mm min^{-1}.

The geometry of the section usually enhances the heat-transfer processes at corners resulting in an increased depth of char and rounding at the corners (see Fig. 8.28). The area of the section lost due to rounding can be calculated from:

$$A = 0.215r^2$$

Where the residual section, i.e. the section unaffected by fire, is not less than 50 mm thick and exposure to fire conditions does not

8. A number underlines twice in this table is restricted in its reduction from previous values by the equivalent value for a plain soffit floor.

Table 8.8 Concrete floors: plain soffit

Nature of construction and materials		Minimum dimensions (mm) excluding any finish for fire resistance of					
		½ h	1 h	1½ h	2 h	3 h	4 h
1. Reinforced concrete (simply supported)							
(a) dense concrete	thickness	75	95	110	125	150	170
	cover	15	20	25	35	45	55
(b) lightweight concrete	thickness	70	90	105	115	135	150
	cover	15	15	20	25	35	45
2. Reinforced concrete (continuous)							
(a) dense concrete	thickness	75	95	110	125	150	170
	cover	15	20	20	25	35	45
(b) lightweight concrete	thickness	70	90	105	115	135	150
	cover	15	15	20	20	25	35
3. Prestressed concrete (simply supported)							
(a) dense concrete	thickness	75	95	110	125	150	170
	cover	20	†25	30	40	55	65
(b) lightweight concrete	thickness	70	90	105	115	135	150
	cover	20	20	30	35	45	60
4. Prestressed concrete (continuous)							
(a) dense concrete	thickness	75	95	110	125	150	170
	cover	20	20	25	35	45	55
(b) lightweight concrete	thickness	70	90	105	115	135	150
	cover	20	20	25	30	35	45

Notes:
1. Cover in this table is the distance from the main tensile reinforcement or tendon to the nearest surface exposed to fire.
2. In the multilayered reinforcements or tendons the effective cover is the average of the values for cover for the individual bars or tendons each value weighted according to the area.
3. When the cover to outermost steel (including links, lacers, etc.) exceeds 40 mm for dense concrete or 50 mm for lightweight concrete, then measures to protect against spalling should be adopted.
4. Density range is dense concrete >2,000 > lightweight concrete ($kg\,m^{-3}$).
5. The numbers in this table accord exactly with Draft CP 110 revised 83/12036 Table 62 and BRE *Guidelines* († *vide* BRE letter to Federation of Concrete Specialists 27 October 1981).

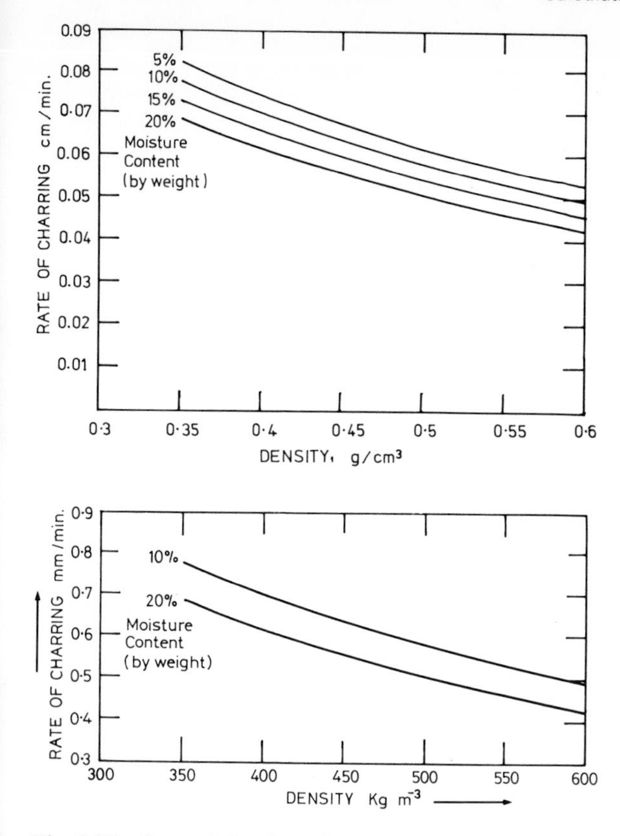

Fig. 8.27 Rate of charring of Douglas fir as function of its density (dry condition) for various moisture contents when exposed to a standard fire (ASTM)

Table 8.9 Notional rate of charring for the calculation of residual section

Species	Charring in 30 min (mins)	Charring in 60 min (mins)
(a) All structural species listed in Table 1 of *CP 112: Part 2: 1971 except those noted in items (b) and (c)	20	40
(b) Western red cedar	25	50
(c) Oak, utile, keruing (gurjun), teak, greenheart, jarrah	15	30

* CP112: Part 2 has been superseded by BS 268

Table 8.10 Names and densities of some structural timbers

Standard name	Botanical species	Other common names	Approximate density at a moisture content of 18 % (kg m^{-3})
Softwoods			
a. Imported			
Douglas fir	*Pseudotsuga menziesii*	B.C. pine	590
Western hemlock (unmixed)	*Tsuga heterophylla*	B.S. hemlock	540
*Western hemlock (commercial)	*Tsuga heterophylla*	Hembal	530
Parana pine	*Araucaria angustifolia*	—	560
Pitch pine	*Pinus palustris* *P. eliotti* *P. caribaea*	Longleaf pitch pine Southern yellow pine Nicaraguan pitch pine Honduras pitch pine	720
Redwood	*Pinus sylvestris*	Baltic redwood	540
Whitewood	*Picea abies* *Abies alba*	Baltic whitewood European whitewood	510
Canadian spruce	*Picea glauca* *Picea sitchensis*	Western white spruce Eastern Canadian spruce Sitka spruce	450
Western red cedar	*Thuja plicata*	B.C. red spruce	380
b. Home-grown			
Douglas fir	*Pseudotsuga menziesii*	—	560

European or Japanese larch	*Larix decidua* *Larix leptolepsis*	Larch	560
Scots pine	*Pinus sylvestris*	Scots fir	540
European spruce	*Picea abies*	Norway spruce	380
Sitka spruce	*Picea sitchensis*	—	400
Hardwoods			
a. Imported			
Abura	*Mitragyna ciliata*	Subaha	590
African mahogany	*Khaya* spp.	Khaya	590
Afrormosia	*Pericopsis elata*	Kokrodua	720
Greenheart	*Ocotea rodiaei*	—	1,060
Gurun/keruing	*Dipterocarpus* spp.	—	720
Iroko	*Chlorophora excelsa*	Mvule	690
Jarrah	*Eucalyptus marginata*	—	910
Karri	*Eucalyptus diversicolor*	—	930
Opepe	*Nauclea diderrichii*	Kusia	780
Red meranti/red seraya	*Shorea* spp.	—	540
Sapele	*Entandrophragma cylindricum*	—	690
Teak	*Tectona grandis*	—	720
b. Home-grown			
European ash	*Fraxinus exelsior*	—	720
European beech	*Fagus sylvatica*	—	720
European oak	*Quercus robur*	—	720

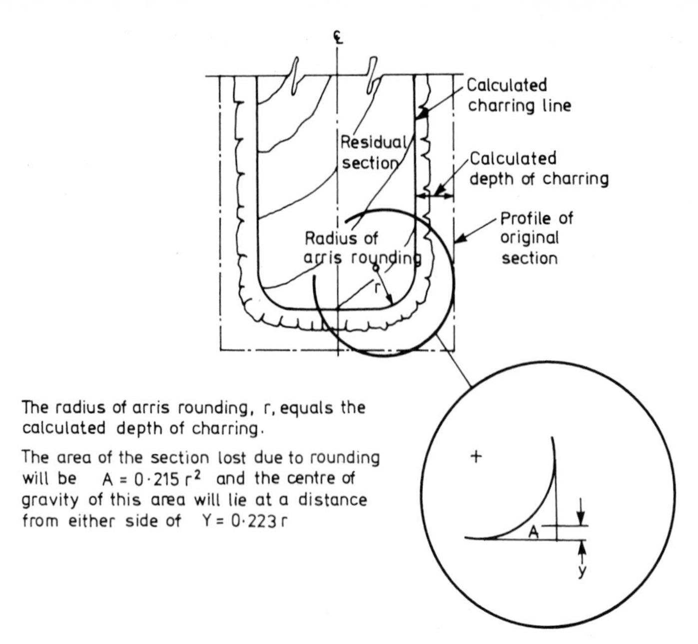

Fig. 8.28 Radius of Arris Rounding

exceed 30 minutes, the additional loss of timber at the corners due to rounding can be ignored.

The thermal conductivity of wood is low hence a very clear interface zone can be identified between the charred and uncharred sections of timber. The charred section acts in two ways:

1. The char insulates the unburned wood, and
2. The char retards the release of flammable volatiles and hence affects the combustion processes.

The general result is that the temperature in the uncharred section is not excessive and a strength reduction of the order of 10 per cent can be anticipated.

The fire resistance of the residual section of structural timber sections can be determined after exposure of the element to heating in the fire-resistance test BS 476: Part 8 and calculation of the loadbearing properties of the residual section can be made. BS 5268: Part 4[16] requires that the residual section, i.e. the section of uncharred timber that would be left after a given period of exposure to fire conditions described in BS 476: Part 8, assuming a steady rate of charring with allowance for accelerated charring at exposed arrises, be computed by subtracting from the appropriate faces the notional amount of charring assumed to occur during the required period of fire exposure.

There are two possible methods which could be adapted to determine the fire resistance of a timber section. These are:

1. Determine the amount of timber available for charring and calculate the time necessary for it to be consumed, or
2. Determine the critical section for structural purposes and add the necessary thicknesses of timber to be consumed by charring in order to achieve the required degree of fire resistance.

For flexural members the loadbearing capacity is calculated using the residual section and stresses of 2.25 times the permissible long-term dry stress for sections having a minimum initial breadth of 70 mm and 2.0 times the permissible long-term dry stress for sections with an initial breadth of less than 70 mm. Deflection of the residual section should not exceed span/30.

WORKED EXAMPLE 6

Calculate the fire resistance of a simply supported beam of home grown Douglas fir spanning 3.5 m and supporting a uniformly distributed load of 4 kN m^{-1}.

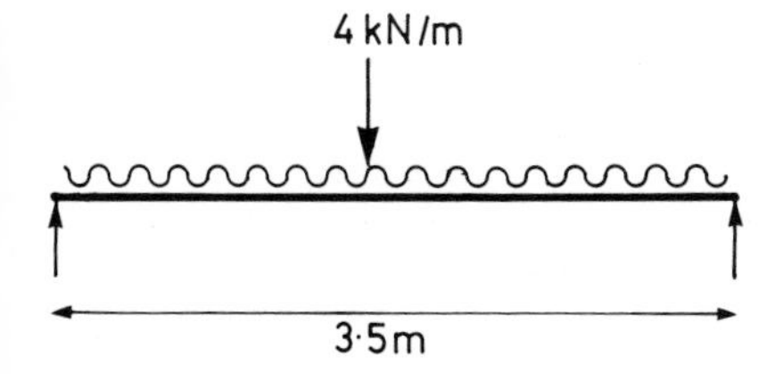

Fig. 8.29 Simply supported timber beam for worked example 8.6

From CP 112: Part 2: 1971, Table 11:

Douglas fir is species group S1, assume Grade 65
Dry stresses (moisture content $\leqslant 18\%$)
Bending and tension parallel to the grain = $f_p = 10.3\ \text{N mm}^{-2}$

BS 5268: Part 2: 1984, Tables 8 and 9:

Douglas fir (British grown), assume Grade M75
Bending parallel to the grain = $f_p = 10\ \text{N mm}^{-2}$

Maximum bending moment is given by:

$$M_a = \frac{WL^2}{8}$$

$$= \frac{4 \times 3.5 \times 3.5}{8}$$

$$= 6.125 \text{ kNm}$$

$$M_a = f_p z$$

$$= f_p \times \frac{bd^2}{6}$$

$$\therefore \quad bd^2 = \frac{6.125 \times 6}{10.3} \times 10^6$$

(Taking $f_p = 10.3$ from CP 112. For comparison f_p from BS 5268: Part 2 can be substituted)

$$= 3{,}567{,}961 \text{ mm}^3$$

Taking $b = 75$ mm, $\quad d = 218$ mm

$\therefore$ take $d = 250$ mm

Maximum permissible stress $f_{Pm} = 10.3 \times 2.25 = 23.18 \text{ N mm}^{-2}$

Method 1

Charring of wood after 30 minutes exposure = 20 mm for each exposed face, Table 8.9, and the beam is exposed on its free faces.

$\therefore$ the uncharred section = 35 × 230 mm

$$\text{stress } f_p = \frac{6.125 \times 6 \times 10^6}{35 \times 230 \times 230}$$

$$= 19.85 \text{ N mm}^{-2}$$

The maximum permissible stress $f_{Pm} = 23.18 \text{ N mm}^{-2}$

$\therefore$ the beam will provide fire resistance for 30 minutes.

Check for deflection For a uniformly-distributed load,

$$\text{maximum deflection} = \frac{5WL^3}{384EI}$$

where:
E = modulus of elasticity
I = moment of inertia

$\therefore$ for rectangular sections, $I = \dfrac{bd^3}{12}$

and $\quad E = 11{,}000 \text{ N mm}^{-2}$

$$\text{Maximum deflection} = \frac{5WL^3}{384EI}$$

$$= \frac{5 \times 4{,}000 \times 3{,}500^3 \times 12}{384 \times 11{,}000 \times 35 \times 230^3}$$

$$= 6 \text{ mm}$$

According to BS 5268: Part 4: 1978, the beam should resist deflection during the fire test, to span/30.

$$\therefore \quad \frac{3{,}500}{30} = 117\,\text{mm}$$

Therefore for fire purposes the beam is satisfactory.

For normal structural purposes, deflection would be limited to 1/360 of the span, i.e. 3,500/360 = 9.7 mm, so even for structural purposes the deflection of the residual section is not excessive.

Method 2

Taking the maximum permissible stress $f_{\text{Pm}} = 23.18\,\text{N mm}^{-2}$

$$bd^2 \text{ for the residual section} = \frac{6.125 \times 6 \times 10^6}{23.18}$$

$$= 1.5854084 \times 10^6\,\text{mm}^3$$

By trial and error the dimensions of b and d can be obtained.

When $b = 35$ and $d = 200$, $bd^2 = 1{,}500{,}000\,\text{mm}^3$
$b = 35$ and $d = 210$, $bd^2 = 1{,}543{,}500\,\text{mm}^3$
$b = 35$ and $d = 212$, $bd^2 = 1{,}573{,}040\,\text{mm}^3$
$b = 35$ and $d = 213$, $bd^2 = 1{,}587{,}915\,\text{mm}^3$

The critical section is $b = 35$ mm and $d = 212$ mm.

$\therefore$ Assuming 20 mm char in 30 minutes, $b = 35 + 40 = 75$ mm

$d = 212 + 40 = 252$ mm

Lie[21] suggested an approximate formula for the calculation of the fire resistance of timber beams and columns. These approximate formulae are:

Beams heated on four sides

$$t_{b4} = 100fB\left(4 - 2\left(\frac{B}{D}\right)\right)$$

Beams heated on three sides

$$t_{b3} = 100fB\left(4 - \frac{B}{D}\right)$$

Columns heated on four sides

$$t_{c4} = 100fD\left(3 - \frac{D}{B}\right)$$

where:

t = fire resistance (mins)
B = breadth of the member (m)
D = depth of the member (m)
f = factor which depends on the load, and for columns also on the effective length.

Values of the f are given in Table 8.11.

Table 8.11 Values of F member type column

Load as % of allowable load	Beam	Column (L/D > /0)	(L/D < /0)
>75	1.0	1.0	1.2
<75750	1.1	1.1	1.3
<50	1.3	1.3	1.5

Applying the approximate formula for a beam heated on three sides to the previous example. ∴ Since $B = 75$ mm and $D = 252$ mm

$$t_{b3} = 100fB\left(4 - \left(\frac{B}{D}\right)\right)$$

Taking $f = 1$,

$$t_{b3} = 100 \times 1 \times 0.075\left(4 - \frac{0.075}{0.250}\right)\right)$$

$= 28$ minutes.

As calculated before the beam has effectively 30 minutes fire resistance using the sections estimated by Method 2 above.

8.11 CONCLUSIONS

The development of methods for calculating fire resistance enables the building designer to adopt a positive approach to fire safety in building design.

The standard time–temperature curve can be used as a basis for calculations predicting fire resistance or alternatively calculations can be based on real fire conditions. The latter requires a knowledge of several factors, i.e. the temperature profile of the real fire, fire loading, ventilation openings for the compartment, heat-transfer mechanisms and temperature rise in constructional components. Increasingly methods of calculating fire resistance are being accepted by statutory bodies as a means of satisfying the requirements of fire safety legislation and it is expected that the acceptance of the calculation of fire resistance as a means of demonstrating compliance with statutory requirements will increase in the future.

APPENDIX 8.12 ESTIMATION OF FIRE RESISTANCE OF A SOLID STEEL COLUMN

```
READY.

 1 REM UNPROTECTED SOLID STEEL
 3 PRINTCHR$(147)
 5 INPUT"TIME STEP";Q
 10 INPUT"AMBIENT TEMPERATURE";TS:S1=TS
 12 PRINT
 13 IF D=5THENOPEN3,4:CMD3:TS=S1
 14 PRINT:PRINT"UNPROTECTED SOLID STEEL":PRINT:PRINT:PRINT"TIME STEP="Q"SEC."
 17 PRINT"AMBIENT TEMPERATURE="TS"K":PRINT
 18 PRINT" AMB.TEMP.,K","FURNACE TEMP.,K","TIME,SEC."
 20 FOR T=0TO3000STEPQ
 30 TF=345*LOG(.133*(T+Q)+1)/LOG(10)+300
 40 DTS=10↑-12*(.7*TF↑4-.9*TS↑4)*Q*.37
 50 PRINTTS,TF,T
 60 TS=TS+DTS
 65 REM IFTS>823THEN75
 70 NEXTT
 75 IFD=5THENPRINT#3:CLOSE3,4
 100 PRINT:INPUT"DO YOU WANT A HARD COPY";A$
 110 IF A$="N"THEN200
 120 IFA$="Y"THEND=5:GOTO5
 200 END
READY.
```

```
UNPROTECTED SOLID STEEL


TIME STEP= 300 SEC.
AMBIENT TEMPERATURE= 300 K
```

AMB.TEMP.,K	FURNACE TEMP.,K	TIME,SEC.
300	856.044542	0
340.916783	958.05692	300
405.028873	1018.18901	600
485.849637	1060.98221	900
578.741804	1094.22946	1200
678.926565	1121.42239	1500
780.585853	1144.42997	1800
876.780557	1164.37036	2100
960.561764	1181.96598	2400
1027.16274	1197.7107	2700
1075.85094	1211.95712	3000

```
UNPROTECTED SOLID STEEL


TIME STEP= 100 SEC.
AMBIENT TEMPERATURE= 300 K
```

AMB.TEMP.,K	FURNACE TEMP.,K	TIME,SEC.
300	698.590933	0
305.89894	797.113634	100
316.063726	856.044542	200
329.640074	898.229754	300
346.106569	931.109802	400
365.095922	958.05692	500
386.324778	980.88842	600
409.559114	1000.69648	700
434.594407	1018.18901	800
461.242925	1033.85116	900
489.324857	1048.02995	1000
518.661745	1060.98221	1100

```
549.071461            1072.90336         1200
580.364425            1083.94552         1300
612.340959            1094.22946         1400
644.789829            1103.85266         1500
677.488117            1112.89491         1600
710.202542            1121.42239         1700
742.692368            1129.49056         1800
774.713885            1137.14639         1900
806.026343            1144.42997         2000
836.398996            1151.37584         2100
865.618782            1158.01393         2200
893.497904            1164.37036         2300
919.880626            1170.46806         2400
944.648509            1176.32729         2500
967.723553            1181.96598         2600
989.0689             1187.40014         2700
1008.6871             1192.64409         2800
1026.61625            1197.7107          2900
1042.92448            1202.61156         3000
```

APPENDIX 8.13 ESTIMATION OF FIRE RESISTANCE OF INSULATED HOLLOW STEEL COLUMN

Estimate the fire resistance for hollow steel column insulated with 25 mm of suitable fire insulation material having a working thermal conductivity of $0.14\ W\ m^{-1}\ K^{-1}$

```
READY.

 1 REM INSULATED STEEL
 3 PRINTCHR$(147)
 5 INPUT"AREA OF INSULATION";A
 6 INPUT"THICKNESS OF INSULATION";P
 7 INPUT"THERMAL CONDUCTIVITY";C
 10 INPUT"AMBIENT TEMPERATURE";TS:S1=TS
 11 INPUT"TIME STEP";Q
 12 PRINT
 13 IF D=5THENOPEN3,4:CMD3:TS=S1
 14 PRINT"INSULATED STEEL":PRINT:PRINT
 15 PRINT"TIME STEP="Q"SEC.:";"AMBIENT TEMP.="TS"K:";"THERMAL CONDUCTIVITY="C
 16 PRINT"THICKNESS OF INSULATION="P"M:";"AREA OF INSULATION="A"M↑2":PRINT:PRI
 19 PRINT"  AMB.TEMP.,K","FURN.TEMP.,K","TIME STEP,SEC"
 20 FOR T=0TO7200STEPQ
 30 TF=345*LOG(.133*(T+Q)+1)/LOG(10)+300
 40 DTS=C*A*(TF-TS)*Q/(3.32E4*P)
 50 PRINTTS,TF,T
 60 TS=TS+DTS
 65 REM  IFTS>850THEN75
 70 NEXTT
 75 IFD=5THENPRINT#3:CLOSE3,4
 100 INPUT"DO YOU WANT A HARD COPY";A$
 110 IF A$="N"THEN200
 120 IFA$="Y"THEND=5:GOTO11
 200 END
READY.

INSULATED STEEL


TIME STEP= 300 SEC.:AMBIENT TEMP.= 300 K:THERMAL CONDUCTIVITY= .14
THICKNESS OF INSULATION= .025 M:AREA OF INSULATION= .6 M↑2


  AMB.TEMP.,K          FURN.TEMP.,K          TIME STEP,SEC
 300             856.044542          0
 316.882316           958.05692          300
 336.349304           1018.18901          600
 357.050943           1060.98221          900
```

378.423315	1094.22946	1200
400.156224	1121.42239	1500
422.054908	1144.42997	1800
443.987259	1164.37036	2100
465.859131	1181.96598	2400
487.60117	1197.7107	2700
509.161122	1211.95712	3000
530.499025	1224.96578	3300
551.58404	1236.93469	3600
572.392276	1248.01781	3900
592.905244	1258.33727	4200
613.108723	1267.99156	4500
632.991913	1277.06126	4800
652.546789	1285.61314	5100
671.767599	1293.70315	5400
690.650461	1301.37864	5700
709.193052	1308.68002	6000
727.394343	1315.64207	6300
745.254394	1322.29493	6600
762.774179	1328.66491	6900
779.95544	1334.77508	7200

INSULATED STEEL

TIME STEP= 100 SEC.:AMBIENT TEMP.= 300 K:THERMAL CONDUCTIVITY= .14
THICKNESS OF INSULATION= .025 M:AREA OF INSULATION= .6 M↑2

AMB.TEMP.,K	FURN.TEMP.,K	TIME STEP,SEC
300	698.590933	0
304.033932	797.113634	100
309.024137	856.044542	200
314.560247	898.229754	300
320.467264	931.109802	400
326.64726	958.05692	500
333.03743	980.88842	600
339.593995	1000.69648	700
346.28467	1018.18901	800
353.084666	1033.85116	900
359.974351	1048.02995	1000
366.937805	1060.98221	1100
373.961869	1072.90336	1200
381.035494	1083.94552	1300
388.149282	1094.22946	1400
395.295154	1103.85266	1500
402.466097	1112.89491	1600
409.655979	1121.42239	1700
416.859398	1129.49056	1800
424.071569	1137.14639	1900
431.28823	1144.42997	2000
438.505568	1151.37584	2100
445.720158	1158.01393	2200
452.928915	1164.37036	2300
460.129045	1170.46806	2400
467.318018	1176.32729	2500
474.493533	1181.96598	2600
481.653495	1187.40014	2700
488.795992	1192.64409	2800
495.919274	1197.7107	2900
503.021741	1202.61156	3000
510.101927	1207.35719	3100
517.158486	1211.95712	3200
524.190183	1216.42002	3300
531.195883	1220.75382	3400
538.174542	1224.96578	3500
545.1252	1229.06258	3600
552.046976	1233.05033	3700
558.939058	1236.93469	3800
565.800701	1240.72089	3900
572.631218	1244.41376	4000
579.429981	1248.01781	4100
586.196412	1251.53719	4200
592.929982	1254.97581	4300
599.630205	1258.33727	4400
606.296638	1261.62497	4500
612.928876	1264.84208	4600
619.526552	1267.99156	4700

```
626.08933      1271.0762     4800
632.616908     1274.09861    4900
639.109012     1277.06126    5000
645.565396     1279.96646    5100
651.985841     1282.8164     5200
658.37015      1285.61314    5300
664.718151     1288.35863    5400
671.029694     1291.05472    5500
677.304646     1293.70315    5600
683.542896     1296.30558    5700
689.744349     1298.86358    5800
695.90893      1301.37864    5900
702.036575     1303.85218    6000
708.127239     1306.28554    6100
714.180889     1308.68002    6200
720.197507     1311.03683    6300
726.177086     1313.35714    6400
732.119631     1315.64207    6500
738.025159     1317.89267    6600
743.893698     1320.10997    6700
749.725284     1322.29493    6800
755.519965     1324.44849    6900
761.277796     1326.57153    7000
766.998841     1328.66491    7100
772.683173     1330.72945    7200
```

In each case the fire resistance afforded to the insulated steel columns is greater than two hours.

REFERENCES

1. Petterson O, *et al.* '*Fire Engineering Design of Steel Structures*', Swedish Institute of Steel Construction, Publication 50, Stockholm 1976.
2. Law M, '*Prediction of fire resistance – Fire resistance requirements for buildings – a new approach*', J.F.R.O. Symposium No. 5, HMSO London 1971
3. Malhotra H L, 'Design of Fire-resisting Structures', Surrey University Press 1982.
4. Petterson O, 'The connection between a real fire exposure and the heating conditions according to standard fire resistance test', *European Convention for Constructional Steelwork*, 1974 Chap. 11. C.E.C.M.-111-74-2E.
5. Witteven J, '*Some aspects with regard to the behaviour and the calculation of steel structures in fire*', J.F.R.O. Symposium No. 2, HMSO, London 1968.
6. Cooke G, *Fire Protection*. Specification 1984.
7. Lie T T, 'Fire resistance of structural steel', *Engineering Journal American Institute of Steel Construction*, Fourth Quarter, Vol. 15, No. 4. 1978.
8. Cooke, G, '*New systems for fire protection: flexible approaches to fire resistance and passive protection*', I.F.S.E.C., London 1985.
9. CP 110, The Structural Use of Concrete, British Standard Code of Practice: Part 1: Design, materials and workmanship, Section 10: Fire resistance, British Standards Institution, London 1972.
10. Lie T T, '*Fire and Buildings*', Applied Science Publishers, 1972.
11. Read R H, *et al.* '*Guidelines for the construction of fire-resisting structural elements*', BRE Report, HMSO, 1981.

12. '*Design and detailing of concrete structures for fire resistance*', Interim guidance by a joint committee of the Institution of Structural Engineers and the Concrete Society, The Institution of Structural Engineers, London 1978.
13. Schedule 8, The Building Regulations 1976, HMSO, London 1976.
14. Forrest J C M and Law M, '*Guidance for the application of tabular data for fire resistance of concrete elements*', The Institution of Structural Engineers, London 1984.
15. FIP/CEB, '*Report on methods of assessment of the fire resistance of concrete structural members*', Cement and Concrete Association, Wexham Springs, Slough 1978.
16. The Structural use of Timber, British Standard Code of Practice BS 5268: Part 4: Section 4.1, 1978.
17. The Structural Use of Timber, British Standard Code of Practice CP 112: Part 2: 1971. British Standards Institution, London.
18. The Structural Use of Timber, British Standard Code of Practice for permissible stress design, materials and workmanship, BS 5268: Part 2: 1984. British Standards Institution, London 1984.
19. Rogowski B F W, '*Charring of timber in fire tests*', Ministry of Technology Fire Officers Committee, Joint Fire Research Organisation Symposium No. 3, HMSO, London 1970.
20. Schaffer E L, '*Charring rate of selected woods – transverse to grain*', U.S. Forestry Service Paper FPL.69, US Dept. Agriculture, 1967.
21. Lie T T, 'A method for assessing the fire resistance of laminated timber beams and columns', *Canadian Journal of Civil Engineering*, **4**, 1977.

CHAPTER 9

Smoke production, movement and control

9.1 INTRODUCTION

In Chapter 1 the historical development of fire safety legislation and fire testing were briefly outlined. It should be noted that the primary occupation of the legislators was concerned with structural fire precautions and smoke was not recognised as a life-threatening agent. An introduction is also given to fire statistics in Chapter 1 and an analysis of current fire statistics will indicate that the majority of fire casualties are a consequence of smoke production and movement during a fire.

The generic title Smoke Production, Movement and Control obscures in part a very complex system which involves the chemical and physical properties of materials, components and building sub-systems. The production of smoke in the fire process is a very complex mechanism and although methods of assessing the propensity of materials to produce smoke have been developed, none so far have been incorporated into fire safety legislation. However, it is known that many materials produce vast quantities of smoke rapidly and it is the rate of smoke production which is important. Perhaps what the professions require is a method of testing, the results of which could be used to grade or rank materials in terms of their potential smoke production under varrious burning regimes.

This Chapter serves as a brief introduction to a very complex area of fire safety engineering combining the chemistry, physics, fluid flow and thermodynamics of smoke production and control.

9.2 SMOKE PRODUCTION

Smoke[1] can be considered in general terms to exist in the form of a cloud of small visible solid and liquid particles that accompany the buoyant plume of hot gases and vapours that arise from a fire bed. In particular terms, smoke consists of the by-products of organic materials that undergo the combustion process. Smoke is created usually during an incomplete combustion process such as:

1. Flaming combustion where a complex reaction series occurs in which the oxidation is too slow to prevent carbon (smoke) particles forming, or
2. A smouldering process where small droplets of tarry substance in a moist form escape when air conditions permit, to produce smoke particles of some 10^{-3} mm in diameter.

The yellowish flames which occur above a solid or liquid fire are due to the presence of small soot particles which either burn away if conditions allow or grow in size and escape from the flame as smoke.

9.2.1 Quantity of smoke

The quantity of smoke release from a flame of a burning material depends on two factors:

1. The chemical nature of the fuel. It has been concluded from investigations that:
   (a) materials or fuels containing within their chemical structure molecules or atoms of oxygen give less smoke than those not containing oxygen
   (b) materials which incorporate within their structure the benzene ring tend to yield large quantities of smoke
2. The fire environment. In this case the quantity of smoke has been shown to depend on:

   (a) the temperature of the combustion and flame zone

   (b) the concentration level of oxygen available to the combustion zone which is in turn directly linked to the ventilation rate of the fire enclosure.

9.2.2 Estimation of quantity of smoke produced

It has been shown that the total volume of air entrained by a fire is so large a component compared with the volume of fuel gases produced that it is possible to equate the production rate of smoke with the rate of air entrainment.[2] The mass rate of air entrainment by a fire and hence the rate of smoke production may be expressed as:

$$\dot{m} = 0.096 P \rho_0 y^{\frac{3}{2}} g^{\frac{1}{2}} \left(\frac{T_0}{T}\right)^{\frac{1}{2}} \qquad \ldots [1]$$

where T_0 = absolute temperature of ambient air (K)
T = absolute temperature of gases in buoyant plume (K)
y = distance between floor and bottom of smoke layer (m)
ρ_0 = density of air at ambient temperature ($kg\,m^{-3}$)
g = acceleration due to gravity = $9.8\,m\,s^{-2}$
$\dot{m}$ = smoke production rate ($kg\,s^{-1}$)
P = perimeter of the fire (m).

Assuming that $\rho_0 = 1.22\,kg\,m^{-3}$ at 300 K

$$T = 1{,}200\,K$$

$$g = 9.81\,m\,s^{-2}$$

$$T_0 = 300\,K.$$

It can be shown that eqn [1] can be written as:

$$\dot{m} = 0.18 P \cdot y^{\frac{3}{2}} \qquad \ldots [2]$$

Thus it is assumed that the smoke production rate depends on the factors expressed in [2].

The volume rate at which smoke is produced depends on the density of air at an elevated temperature and can be easily calculated using the fact that

$$\rho \propto \frac{1}{T}$$

or $$\frac{\rho_1}{\rho_2} = \frac{T_2}{T_2} \qquad \ldots [3]$$

The perimeter P of the fire may for design and calculation purposes be taken as follows for a range of fires considered in earlier Chapters:

$\dot{Q} = 5.0$ MW (large fire) $\quad P = 4 \times 3 = 12$ m

$\dot{Q} = 1.25$ MW (small fire) $\quad P = 4 \times 1.5 = 6$ m

$\dot{Q} = 0.5$ MW (very small fire) $\quad P = 4 \times 1.0 = 4$ m

WORKED EXAMPLE 1

Calculate

1. Initial theoretical smoke production rate for a small fire in a partially-ventilated enclosure 8 m × 5 m × 3 m, a sketch of which is given in Fig. 9.1.
2. Time taken for the smoke layer to move downwards to a level 1.5 m above the floor.

Before completing this calculation it is necessary to consider the analysis required to generate the equations necessary for solutions.

Analysis

As shown in Fig. 9.1, let the height of the base of the smoke layer above the floor be y, the floor area be A and enclosure height be h. Thus at any time t

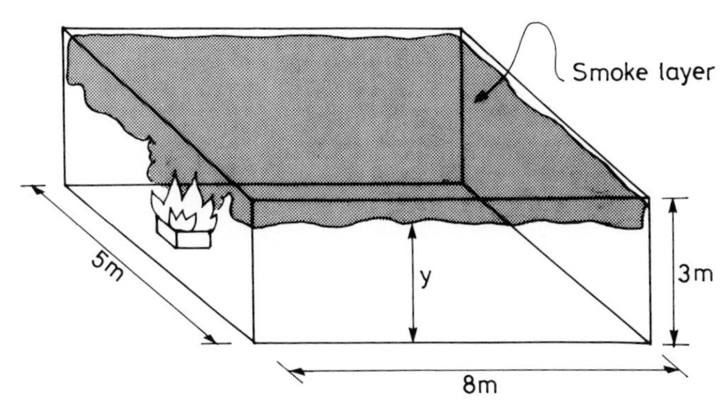

Fig. 9.1 Height of smoke layer above floor level

after the start of smoke production the smoke mass production rate can be expressed as:

$$\dot{m} = \frac{\mathrm{d}}{\mathrm{d}t}[\rho^1 A(h - y)] = -\rho^1 A \frac{\mathrm{d}y}{\mathrm{d}t} \qquad \ldots [4]$$

where ρ^1 = density of the smoke at ceiling level where the temperature may be assumed to be 400–500 °C.

Now eqn [4] into eqn [1] gives:

$$0.096 P \rho_0 g^{\frac{1}{2}} \left(\frac{T_0}{T}\right)^{\frac{1}{2}} y^{\frac{3}{2}} = -\rho^1 A \frac{\mathrm{d}y}{\mathrm{d}t}$$

Rewriting letting $K = 0.096\rho_0 \left(\frac{T_0}{T}\right)^{\frac{1}{2}} \frac{1}{2} - g^{\frac{1}{2}}$

i.e.,
$$t = -\frac{\rho^1 A}{KP}\int \frac{dy}{y^{\frac{1}{2}}}$$

and integrating and using boundary conditions when $t=0$, $y=h$

i.e.
$$t = \frac{\rho^1 A}{KP}\left[\frac{+2}{y^{\frac{1}{2}}}\right] + \text{const}$$

where
$$\text{const} = \frac{-2\rho^1 A}{KPh^{\frac{1}{2}}}$$

$$\therefore \quad t = \frac{2\rho^1 A}{KP}\left[\frac{1}{y^{\frac{1}{2}}} - \frac{1}{h^{\frac{1}{2}}}\right]$$

$\therefore$ since $\rho^1 = \rho_0 \dfrac{300}{673}$

(assuming gas temperature = 673 K at ceiling)

then
$$t = \frac{2\rho_0\left(\frac{300}{673}\right)A}{0.096\rho_0\left(\frac{300}{1{,}200}\right)^{\frac{1}{2}} g^{\frac{1}{2}}P}\left[\frac{1}{y^{\frac{1}{2}}} - \frac{1}{h^{\frac{1}{2}}}\right]$$

at $T = 1{,}200$ K (fire temperature).

$$\therefore \quad t = \frac{A\left(\frac{300}{673}\right)\times 2}{0.096 g^{\frac{1}{2}}P}\left[\frac{1}{y^{\frac{1}{2}}} - \frac{1}{h^{\frac{1}{2}}}\right]$$

$$\therefore \quad t = \frac{4A \times 0.45}{0.096 g^{\frac{1}{2}}P}\left[\frac{1}{y^{\frac{1}{2}}} - \frac{1}{h^{\frac{1}{2}}}\right]$$

$$\therefore \quad t = \frac{20A}{Pg^{\frac{1}{2}}}\left[\frac{1}{y^{\frac{1}{2}}} - \frac{1}{h^{\frac{1}{2}}}\right] \qquad \ldots [5]$$

Solution

Here $A = 40\ \text{m}^2$
$P = 6$ m (assuming a small fire)
$g = 9.81\ \text{m s}^{-2}$
$h = 3.0$ m.

Then

1.
$$\dot{m} = 0.18 P y^{\frac{3}{2}}$$

$\therefore$ When $y = 3.0$ m
$$\dot{m} = 0.18 \times 6 \times 3^{\frac{3}{2}}$$
$$= 5.6\ \text{kg s}^{-1} \text{ (approx.)}.$$

2. Using $$t = \frac{20A}{Pg^{\frac{1}{2}}}\left[\frac{1}{1.5^{\frac{1}{2}}} - \frac{1}{3.0^{\frac{1}{2}}}\right]$$

$$= \frac{20 \times 40}{6 \times 3.13}[0.816 - 0.577]$$

$\therefore$ $$t \simeq 10 \text{ seconds.}$$

So after a time of 7 seconds or so the smoke layer from this small fire is estimated to have dropped to a level 1.5 m above the floor. The reader must appreciate that in generating this simple mathematical model several assumptions were made which may be inadequate for a real fire situation within a ventilated enclosure which suggests a longer time than 10 seconds is available for occupants to make good an escape.

9.3 VISIBILITY AND SMOKE DENSITY

9.3.1 Obscuration and optical density

Smoke due to its particulate nature can reduce drastically the ability of a person to see when trying to escape from a fire zone. This reduction in the visibility depends on the nature of the smoke and the type and level of illumination of the escape route. Smoke density can be measured in terms of reduction of the intensity of a light beam as it passes through a smoke-laden atmosphere, see Fig. 9.2.

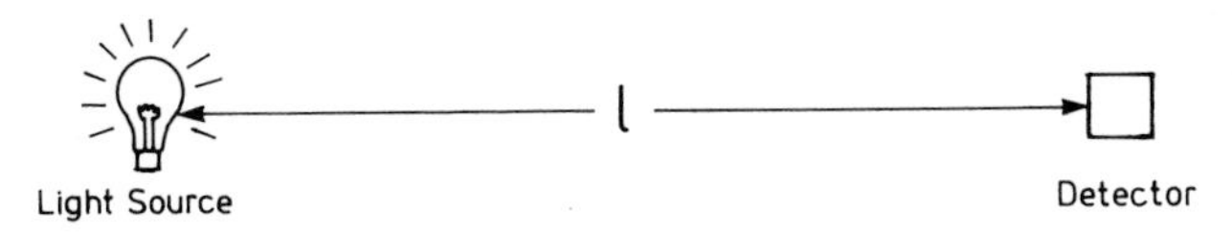

Fig. 9.2 Basic elements of optical smoke detector

This type of objective measurement can be expressed as either light obscuration or optical density.

1. Light Obscuration S is defined as:

$$S = 100 \times \left(\frac{I_0 - I}{I_0}\right)$$

when I_0 is the intensity of incident light with no smoke present. I is the intensity measure at same distance from the light source with smoke present (Fig. 9.2).

2. Optical density D is defined as:

$$D = -10\log_{10}\left(\frac{I}{I_0}\right) \text{ and is given in db(s)}$$

or $D = 10\log_{10}\left(\frac{I_0}{I}\right)$

This logarithmic scale can be shown to relate to a person's subjective visibility.

Assuming that Beer's law holds:

i.e. $I = I_0\,\mathrm{e}^{-kc\ell}$

where k is some constant depending on the smoke source,
c is the concentration of smoke particles,
ℓ is the path length traversed by light in Fig. 9.2.

$$\therefore \quad \frac{I}{I_0} = \mathrm{e}^{-kc\ell}$$

$$\frac{I_0}{I} = \mathrm{e}^{kc\ell}$$

$$\therefore \quad \ln\left(\frac{I_0}{I}\right) = kc\ell$$

$$\therefore \quad \frac{1}{2.302}\cdot\ln\left(\frac{I_0}{I}\right) = \frac{kc\ell}{2.302} = \log_{10}\left(\frac{I_0}{I}\right)$$

Multiplying by 10

$$\therefore \quad \frac{10kc\ell}{2.302} = 10\log_{10}\left(\frac{I_0}{I}\right) = D \text{ (optical density)}$$

$\therefore \quad D \propto c$

and $\quad D \propto \ell$

i.e. $\quad \dfrac{D}{\ell} = \dfrac{10}{2.302}kc$

Here D/ℓ is expressed as optical density per metre.

This relationship relates the D/ℓ to smoke concentration and thus to the visibility.

From experimental work,[3] it has been shown that it is possible to relate optical density per metre D/ℓ to visibility,

i.e. For front illumination of a sign

$$\text{Visibility (m)} = \frac{10}{\dfrac{D}{\ell}}$$

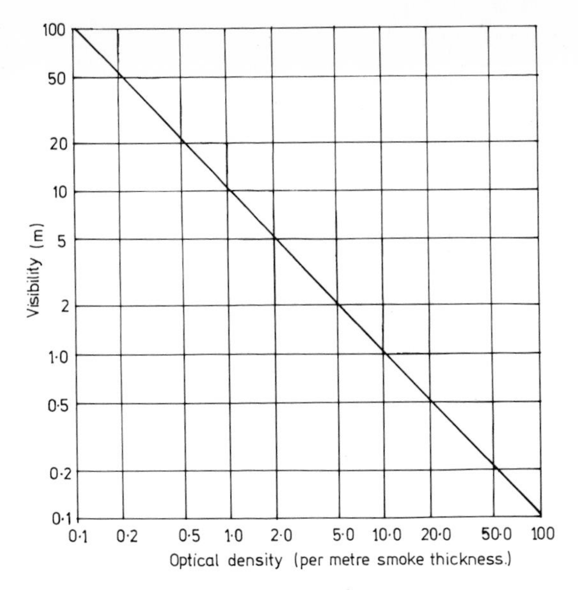

Fig. 9.3 Comparison of visibility measurements (front illumination)

For back illumination of a sign

$$\text{Visibility (m)} = \frac{25}{\frac{D}{\ell}}$$

The experimental results are expressed graphically for the front illumination case, see Fig. 9.3.

Thus it follows that if D/ℓ decreases as it must do if the smoke concentration c is decreased then the visibility will increase. Such a decrease in smoke concentration can be achieved by diluting the smoke with fresh air. This will be discussed later under smoke control tactics.

WORKED EXAMPLE 2

Calculate the smoke dilution by fresh air necessary to increase the visibility in an egress route from 2 m to 5 m.

Solution

Assuming that signs are front illuminated then referring Fig. 9.3:

$$\text{Visibility} = \frac{10}{D/\ell}$$

$\therefore \quad D/\ell_1 = 5\,\text{dB}$

$\quad\quad D/\ell_2 = 2\,\text{dB}$

Since $D_\ell \propto$ concentration of smoke particles

$$\therefore \quad \frac{c_1}{c_2} = \frac{D/\ell_1}{D/\ell_2} = \frac{5}{2}$$

$\therefore \quad c_1 = \tfrac{5}{2}c_2$

$\therefore$ The smoke must be diluted by 2.5 times its volume of fresh *air*.

9.3.2 Smoke loading and visibility

The knowledge of the propensity of a material to produce smoke is important and can be measured experimentally under controlled conditions in a smoke density chamber[4] where the smoke is collected in a fixed volume and the resulting observation measured.

This is expressed in terms of the specific optical density D_m which is defined as:

$$D_m = \frac{1}{10} \cdot \frac{D}{\ell} \cdot \frac{V}{A_s} \quad \text{(bels)}$$

where V = volume of chamber (m^3)
A_s = exposed surface area of sample.

It has been noted that D_m is dependent on the thickness of the sample and an alternative method of expressing and measuring the smoke production potential[5] is given by the quantity Standard Optical Density, D_0, where:

$$D_0 = \frac{D}{\ell} \cdot \frac{V}{W_L} \quad (\text{db m}^{-1}\,.\,\text{m}^3\,\text{g}^{-1})$$

where W_L is the mass of the material that has been burnt away during the test (measured in grams).

Values of D_0 as determined at the Fire Safety Engineering Unit of Edinburgh University are given in Table 9.1 for flaming and non-flaming combustion. Using this information it is possible to get an approximate value for the 'smoke load' and the visibility potential for a compartment housing a given fire load during a fire situation.

Table 9.1

Material	Standard optical density {db m^{-1} m^3 g^{-1}} OR {obscura m^3 g^{-1}}	
	Flaming	Non-flaming
Fire insulating board	0.6	1.8
Chipboard	0.37	1.9
Hardboard	0.35	1.7
Birch plywood	0.17	1.7
External plywood	0.18	1.5
Cellulose	0.22	2.4
Rigid PVC	1.7	1.8
Rigid PVF	4.2	1.7
Flexible PVF	0.96	5.1
Plasterboard	0.042	0.39

WORKED EXAMPLE

Determine the visibility in a room of 200 m^3 volume after a 2.0 kg block of rigid polyurethane has undergone flaming combustion to leave behind 20 per cent by weight of residual char.

Solution

20 per cent by weight left, ∴ 80 per cent burnt away
∴ 1.6 kg burnt away, i.e. 1600 g burnt away.
Consulting Table 9.1, $D_0 = 4.2$

Using the expression $$D_0 = \frac{D}{\ell} \cdot \frac{V}{W_L}$$

$$4.2 = \frac{D}{\ell} \cdot \frac{200}{1{,}600}$$

∴ $$\frac{D}{\ell} = 33.6$$

From the graph Fig. 9.3 or the expression for visibility,

i.e., $$\text{Visibility} = \frac{10}{D/\ell} = \frac{10}{33.6} \simeq 0.3\ \text{m}$$

(assuming front illumination).

Table 9.2 Some toxic species in fire gases

Species	TLV (ppm)	STEL (ppm)	Comments
Carbon monoxide	50	400	0.3% fatal in 30 minutes
Carbon dioxide	5,000	1,500	12% causes immediate unconsciousness and rapid death
Hydrogen cyanide	10	–	280 ppm immediately fatal
Ammonia	25	35	irritant in 500 ppm – 0.2% is fatal
Nitrogen dioxide	3	5	250 ppm fatal in a few minutes

TLV = Threshold Limit Value (for an 8-hour time weighted average)
STEL = Short Term Exposure Limit
(Figures taken from H.S.E. Guidance Note EH.15/79)

9.3.3 Toxic combustion products

Many of the toxic products like smoke are the result of incomplete combustion. Table 9.2 lists the important species, indicating just how lethal some of the toxicants are.

The most important toxic product from any fire situation is carbon monoxide, the yield of which has been shown to depend on the fuel and level of ventilation. Rabash and Stark[6] showed that the production of carbon monoxide in small-scale enclosure fires correlated well with $A\sqrt{H/W}$ where $A\sqrt{H}$ is the opening factor for the enclosure and W was the fire load.

In addition to carbon monoxide, hydrogen cyanide can be produced from materials containing nitrogen within their chemical structure. Materials such as nylon and polyurethane would then be expected to yield hydrogen cyanide during combustion. This has been shown to be the case both experimentally during laboratory tests and also in full-scale experimental fires.

Hydrogen cyanide is approximately ten-times more toxic than carbon monoxide. However, since the yield of the HCN is usually very low even for fires involving large quantities of polyurethane, it follows that carbon monoxide remains the major hazard.

As stated, some materials produce large quantities of smoke and toxic gases, see Table 9.2. It thus follows that these materials should not be used in situations when long egress or escape times are necessary or where the occupants suffer from respiratory diseases.

So if the identification of such hazardous materials can be made easy by direct identification or by reference to tests in order to eliminate their use, then the chance of excessive smoke and toxic gas production can be minimised to create a less dangerous environment in the event of a fire.

9.4 FLUID FLOW

In order that the reader can acquaint himself more completely with this subject it is necessary to have a brief revision of the elementary aspects of fluid flow. The following paragraphs deal with the relevant aspects of this subject.

9.4.1 Bernoulli flow equations

An ideal fluid is considered to be non-viscous and incompressible. Such a flowing fluid can be pictured as shown in Fig. 9.4 using streamlines which in the non-turbulent case indicates the direction of the fluid particles as they move.

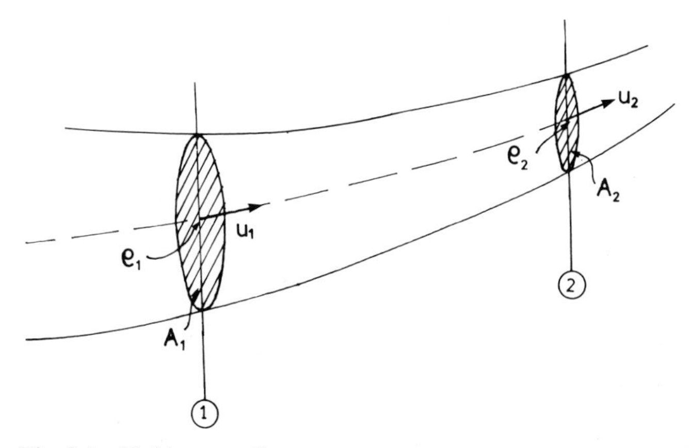

Fig. 9.4 Fluid stream lines

The conservation of mass yields the following continuity equation:

$$\rho_1 u_1 A_1 = \rho_2 u_2 A_2 = \dot{m}$$

or when $\rho_1 = \rho_2$ then $A_1 u_1 = A_2 u_2 = \dot{Q}$

where:

$\dot{m}$ = mass flow rate of fluid in tube
$\dot{Q}$ = volume flow rate
A = transverse area of the streamtube
ρ = density of the fluid.

Considering now the energy of the flowing fluid as it flows along the streamtube, which must be conserved also.

This is adequately expressed by the Bernoulli Equation which in its elementary form considers the kinetic energy of translation, the pressure energy and the potential energy at any point in the fluid to remain constant.

Referring to the two reference points, Fig. 9.4, the Equation states the energy conservation as follows:

$$p_1 + \tfrac{1}{2}\rho_1 u_1^2 + \rho_1 g h_1 = p_2 + \tfrac{1}{2}\rho_2 u_2 L + \rho_2 g h_2 \qquad \ldots[6]$$

where p = pressure in fluid
h = height of point above a datum.

It should be noted that each of the terms in eqn [6] has the dimension of pressure which is the energy per unit volume of the flowing fluid. Equation [6] has no term relating to thermal energy, conversion of energy to heat via viscosity or to work done by the fluid itself. Thus eqn [6] as given refers to a situation where these factors are not present or where they can be neglected.

Measuring fluid flow

The measurement of fluid flow is an important aspect of fluid mechanics and, as such, various methods by which it can be determined will now be discussed.

Venturi meter

A sketch of a Venturi meter is shown in Fig. 9.5. From the application of Bernoulli's Equation it is clear that when a fluid that is moving horizontally it speeds up causing a localised decrease in fluid pressure. This change in pressure will be picked up by the manometer which registers it as a difference in height h of the manometer fluid, i.e., using eqn [6]

$$p_1 + \tfrac{1}{2}\rho u_1^2 = p_2 + \tfrac{1}{2}\rho u_2^2$$

and the fact that $A_1 u_1 = A_2 u_2$:

and for the manometer

$$p_1 + \rho g h_1 = p_2 + \rho g(h_1 - h) + \rho_G g h$$

where ρ_G = density of manometer fluid.

Rearranging these equations,

$$\therefore \quad \dot{Q} = u_2 A_2 = A_2 \left[\frac{2gh(\rho_G - \rho)}{\rho\left[1 - \left(\frac{A_2}{A_1}\right)^2\right]} \right]^{\frac{1}{2}}$$

As an illustration a worked example will be given in the next paragraph.

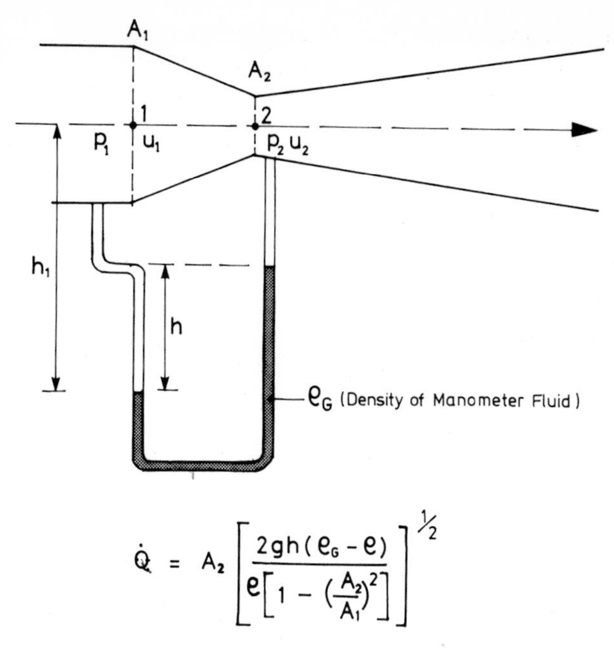

$$\dot{Q} = A_2 \left[\frac{2gh(\varrho_G - \varrho)}{\varrho\left[1 - \left(\frac{A_2}{A_1}\right)^2\right]} \right]^{1/2}$$

Fig. 9.5 Venturi meter

WORKED EXAMPLE 4

The rate of air flow in a circular pipe 300 mm diameter is measured by means of a Venturi meter of throat diameter 150 mm (see Fig. 9.6). If the manometer uses a liquid of density 800 kg m^{-3}, determine the rate of air flow when each reading S is 200 mm, assuming density of air to be 1.3 kg m^{-3}.

Solution

Using the expression derived in previous paragraphs:

$$A_1 = \frac{\pi \times 300^2}{4} \text{mm}^2 \qquad A_2 = \frac{\pi \times 150^2}{4} \text{mm}^2$$

$$\therefore \quad \frac{A_2}{A_1} = \frac{1}{4} \qquad \left(\frac{A_2}{A_1}\right)^2 = \frac{1}{16}$$

$$h = \text{S} \sin 30^\circ$$

$$\therefore \quad h = 2 \times 10^{-1} \sin 30$$

$$= \underline{10^{-1} \text{ m}}$$

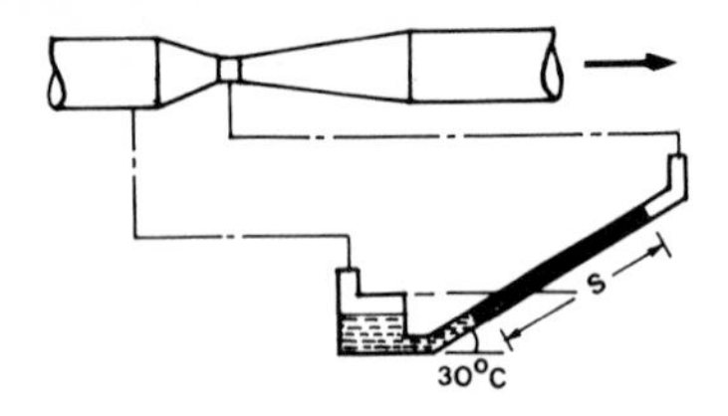

Fig. 9.6 Venturi meter and inclined manometer for worked example

$$\therefore \quad \dot{Q} = \frac{\pi(1.5)^2 \times 10^{-2}}{4} \sqrt{\frac{2 \times 10 \times 10^{-1}(800 - 1.3)}{1.3(1 - \frac{1}{16})}}$$

$$= \frac{\pi \times 2.25 \times 10^{-2}}{4} \cdot 4 \sqrt{\frac{2 \times 798.7}{1.3 \times 15}}$$

$$= \pi \times 2.25 \times 10^{-2} \times 9.05$$

$$\dot{Q} = \underline{0.639 \text{ m}^3 \text{ s}^{-1}} \quad \text{(Ans)}$$

Pitot-static tube

The pitot-static tube is an alternative method to the Venturi meter and in its simplest form consists of a thin bent tube inserted into the fluid as shown in Fig. 9.7. When the fluid flowing along streamline passing through (1) is brought to rest at (2) the tip of the tube, using Bernoulli's Equation:

$$p_1 + \tfrac{1}{2}\rho u_1^2 = p_2 + 0$$

$$p_2 - p_1 = \tfrac{1}{2}\rho u_1^2$$

$$\therefore \quad u_1 = \left[\frac{2(p_2 - p_1)}{\rho}\right]^{\frac{1}{2}}$$

If an inclined monometer is used to record the pressure difference $p_2 - p_1$ then:

$$u_1 = \left[\frac{2gh(\rho_G - \rho)}{\rho}\right]^{\frac{1}{2}}$$

where $\rho_G =$ density of fluid in manometer.

Again to give the reader a better understanding of the pitot-static tube, a worked example is presented below.

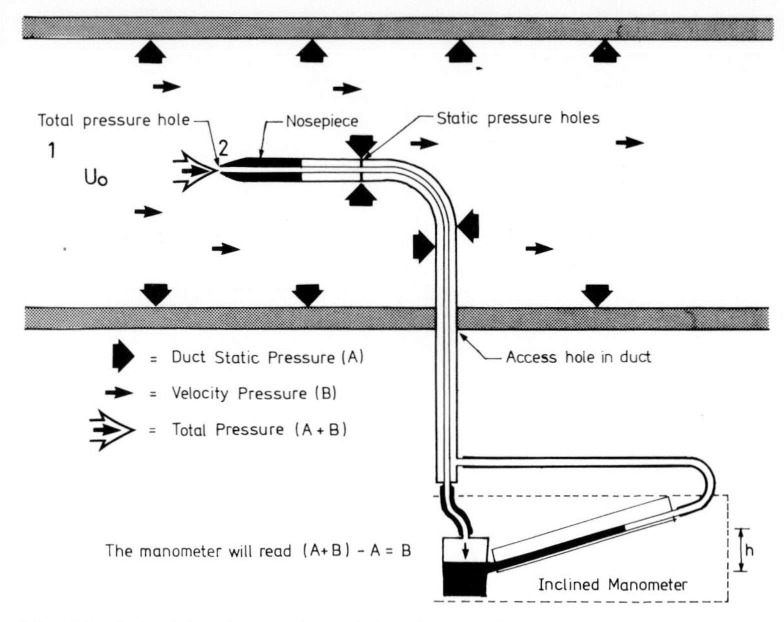

Fig. 9.7 Principle of operation of the pitot-static tube

WORKED EXAMPLE 5

Derive an expression using first principles for the velocity of fluid flow using the pitot tube arrangement shown in Fig. 9.8 and estimate the length L of the included oil column in the inclined manometer given that the density of oil is $800\,\mathrm{kg\,m^{-3}}$ and the velocity and density of air are $15\,\mathrm{m\,s^{-1}}$ and $1.2\,\mathrm{kg\,m^{-3}}$, respectively.

Assume that g (acceleration of gravity) $= 10\,\mathrm{m\,s^{-2}}$.

Solution

For first part see notes in previous paragraph.

$$u_1 = \left[2gh\frac{(\rho_G - \rho)}{\rho}\right]^{\frac{1}{2}}$$

Here $h = L \sin 10^\circ$

$$h = L \times 0.174$$

$$\therefore \quad u_1 = \left[\frac{2 \times 10 \times L \times 0.174(\rho_G - \rho)}{\rho}\right]^{\frac{1}{2}}$$

$$= \sqrt{\frac{3.48L \times 798.8}{1.2}}$$

$$u_u = 48.13\sqrt{L}$$

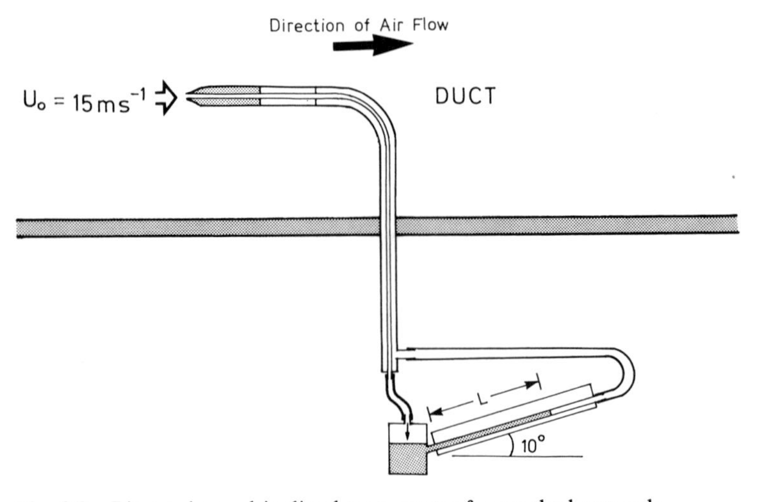

Fig. 9.8 Pitot-tube and inclined manometer for worked example

$\therefore$ where $u_1 = 15\,\mathrm{m\,s^{-1}}$

$$\sqrt{L} = \frac{15}{48.13}$$

$$\therefore \quad L = \left(\frac{15}{48.13}\right)^2 = 0.097\,\mathrm{m}$$

$$L = \underline{9.7\,\mathrm{cm}} \quad \text{or 97 mm } (Ans)$$

Orifice flow meter

It is now necessary to consider the flow and a fluid through an opening or orifice as shown in Fig. 9.9.

Using Bernoulli's Equation

$$p_1 + \tfrac{1}{2}\rho u_1^2 = p_2 + \tfrac{1}{2}\rho u_2^2$$

Since $u_1 = 0$ in this case

$$\therefore \quad p_1 - p_2 = \tfrac{1}{2}\rho u_2^2$$

$\therefore$ Theoretical flow rate $\dot{Q}_{\mathrm{Th.}} = A\left[\dfrac{2}{\rho}\cdot(p_1 - p_2)\right]^{\frac{1}{2}}$

$\therefore$ Actual flow rate is $\dot{Q} = CQ_{\mathrm{Th.}}$, where C = discharge coefficient.

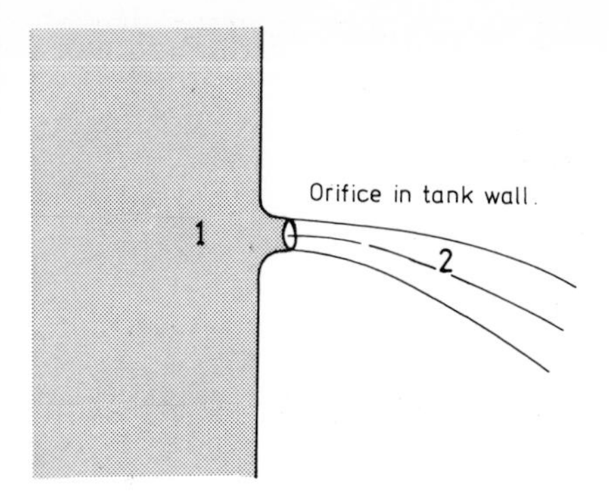

Fig. 9.9 Orifice flow

It must be noted that C can be as large as 0.98 for a well rounded entrance as shown in Fig. 9.9 or as small as 0.6 for sharp or square-edged orifices.

WORKED EXAMPLE 6

Calculate the air-leakage rate through an orifice having dimensions (800 mm × 1,000 mm) in an external wall of a building when a pressure difference between the internal and external environment is 50 Pa.

Solution

$$\dot{Q} = CA\left[\frac{2\,\Delta p}{\rho}\right]^{\frac{1}{2}}$$

Assume in this case $C = 0.6$ (square-edged orifice) and density of air $= 1.3\ \text{kg m}^{-3}$

$$\dot{Q} = 0.6 \times 8 \times 10^{-1}\left[\frac{2 \times 50}{1.3}\right]^{\frac{1}{2}}$$

$$= 4.8 \times 10^{-1}\sqrt{\frac{100}{1.3}}.$$

$$\dot{Q} = \underline{4.2\ m^3\ s^{-1}} \quad (Ans)$$

9.4.3 Fluid flow over surfaces

As an application of the Bernoulli's Equation the flow of air or a liquid around an aerofoil and a flat plate will now be considered briefly.

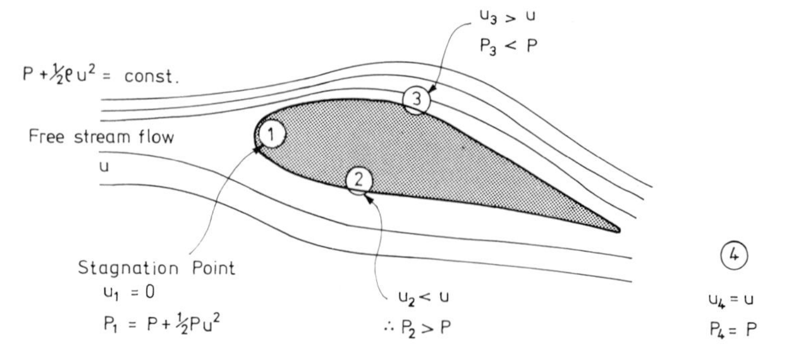

Fig. 9.10 Streamlines around aerofoil

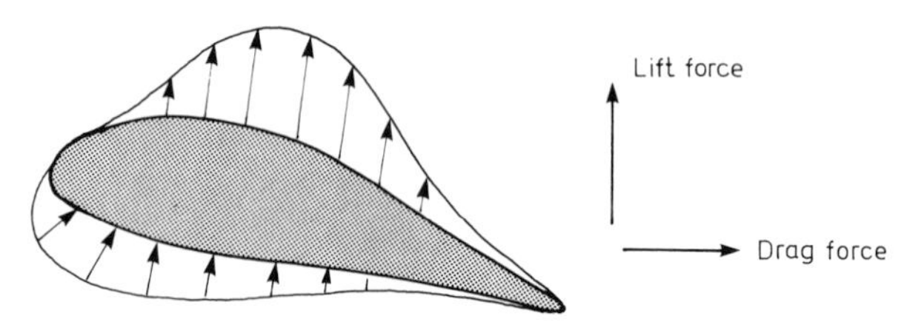

Fig. 9.11 Variation of lift forces over aerofoil

Aerofoil flow

As shown in Fig. 9.10 the streamlines become crowded above and separated below the aerofoil, creating an increase in local pressure below and decrease above the aerofoil.

The forces thus produced are as shown in Fig. 9.11 where the resultant force consist of a lift and drag component.

Flat plate flow

As shown in Fig. 9.12, the velocity of the fluid is zero at the surface and increases in the vertical direction to a maximum (free-stream velocity). This is shown as O BQ in Fig. 9.12. This velocity variation takes place in a 'Boundary Layer'.

The laws governing the velocity distribution across the boundary layer are different in the streamlined section from that of the turbulent region. It has been assumed that within this layer the fluid flow is parallel to the plate.

Further information regarding the significance and properties of the boundary layer are to be found in texts concerned solely or mainly with fluid mechanics. A simple reference will be made later in this Chapter to air flow over buildings and land surfaces.

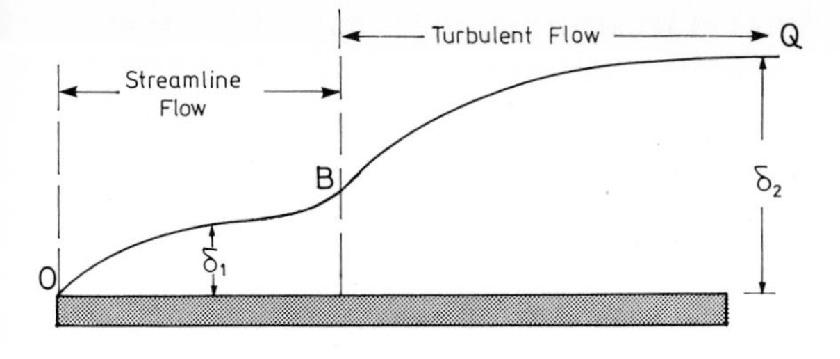

Fig. 9.12 Transition from streamlined to turbulent flow

9.5 SMOKE MOVEMENT

9.5.1 Introduction

In fire safety terms, smoke is always associated with the toxic and hazardous products of combustion which are driven away from the fire bed by the buoyant forces generated by the gas density changes within the fire zone. As discussed in Chapter 3, the convection process entrains air into the combustion zone and into the buoyant fire-gas plume. Since the temperature of this plume is not high enough to cause complete mixing of the oxygen in the entrained air with the volatiles, the reaction is chemically incomplete. It is by this mechanism that smoke and toxic products such as carbon monoxide are produced. Thus it is this unreacted air which has been entrained with the smoke which is considered to form the major component of the fire gas plume.

9.5.2 Forces available for smoke movement

The total force that acts on any discrete compartment within a building consists of four components.[7] Namely, the pressue due to:

1. Stack effect
2. Wind pressure effect
3. Mechanical air handling systems
4. Fire stack effect,

i.e. $P_{\text{total}} = (p_1) + (p_2) + (p_3) + (p_4)$ ...[7]

The total force produced by the summation of these pressures may yield an overall positive or negative pressure compared to the spaces surrounding the fire enclosure or space.

The following paragraphs describe and explain each of the four pressure components listed in [7].

The stack effect

It is well known and appreciated that the pressure in a fluid increases with depth. Consider the case of a cylindrical section of a fluid, Fig. 9.13, where $\delta p = -\rho g\,\delta h$ and the increase in height is measured as δh.

$$\therefore \quad \mathrm{d}p = -\rho g\,\mathrm{d}h$$

$$\mathrm{d}h = \frac{-\mathrm{d}p}{\rho g}$$

Using the Gas Laws it follows that:

$$\therefore \quad \frac{p_1}{\rho_1} = \frac{p_0}{\rho_0}$$

$$\therefore \quad g\rho = \frac{p\rho_0 g}{p_0}$$

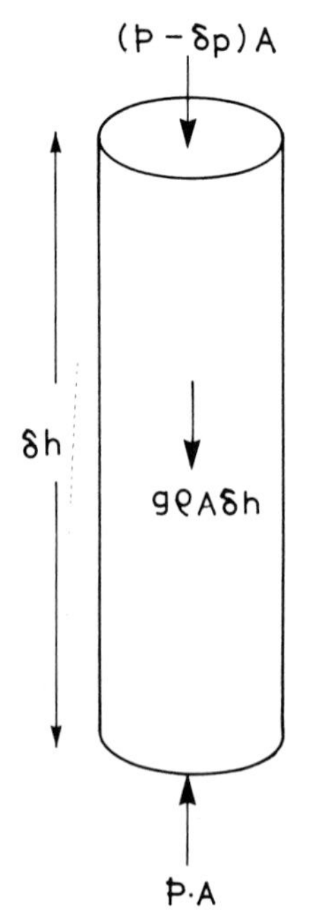

Fig. 9.13 Cylindrical section through homogeneous fluid

where p_0 is pressure at reference level.

$$\therefore \quad dh = \frac{-dp \cdot p_0}{pg\rho_0}$$

This is now integrated

$$h = -\frac{p_0}{\rho_0 g} \ln p + \text{const}$$

Using the fact that when $h = 0$, $p = p_0$

$$\therefore \quad \text{const} = \frac{-p_0}{\rho_0 g} \ln p_0$$

$$\therefore \quad h = \frac{-p_0}{\rho_0 g} \ln\left(\frac{p}{p_0}\right)$$

$$\therefore \quad e^{\frac{-\rho_0 gh}{p_0}} = \frac{p}{p_0}$$

$$\therefore \quad p = p_0 \, e^{\frac{-\rho_0 gh}{p_0}}$$

This can be expanded and the higher order terms ignored.

$$p(h) = p_0 \left[1 - \frac{h\rho_0 g}{p_0} + (\quad)^2 + \cdots \underset{(\leftarrow ignored \rightarrow)}{}\right]$$

$$\therefore \quad p(h) = p_0 - h\rho_0 g$$

This may be expressed in graphical terms as follows (Fig. 9.14), where the slope is $-(\rho_0 g)$.

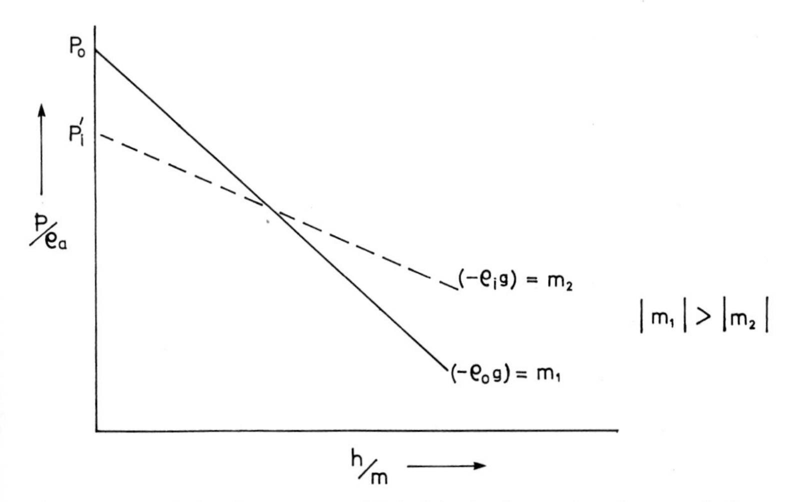

Fig. 9.14 Variation in pressure with height for internal and external air

A similar argument holds for the internal air but in this case since usually:

$T_i > T$ then $\rho_0 > \rho_i$

which means that

$|-\rho_0 g| > |-\rho_i g|$

This fact can now be exploited to consider the variation in pressure in a vertical shaft ventilated at top and bottom, Fig. 9.15. As can be seen from Fig. 9.15 the change in external pressure is more rapid than that inside. This creates a situation where a cross-over occurs where the pressure inside equals external pressure at the same height above the reference level taken in this case as the bottom of the shaft. The plane containing this neutral point is referred to as the Neutral Pressure Plane (NPP).

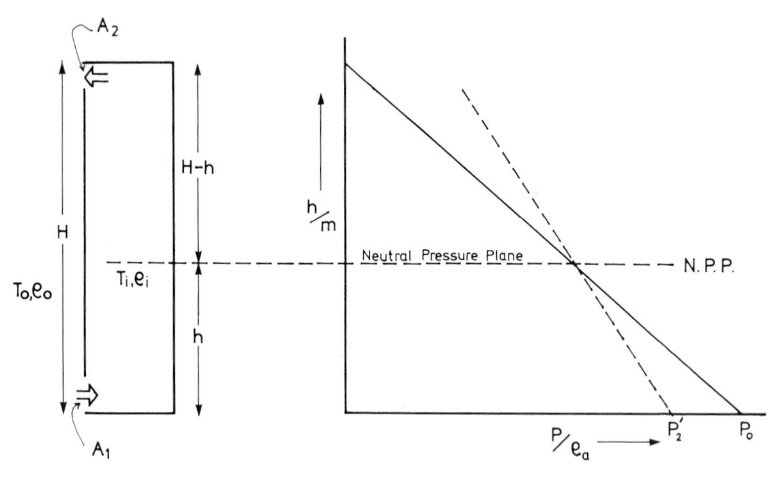

Fig. 9.15 Location of neutral pressure plane

Since the only openings are to top and bottom air will move in at the bottom and out at top. Also if leaks occur in the facade of the shaft, air can be drawn in below NPP and will be forced out above the NPP. The pressure difference of p_i over p_0, i.e. Δp can be calculated for any height h above the NPP relative to barometric pressure at NPP,

i.e. $\Delta p = gh(\rho_0 - \rho_i)$

Since $pV = MRT$

Then $\dfrac{p}{RT} = \dfrac{M}{V}$

Assuming gas at atmospheric pressure:

$$\rho_0 = \frac{p}{RT_0} \quad \text{and} \quad \rho_i = \frac{p}{RT_i}$$

$$\therefore \quad \Delta p = \frac{ghp}{R}\left[\frac{1}{T_0} - \frac{1}{T_i}\right]$$

If $g = 9.8 \text{ m s}^{-2}$
$R = 287 \text{ J kg}^{-1} \text{ K}^{-1}$
$p = 10^5 \text{ Pa}$

$$\Delta p = 3{,}420\,\text{h}\left[\frac{1}{T_0} - \frac{1}{T_i}\right]\text{Pa} \qquad [7(a)]$$

$$\Delta p = 3{,}420\,\text{h}\left[\frac{T_i - T_0}{T_i T_0}\right]\text{Pa} \qquad [7(b)]$$

Estimating the position of the NPP using the mass flow conservation law of fluid flow,

$$\dot{m} = \rho\, CA\left[\frac{2}{\rho}\cdot\Delta p\right]^{\frac{1}{2}}$$

i.e. $\dot{m}_{\text{in}} = \dot{m}_{\text{out}}$

Referring to Fig. 9.15, where h is the distance of opening A from NPP,

$$\rho_1\, CA_1\left[\frac{2}{\rho_1}\cdot\Delta p_1\right]^{\frac{1}{2}} = \rho_2\, CA\left[\frac{2}{\rho_2}\cdot\Delta p_2\right]^{\frac{1}{2}}$$

$$\frac{\Delta p_1}{\Delta p_2} = \left(\frac{A_2}{A_1}\right)^2 \cdot \frac{\rho_2}{\rho_1}$$

Using

$$\frac{\Delta p_1}{\Delta p_2} = \frac{h}{H - h}$$

and

$$\frac{\rho_2}{\rho_1} = \frac{T_0}{T_i}$$

$$\therefore \quad \frac{h}{(H-h)} = \left(\frac{A_2}{A_1}\right)^2 \cdot \left(\frac{T_0}{T_i}\right) \qquad \ldots[8]$$

$$\therefore \quad h = \frac{H}{\left[1 + \left(\frac{A_1}{A_2}\right)^2 \cdot \left(\frac{T_i}{T_0}\right)\right]}$$

Then if $$A_1 = A_2$$

$$\frac{h_1}{H - h_1} = \frac{T_0}{T_i}$$

When $$T_0 = T_i \quad h_1 = \frac{H}{2} = h_2$$

while if $$T_i = 3T_0 \quad \frac{h_1}{H - h_1} = \frac{1}{3}$$

$$4h_1 = H, \quad h_1 = \frac{H}{4} \quad \text{and} \quad h_2 = \tfrac{3}{4}H$$

WORKED EXAMPLE 7

Estimate the steady state in and out mass flow rate of air in a shaft 10 m tall, open at top and bottom given that the internal and external temperatures are 7 and 17 °C, respectively, and that the inlet area is 2 m^2 and outlet area is 1 m^2. Assume that both openings have a discharge coefficient equal to 0.75 and density of air at 7 °C is 1.26 kg m^{-3}.

Solution

Using the expression [8]:

$$\frac{h_1}{h_2} = \left(\frac{A_2}{A_1}\right)^2 \cdot \frac{T_0}{T_i}$$

$$\frac{h_1}{h_2} = \left(\frac{1}{4}\right) \cdot \frac{280}{290} = 0.24$$

$\therefore$ $h_1 = 0.24h_2$ since $h_1 + h_2 = 10$ m

then $1.24h_2 = 10$ m

$\therefore$ $h_2 = 8.06$ m

and $h_1 = 1.9$ m

Calculation of mass flow rate,

i.e. $$\dot{m} = \rho\, CA \left[\frac{2}{\rho} \cdot \Delta p\right]^{\frac{1}{2}}$$

$\therefore$ in flow rate $$\dot{m}_{\text{in}} = \rho_1\, CA_1 \left[\frac{2}{\rho} \cdot \Delta p_1\right]^{\frac{1}{2}}$$

$$m_{\text{in}} = CA_1 [2\rho_1\, \Delta p_1]^{\frac{1}{2}}$$

Using eqn [7a]

$$\therefore \quad \Delta p_1 = 3{,}420 \cdot h_1 \cdot \left[\frac{T_i - T_0}{T_i T_0}\right]$$

$$\therefore \quad \Delta p_1 = \frac{3{,}420 \times 1.94 \times 10}{280 \times 290} = \underline{0.817 \text{ Pa}}$$

$$\therefore \quad \dot{m} = 0.75 \times 2\sqrt{2 \times 1.26 \times 0.817}$$

$$= \underline{2.15 \text{ kg s}^{-1}} \quad \text{(Ans)}$$

A similar calculation can be applied to obtain $\dot{m}_2$, but in any case, $\dot{m}_1 = \dot{m}_2$.

Wind pressure effect, P_{WIND}

The flow of air over a body or building is shown in Fig. 9.16. The pressures over buildings can be represented by a dimensionless

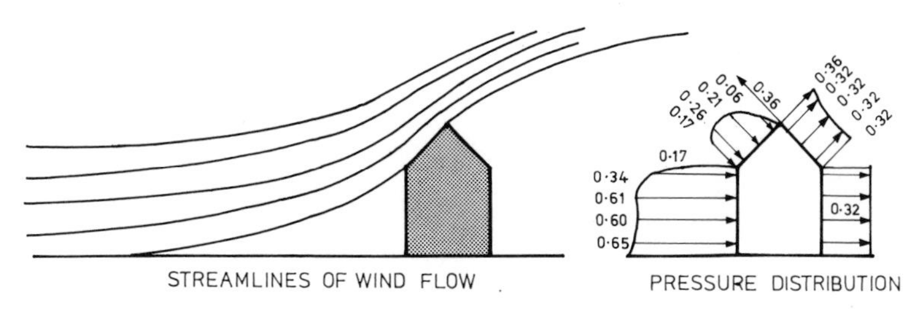

Fig. 9.16 Air flow around buildings

coefficient C_p which is defined as follows:

$$C_p = \frac{p}{\frac{1}{2}\rho v^2}$$

where $\frac{1}{2}\rho v^2 = p_s =$ dynamic pressure due to wind
and $p =$ wind pressure at a point on the surface of a building,
i.e. $p = C_p p_s = C_p(\frac{1}{2}\rho v^2)$

WORKED EXAMPLE 8

Given that the incident wind speed at a certain level above the ground is 20 m s^{-1}, evaluate the wind pressure at a point on the surface of a building when the wind pressure coefficient is known to be 0.7.

Assume that density of air at 20 °C is 1.26 kg m^{-3}.

Solution

$$p_s = \tfrac{1}{2}\rho v^2$$
$$= \tfrac{1}{2} \times 1.26 \times (20)^2$$
$$= 0.63 \times 400 = \underline{253\ Pa}$$
$$\therefore \quad = C_p p_s = 0.7 \times 253 = \underline{177.1\ \text{Pa}} \quad (Ans)$$

Variation of wind speed with height

It is well known that the wind speed varies with height as shown in Fig. 9.17.[8] This variation can be described in mathematical terms by:

$$\frac{V_h}{V_G} = \left(\frac{h}{h_G}\right)^{\alpha}$$

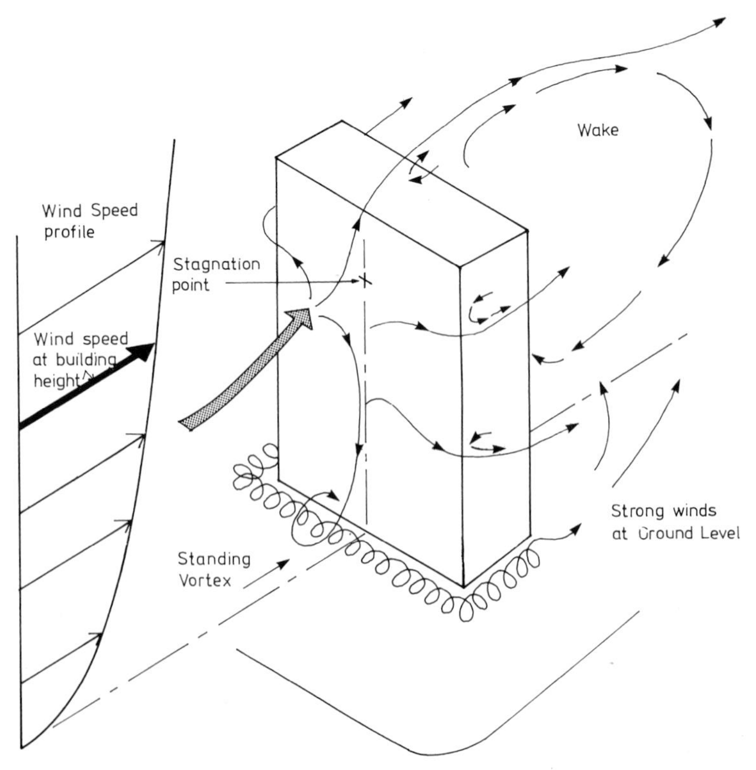

Fig. 9.17 Flow around a building in a boundary layer

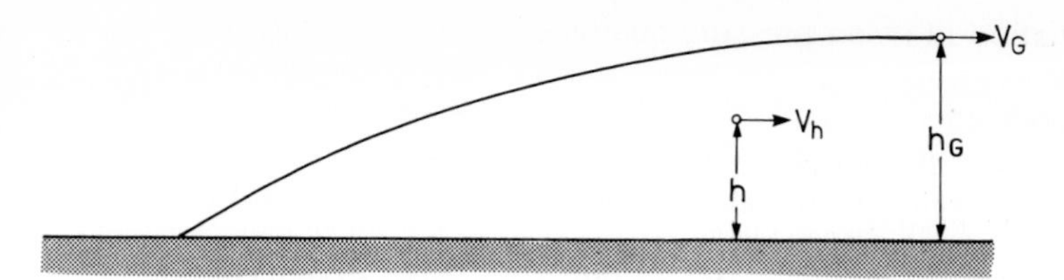

Fig. 9.18 Variation in wind speed with boundary layer thickness

where V_G is the mean gradient wind speed at the edge of the boundary layer of thickness h_G (see Fig. 9.18). Since the type of terrain and ground characteristics vary, different values are used for α.

WORKED EXAMPLE 9

Evaluate the mean wind speed at a height of 50 m for a suburban area in well-wooded surroundings given that the mean gradient wind speed is $80\ \text{m s}^{-1}$

Solution

From Table 9.3, $h = 400\ \text{m}$

and $\alpha = 0.25$

$\therefore$ using above equation $V_{50} = 80\left(\dfrac{50}{400}\right)^{0.25}$

$= \underline{47.7\ \text{m s}^{-1}}$ (*Ans*)

Table 9.3

Type of terrain	h_G (m)	index (α)	$V_h = V_G\left(\dfrac{h}{h_G}\right)^{\alpha}$
Open country low scrub	300	0.15	$V_h = V_G\left(\dfrac{h}{h_G}\right)^{0.15}$
Small towns suburban areas	400	0.25	$V_h = V_G\left(\dfrac{h}{h_G}\right)^{0.25}$
City centres with many tall buildings	500	0.36	$V_h = V_G\left(\dfrac{h}{h_G}\right)^{0.36}$

Note: V_G = free stream velocity at the edge of the boundary layer.

Variation of mean pressure coefficient with wind direction

Experiments[8] using models in low-velocity wind tunnels and full-scale tests indicate that the wind pressure varies with time and the mean pressure coefficient C_p varies with wind direction, Fig. 9.19 and Fig. 9.20, respectively.

As the air flows over and around a building, sections of the building surface suffer a suction effect which is due to the localised

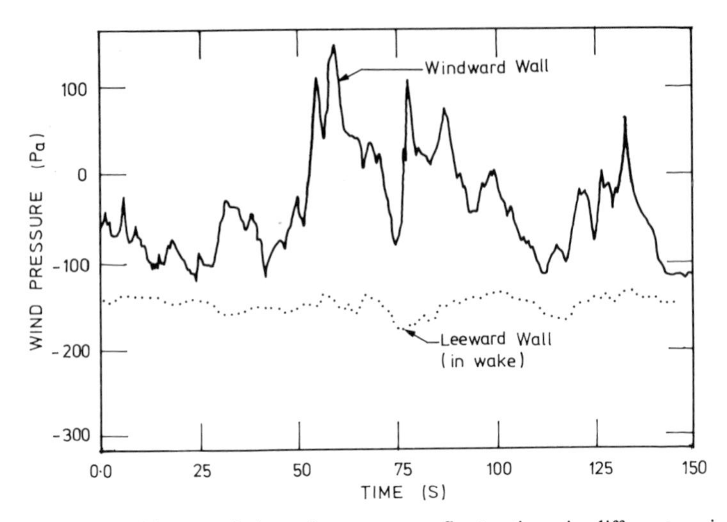

Fig. 9.19 Characteristic surface pressure fluctuations in different regions of flow

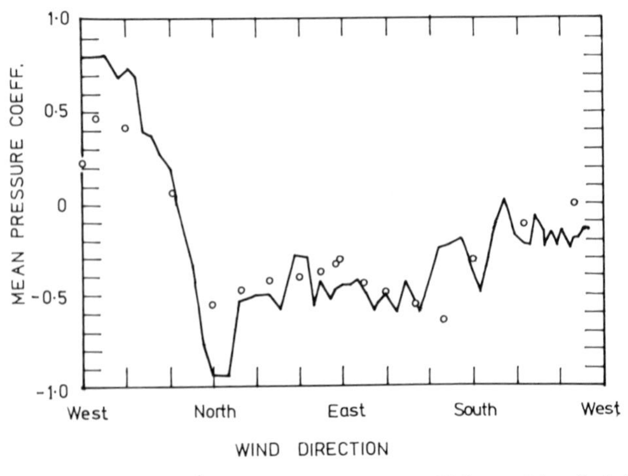

Fig. 9.20 Variation of mean pressure coefficient with wind direction

Table 9.4 Pressure coefficients C_{pe} for vertical walls of rectangular clad buildings

Building height ratio	Building plan ratio	Side elevation	Plan	Wind angle α	C_{pe} for surface A	B	C	D	Local C_{pe}
$\frac{h}{w} < \frac{1}{2}$	$1 < \frac{l}{w} < \frac{3}{2}$	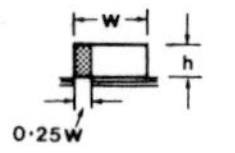 0·25w	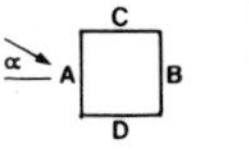	0°	+0·7	−0·2	−0·5	−0·5	−0·8
				90°	−0·5	−0·5	+0·7	−0·2	
	$\frac{3}{2} < \frac{l}{w} < 4$		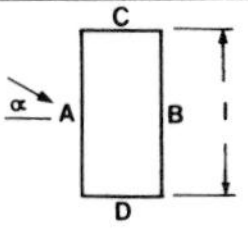	0°	+0·7	−0·25	−0·6	−0·6	−1·0
				90°	−0·5	−0·5	+0·7	−0·1	
$\frac{1}{2} < \frac{h}{w} < \frac{3}{2}$	$1 < \frac{l}{w} < \frac{3}{2}$		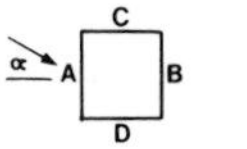	0°	+0·7	−0·25	−0·6	−0·6	−1·1
				90°	−0·6	−0·6	+0·7	−0·25	
	$\frac{3}{2} < \frac{l}{w} < 4$		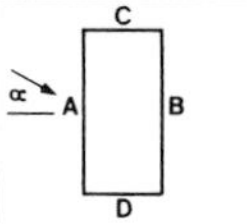	0°	+0·7	−0·3	−0·7	−0·7	−1·1
				90°	−0·5	−0·5	+0·7	−0·1	
$\frac{3}{2} < \frac{h}{w} < 6$	$1 < \frac{l}{w} < \frac{3}{2}$	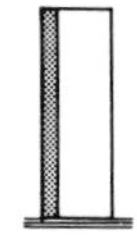	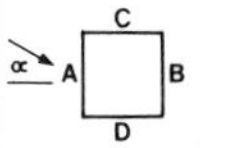	0°	+0·8	−0·25	−0·8	−0·8	−1·2
				90°	−0·8	−0·8	+0·8	−0·25	
	$\frac{3}{2} < \frac{l}{w} < 4$	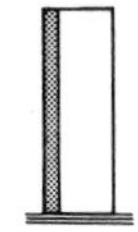	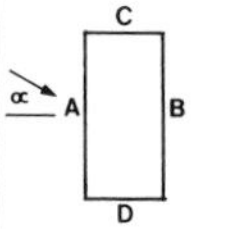	0°	+0·7	−0·4	−0·7	−0·7	−1·2
				90°	−0·5	−0·5	+0·8	−0·1	

l = length of major face of building.
W = width of building (length of minor face).

pressure decrease due to parallel fluid flow (Bernoulli's Equation) Fig. 9.16. Table 9.4 gives a summary of the pressure coefficients C_{pe} for the vertical walls of rectangular buildings. Thus the total wind load on a structure may be determined by the vectorial summation of the loads of the various wall and roof surfaces.

Referring specifically to the fire situation, wind can disturb the air inside a building since the pressure so developed on leaky external surfaces can give rise to horizontal air movement into and out of a building. The magnitude of this leakiness is assumed to be 1 or 2 per cent of overall facade surface area. It has been estimated that even for a modest wind speed an air flow of 1 to 2 m^3 per square metre per second can be expected. This air can be either dissipated by leakage within the building or it may create pressure differences across less leaky barriers. Such wind-induced pressures may help to negate the pressures due to the stack effect or it can assist in the transmission of smoke from one space to another.

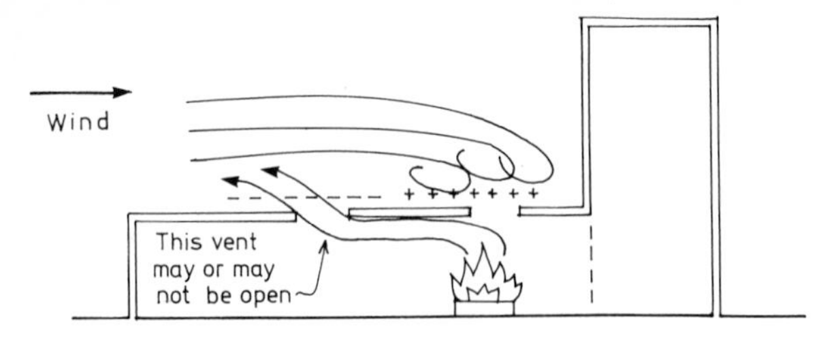

Fig. 9.21 Shortcomings in a natural ventilation system

It must be noted that during a fire the variation in the pressure differences over a roof plays an important role where vents are used to remove smoke from a fire zone. If the case depicted in Fig. 9.21 occurs, the wind-pressure effect will tend to defeat the action of the stack effect in the vent region over the fire bed.

Thus it must be emphasised that natural venting systems are impossible to design with any certainty so that the situation shown in Fig. 9.21 will never occur.

WORKED EXAMPLE 10

Compare the wind pressure produced by a wind 30 m above ground level having a speed of $5\,\mathrm{m\,s^{-1}}$ with that produced by the stack effect in a building closed at the top in which the air temperature is 17 °C while that outside is 7 °C.

Wind pressure $p_w = \frac{1}{2}\rho v^2$

$= \frac{1}{2} \times 1.26 \times 5^2$

$= \underline{15.75\ Pa}$ (*Ans*)

Stack effect $p_s = gh(\rho_o - \rho_i)$ $\rho_i = 1.26 \times \left(\frac{280}{290}\right)$

$p_s = 10 \times 30 \times 0.043 = \underline{13\ \mathrm{Pa}}$ (*Ans*)

Thus in this case the stack-effect pressure compares with the wind pressure at this height above ground level.

Mechanical systems, P_{MECH}

It is usual in this case that the air flow rate into a building is greater than the outflow thus creating a small residual pressure inside the building. The flow of air and smoke when such a system is working can be determined, but in the case of a fire when it is normal to close down the mechanical systems, the direction of air and smoke flow will become random depending on the other forces present at the time.

Thus if such a mechanical system could in the event of a fire be switched to an emergency mode, this would help to control smoke movement and be designed so as to help keep egress routes relatively smoke-free for a reasonable period of time for the safe escape of occupants. If sophisticated air-conditioning systems are used in buildings, they will be able to accommodate the problems of stack and wind pressure and keep the internal environment in a fixed state. The actual pressures developed by such mechanical systems depend on the leakage and infiltration filtration rates of the building and are calculable by conventional techniques.

Fire stack pressure, $P_{\text{FIRE STACK}}$

This pressure developed in a fire situation is really an enhanced stack-pressure effect due to the very hot gases produced by the combustion processes during a fire.

It has been shown in the earlier section dealing with the environmental stack pressure effect that:

$$P_{\text{stack}} = gh(\rho_0 - \rho_i) = \Delta p_i$$

and

$$\Delta p_i = \frac{gph}{R} \cdot \left[\frac{1}{T_0} - \frac{1}{T_i}\right]$$

or

$$\Delta p_i = 3.42 \times 10^{-2} \times ph \cdot \left[\frac{1}{T_0} - \frac{1}{T_i}\right]$$

Assuming $p = 10^5$ Pa

$$\therefore \quad \Delta p_i = 3{,}420h \cdot \left[\frac{1}{T_0} - \frac{1}{T_i}\right]$$

where h is the distance of point above or below NPP.

As an exercise the reader may like to verify the following tables giving the fire stack pressure at various heights above NPP for different internal and external temperature differences.

(a) $\Delta T = 200$

	(h/m) =	1	2	5	10	20
$T_0 = 300$ K	$\left(\frac{P_{\text{STACK}}}{Pa}\right)$ =	4.55	9.6	22.75	45.5	91

(b) $\Delta T = 500$

	(h/m) =	1	2	5	10	20
$T_0 = 300$ K	$\left(\frac{P_{\text{STACK}}}{Pa}\right)$ =	7.1	14.2	35.5	71.0	142

A graph showing the variation of P_{STACK} with height h above the NPP is given in Fig. 9.22.

It is now useful to consider how the NPP varies during a fire situation. As previously discussed and proved in an earlier paragraph,

$$\frac{h_1}{h_2} = \frac{h}{H-h} = \frac{T_0}{T_i}\left(\frac{A_2}{A_1}\right)^2$$

describes the way in which the NPP varies with:

1. Ratio of area of inlet to area of outlet (A_1/A_2)
2. Difference in T_0 and T_i, i.e. T_0/T_i.

Consider the following situation:

(a) when $A_1 = A_2$ and letting $T_i = T_F = 900$ K

$$\text{then} \quad \frac{h}{H-h} = \frac{1}{3}$$

$$3h = H - h$$

$$\therefore \quad h = \frac{H}{4}$$

(b) when $A_1 = 2A_2$ and $T_F = 900$ K

$$\therefore \quad \frac{A_2}{A_1} = \left(\frac{1}{2}\right)$$

$$\frac{h}{H-h} = \frac{1}{3} \times \left(\frac{1}{2}\right)^2 = \frac{1}{12}$$

$$12h = H - h$$

$$\therefore \quad h = \frac{H}{13}$$

(c) Consider a case of an open door, or window.

As shown in Fig. 9.24, let door width be d.

Assuming for simplicity that the velocity profile for gas entering and leaving the enclosure is constant across the area above and below the NPP.

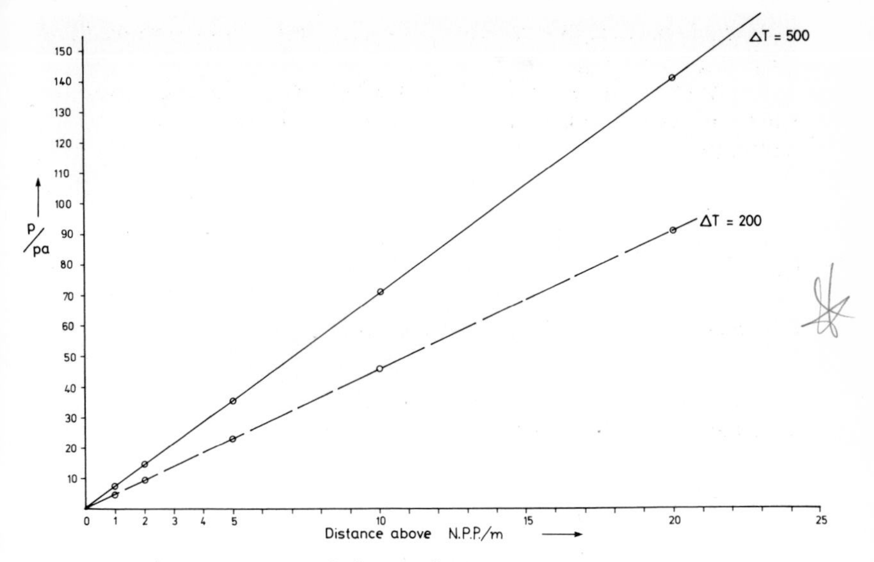

Fig. 9.22 Pressure developed above a fire

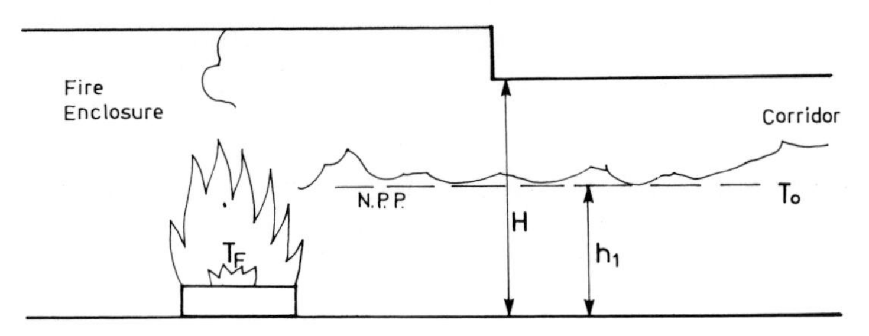

Fig. 9.23 Location of neutral pressure plane in an enclosure

Assuming door width $= d$

$$\therefore \qquad A_1 = h_1 d \qquad \text{and} \qquad A_2 = h_2 d$$

where $\qquad h_2 = H - h$

Now since $\dfrac{h_1}{h_2} = \dfrac{T_0}{T_f}\left(\dfrac{A_2}{A_1}\right)^2 = \dfrac{T_0}{T_F}\left(\dfrac{h_2^2 d^2}{h_1^2 d^2}\right)$

$$\therefore \qquad \frac{h_1^3}{h_2^3} = \frac{T_0}{T_F}$$

Fig. 9.24 Location of neutral pressure plane in a doorway

If it can be assumed that $T_F = 900$ K and $T_0 = 300$ K an estimate of the location of the NPP can be made,

i.e. $$\frac{h_1}{h_2} = \left(\frac{300}{900}\right)^{\frac{1}{3}} = (0.33)^{\frac{1}{3}} = 0.69$$

$$h_2 = \frac{h_1}{0.69} = 1.44h_1$$

$$\therefore \quad 2.44h_1 = H = 2\text{ m} \quad \text{(say)}$$

$$h_1 = 0.82\text{ m} \quad \text{(approximately).}$$

Thus the NPP should be just below the half height of the opening.

This approximate calculation would seem to be in agreement with the kind of smoke profiles produced during test fires using an open door or open door/corridor configuration, see Fig. 9.23 and Fig. 9.24.

WORKED EXAMPLE 11

Estimate the pressure due to fire in a compartment at the top of a 2 m door assuming that the clearance at top and bottom of door is the same. Assuming the temperature of fire to be 900 K.

Solution

Take:

$T_F = 900$ K

$T_0 = 300$ K

Calculation $A_1 = A_2$

$$\frac{h_2}{h_1} = \left(\frac{A_1}{A_2}\right)^2 \frac{T_F}{T_0}$$

$$\frac{h_2}{h_1} = 3$$

$$h_2 = 3h_1$$

$$\therefore \quad 4h_1 = 2\text{ m}, \qquad h_1 = 0.5$$

$$h_2 = \underline{1.5\text{ m}} \quad (Ans)$$

Inserting this value into equation for the stack-induced pressure

$$\Delta p = 3{,}420 \times 1.5\left[\frac{1}{300} - \frac{1}{900}\right] \text{ Pa}$$

$$= \frac{3{,}420 \times 1.5 \times 2}{900}$$

$$= \Delta p = \frac{34.20}{3.00} = \underline{11.4\text{ Pa}} \quad (Ans)$$

9.6 SMOKE CONTROL

9.6.1 Introduction

Having discussed smoke production and movement, consideration must also be given briefly to methods of smoke control. In Chapter

12 various aspects of smoke control are discussed in the context of the provision of effective means of escape from a building in the event of a fire, and much of what is discussed there could be applied in principle to a building as a whole. In general the decision to employ some method of smoke control will be in response to a perceived threat from smoke in the event of a fire. The method of smoke control employed, if any, will be determined principally by the building occupancy.

Building occupancy is discussed in Chapter 13 and in the context of this Chapter has two components: people and building function. These two components will determine, in essence, priorities which will influence the decision to employ a method of smoke control and the method employed. As illustrated below, both components are present in all of the broad categories illustrated, but will assume different priority levels perhaps for each.

Components	*Building type*	e.g.
People	Public Assembly Institutional	Leisure Centre Health Care
Building Function	Commercial Industrial	Office Factory

Thus in public assembly and institutional type buildings the life safety of the occupants will be the dominant consideration whereas in industrial type buildings where few people may be employed, the avoidance of damage to contents and limiting the fire damage to the building may be the dominant factors.

9.6.2 Methods of smoke control

Smoke-control systems may be classified as 'natural' or 'powered' systems. Natural systems of smoke control rely on environmental forces as discussed previously to move smoke away from occupied spaces, whereas powered systems will utilise the energy developed by mechanical systems to move the smoke along predetermined routes.

Three broad categories of methods of smoke control will be considered. These are:

1. Dilution
2. Removal
3. Containment.

Dilution

The relationship between optical density, D/ℓ, and the product of the length of the light transmission path, ℓ, and the concentration of the smoke, c, was established earlier:

$$D = \ell CB$$

where B is a constant depending upon the nature of the smoke.
Similarly the relationship between optical density and visibility was also discussed. Simply stated (Section 9.3.1)

1. For front illumination,

$$\text{visibility (in metres)} = \frac{10}{\text{optical density per metre}}$$

2. For rear illumination,

$$\text{visibility (in metres)} = \frac{25}{\text{optical density per metre}}$$

It follows that because the concentration of smoke affects optical density it also significantly affects visibility. Thus if the smoke produced in a fire can be sufficiently diluted such that escape routes and other critical building spaces can be safely used and the threat from smoke considerably reduced, if not eliminated, dilution can be considered as an effective means of smoke control. The utilisation of this concept (of dilution) as a means of smoke control may arise naturally from the geometry of the building spaces under consideration. If the building space is large enough (Fig. 9.24) the smoke may accumulate at the higher levels and some considerable time may elapse before any direct threat from the smoke would be encountered.

Open corridors and pedestrian walkways, Fig. 9.24, are other situations where dilution may be employed.

It should be pointed out that whilst valid as a concept, dilution as an effective means of smoke control in the practical sense has a limited number of applications.

Fig. 9.25 Large indoor sports area with domed roof structure

Removal

The removal of smoke may be by natural ventilation, mechanical extraction or a combination of both. In Chapter 12 the reliability of natural ventilation systems is questioned and it is not intended to discuss this aspect further, but the reader should understand that natural ventilation systems may not operate when, or as effectively, as required.

Cross-ventilation is a smoke control technique which has been used in the United Kingdom for many years to protect internal common-access corridors in blocks of flats and maisonettes, Fig. 9.26. The ventilated lobby, Fig. 9.39 is also a method of smoke control which is used to prevent or retard the passage of smoke into the stairwell. A high-level vent is incorporated so that the gas pressure generated in the lobby does not exceed ambient. Both methods of smoke control outlined above may be employed in two different buildings adjacent to each other on the same site and having exactly the same orientation. For different reasons neither may prove to be an effective method of smoke control.

Earlier the forces available to move smoke around buildings were

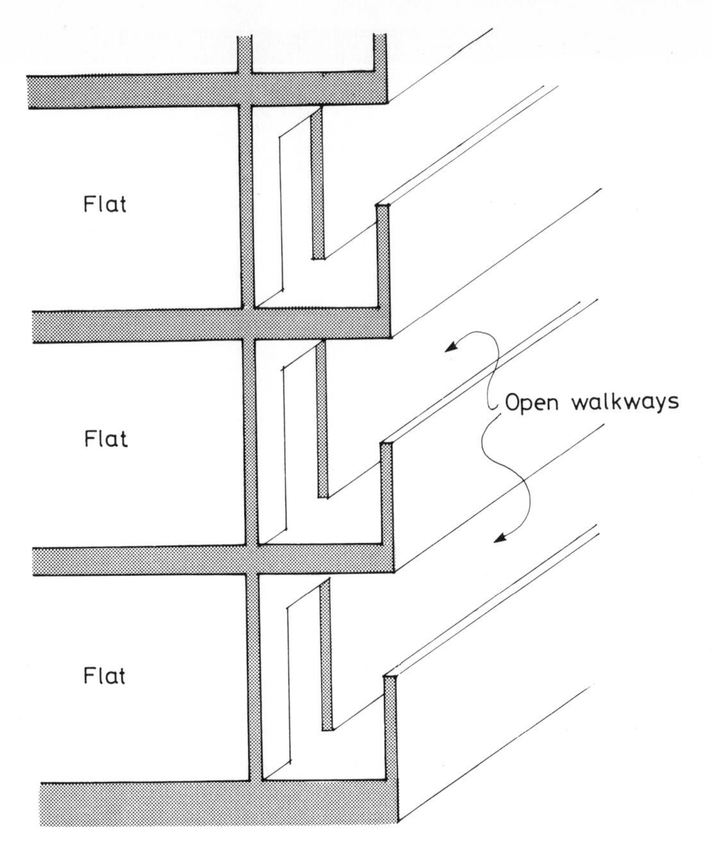

Fig. 9.26 Block of flats with open linked walkways

considered. The important available force for natural smoke-control systems is that generated by the wind. If it is to be assumed that a smoke-control system should be available and effective at all material times, then any reliance on wind, which is both unpredictable and uncontrollable, as a generating force to move smoke away from threatened to external spaces is questionable.

In Fig. 9.27, if the velocity of the prevailing wind is sufficient, positive and negative pressure coefficients will be generated such that the smoke generated from a fire would be both diluted and dispersed to the outside. If the direction of the winds is sufficiently changed, Fig. 9.28, neither dilution or dispersal of the smoke may occur. Similarly with the ventilated lobby, Fig. 9.39, the sind orientation and velocity may be such that ventilation of the lobby will not be achieved. According to some estimates,[7] the wind velocity will be too low for considerable periods of time to have any real effect on smoke dilution or movement away from the threatened spaces.

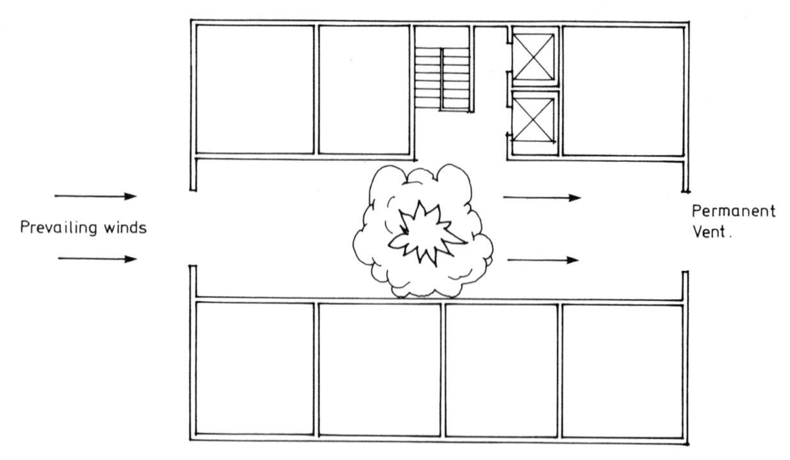

Fig. 9.27 Situation where smoke may be removed by cross ventilation

Suffice to say at this juncture that in 1982 a review of the effectiveness of cross ventilation was initiated.[8] Roof venting is also extensively used as a method of smoke control,[9] Fig. 9.38. Notwithstanding the unpredictable and uncontrollable nature of the weather conditions, it is possible to calculate the vent areas necessary to vent a fire of known size where natural ventilation is employed.

WORKED EXAMPLE 12

Calculate:

1. Vent area required for the building shown in Fig. 9.29
2. Size of the individual vents
3. Comment on the distribution of the vents given:

Density of gases

K	ρ (kg m^{-3})
273	1.293
280	1.26
290	1.22
300	1.18
546	0.65
700	0.50
1,100	0.32

Data:

$T_0 = 290$ K $\quad P = 12$ m

$T_f = 1{,}100$ K $\quad y = 8$ m

$T_c = 546$ K $\quad d = 2$ m

$$\rho_c = \rho_0 \cdot \frac{T_0}{T_c} = 1.22 \times \frac{290}{546} = 0.65 \text{ kg m}^{-3}$$

Following the procedure outlined by Morgan and Marchant:[10]

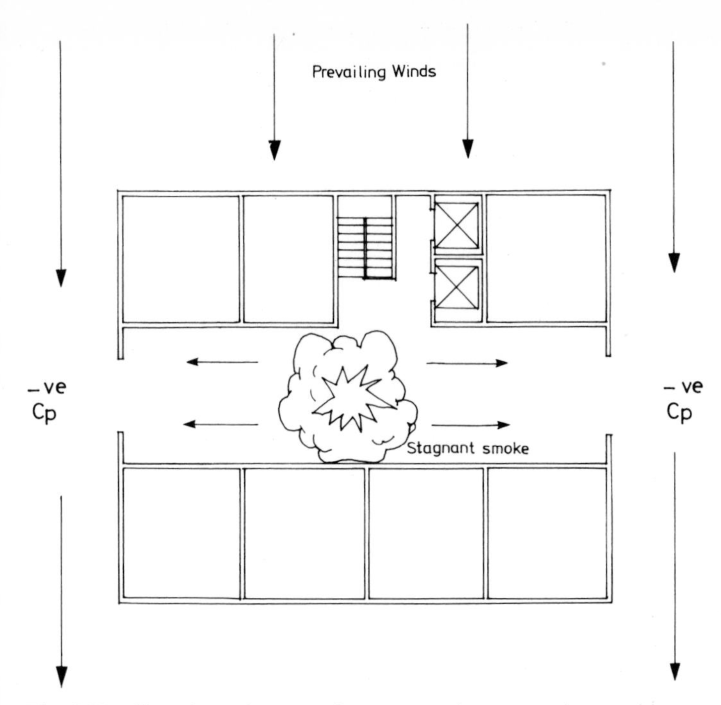

Fig. 9.28 Situation where smoke may not be removed

1. The mass flow rate of gases from the fire is given by the expression[11]

$$\dot{M}_f = 0.096P(y)^{\frac{3}{2}}\rho_0(gT_0/T_f)^{\frac{1}{2}}$$

$$= 0.096 \times 12 \times (8)^{\frac{3}{2}} \times 1.22\left(\frac{9.80 \times 290}{1{,}100}\right)^{\frac{1}{2}}$$

$$= 51.12\ \text{kg s}^{-1}$$

The volume rate of flow at the ceiling (V_c) is given by:

$$V_c = \frac{\dot{M}_f}{\rho_c} = \frac{51.12}{0.65} = 78.65\ \text{m}^3\ \text{s}^{-1}$$

The total area of vent required can be calculated from the expression[12]

$$M_V = \frac{C_V A_V \rho_0 \{2.g.d.(T_g - T_0)T_0\}^{\frac{1}{2}}}{T_0 + (T_g - T_0)}$$

$$= \frac{0.6 \times A_{Vm} \times 1.22 \times 2.98 \times 2(546 - 290)290^{\frac{1}{2}}}{290 + (546 - 290)}$$

$$\dot{M}_V = 2.29A_{Vm}$$

$\therefore$ Total area of vent required, $A_V = \frac{51.12}{2.29}$

$$= 22.32\ \text{m}^2$$

2. Checking on the flow rate and vent area so that in conditions of zero outside wind pressure the gas will flow at less than the critical flow rate, V_0, for a given depth of smoke layer as defined by Spratt and Heselden[13]

$$V_0 = \frac{2gd^5(T_g - T_0)T_0^{1/2}}{T_g}$$

$$= \frac{2 \times 9.9 \times 2^5(546 - 290)290^{\frac{1}{2}}}{546}$$

$$= 17.67 \text{ m}^3 \text{ s}^{-1}$$

N.B. T_g is the temperature of the gases in the mall. T_g is the temperature of the gases in the rising plume and in this example is taken as equivalent to T_c the temperature of the gases at ceiling level.

Volume rate of flow at the ceiling ($T_c = 546$ K) $= 78.65 \text{ m}^3 \text{ s}^{-1}$. Reducing to ambient temperature,

$$\text{volume rate of flow at ceiling} = \frac{78.65 \times 290}{546}$$

$$= 41.77 \text{ m}^3 \text{ s}^{-1}$$

It can be seen that the volume rate of flow at ceiling level reduced to ambient temperature ($41.77 \text{ m}^3 \text{ s}^{-1}$) is over twice the critical flow rate ($17.67 \text{ m}^3 \text{ s}^{-1}$). Therefore the total vent area of 22.32 m^2 should be divided into three separate areas of 7.4 m^2 giving an individual vent size of $2.72 \text{ m} \times 2.72 \text{ m}$ so that V_0 is not exceeded through any one vent, i.e.

$$A_V = \frac{22.32}{3}$$

$$= 7.4 \text{ m}^2$$

A check can be made on the vent area, and screen depth (if screens are to be employed) using:

$$A_{Vm} = 2.4d^2$$

Taking $d = 2$ m

$$A_{Vm} = 9.6 \text{ m}^2$$

$\therefore$ Vent size $= 7.4 \text{ m}^2$ is within limits.

3. Obviously the ideal location for a vent would be directly over the fire but it is not possible to predict exactly where a fire will occur. It is suggested therefore that the building be divided into three equal areas on plan incorporating 3 m drop screens from the ceiling and the vents located as shown in Fig. 9.29. It should be noted that this is a rather simple example which ignores wind effects completely.

The reader should pause and consider the critical flow rate V_0 mentioned earlier. This flow rate is determined in idealised conditions Fig. 9.31 so that the smoke exhausts through the vent without disturbing the bottom of the smoke layer, i.e. steady-state conditions exist where the rate of smoke production is equal to the flow rate of smoke through the vent. If the critical flow rate is exceeded the bottom of the smoke layer is disturbed by entrained air and the vent operates inefficiently exhausting some smoke and much entrained air, Fig. 9.31.

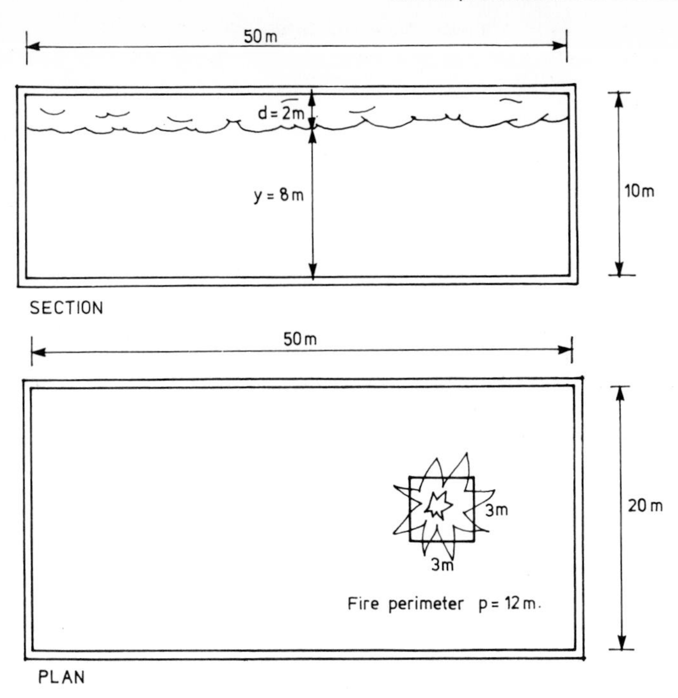

Fig. 9.29 Section and plan of single-storey building

Venting will not be effective unless it is possible for cold air to flow into the building to replace the hot gases which are required to flow out through the roof vents. Thus adequate inlets for cold air must be provided. Vents can be operated automatically by an alarm system or by a smoke detector. They can also be power assisted, i.e. have mechanical-extraction systems incorporated. It must be stressed that this example is for an ideal situation where external wind effects have been ignored.

The calculations to prevent smoke-logging using mechanical systems differ from those for natural ventilation in that rate of smoke production for a given fire in a given volume is matched by the extraction rate of the mechanical system totally or partially.

Options: dilution or removal?

Consider a modern building which incorporates an atrium, Fig. 9.32. Should a fire occur, several alternatives are available for coping with the smoke produced from the fire.

1. The concept of dilution could be employed and provided the atrium volume was sufficient, smoke could be encouraged into it, filling the atrium while the remainder of the building is evacuated. This option ignores the psychological impact of dense black smoke visible to persons on upper floors of the building

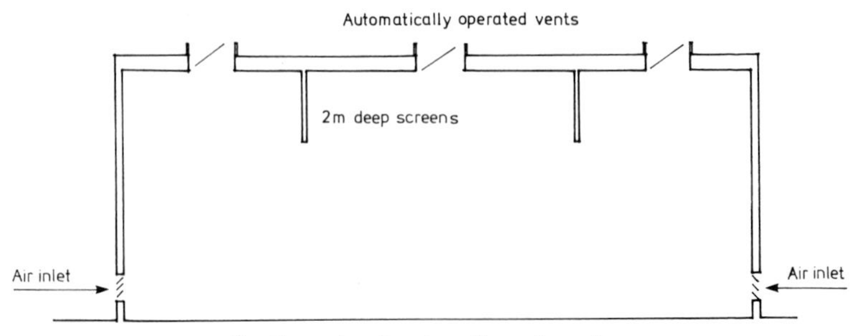

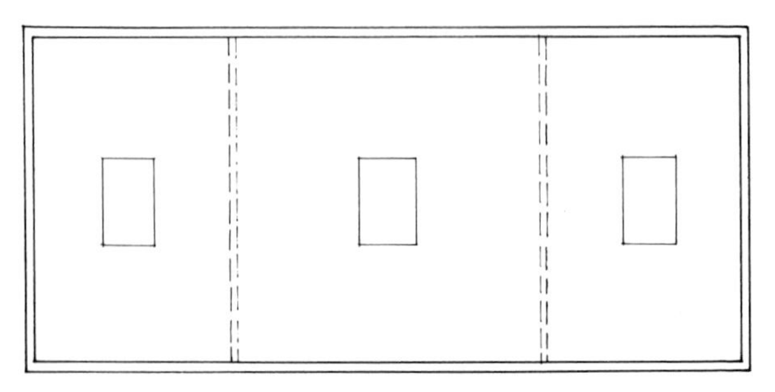

Fig. 9.30 Section and plan showing location of vents

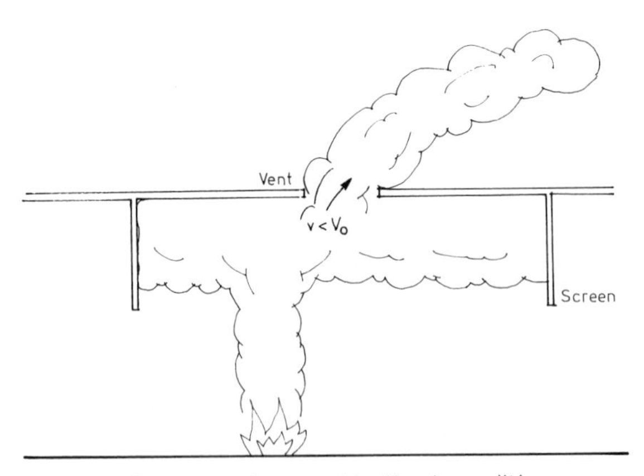

Fig. 9.31 Elevation showing idealised conditions ($V_0 < 17.67\,m^3\,s^{-1}$)

flowing into the rapidly smoke-logging atrium. For the latter reason it may not be an acceptable option unless the upper floors could be effectively screened from the atrium by automatically operating shutters.

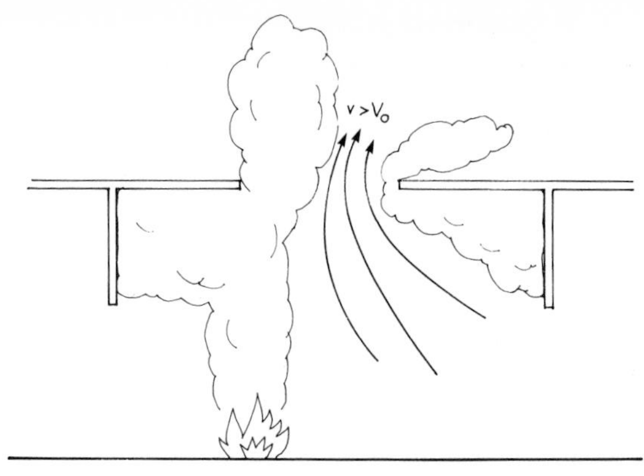

Fig. 9.32 Critical flow rate exceeded resulting in inefficient venting ($V_0 > 17.67\,\mathrm{m^3\,s^{-1}}$)

2. Option two could be similar to option one except that the upper floors could be evacuated at the pace of the base of the descending smoke layer. The rate R at which the base of the smoke layer falls can be simply estimated from:

$$R = \frac{V}{A} \quad \mathrm{m\,s^{-1}}$$

where V is the volume of smoke being produced and A is the area of the atrium. Thus it is possible to estimate the time taken for the descending base of the smoke layer to reach each floor and the time for complete smoke-logging of the atrium to occur.

3. Option three would incorporate a mechanical extraction system to control smoke in the atrium. The following worked example is based on option three.

WORKED EXAMPLE 13

The atrium in the building as shown in Fig. 9.33 is 17 m high and has an opening area of 120 m^2. If the fans located at the head of this atrium extract smoke at ambient temperature at a rate of $25.0\,\mathrm{m^3\,s^{-1}}$:

Estimate (a) the depth of the layer of smoke at the top of the atrium at equilibrium
(b) the time taken for this equilibrium situation to occur.

Solution

Assuming the fire remains small for this period, having a heat output of 0.5 MW, and that smoke is extracted at or around ambient temperature:

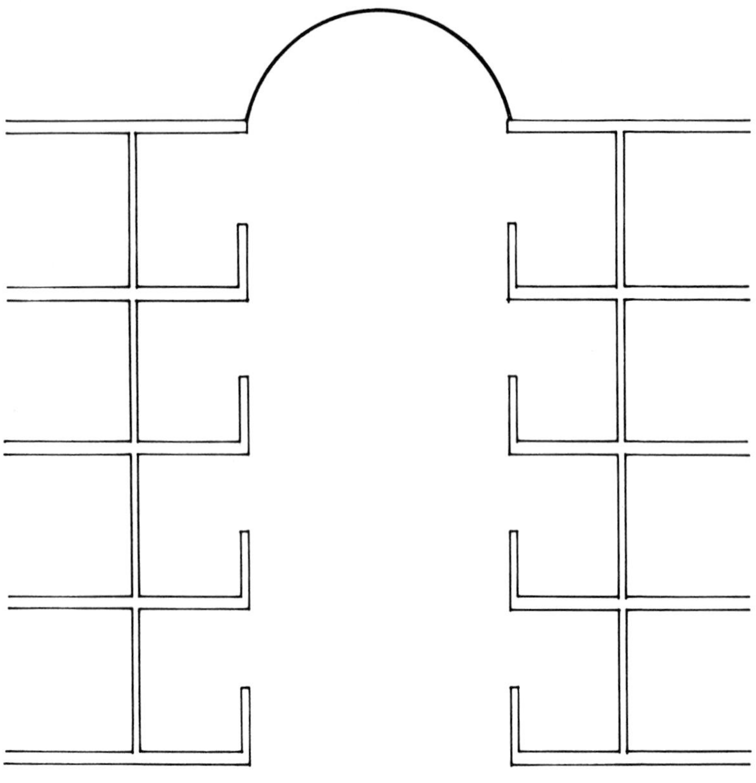

Fig. 9.33 Atrium space which may be utilised for smoke control

(a) Using $\dot{m} = 0.188Py^{\frac{3}{2}}$ kg s^{-1}

To convert to volume of gas at ambient

$$\dot{v} = \frac{0.188Py^{\frac{3}{2}}}{\rho} = 25 \text{ m}^3 \text{ s}^{-1}$$

For a small fire, 0.5 MW; $P \sim 4$ m

and using $\rho = 1.2$ kg m^{-3}

then $$\frac{0.188 \times 4 \times y^{\frac{3}{2}}}{1.2} = 25$$

$$y^{\frac{3}{2}} = \frac{32}{4 \times 0.188}$$

$$y = \underline{12.3 \text{ m}} \quad (Ans)$$

Thus at equilibrium the smoke level will be approximately 12.3 m above the floor.

(b) This situation requires some additional analysis and computation to

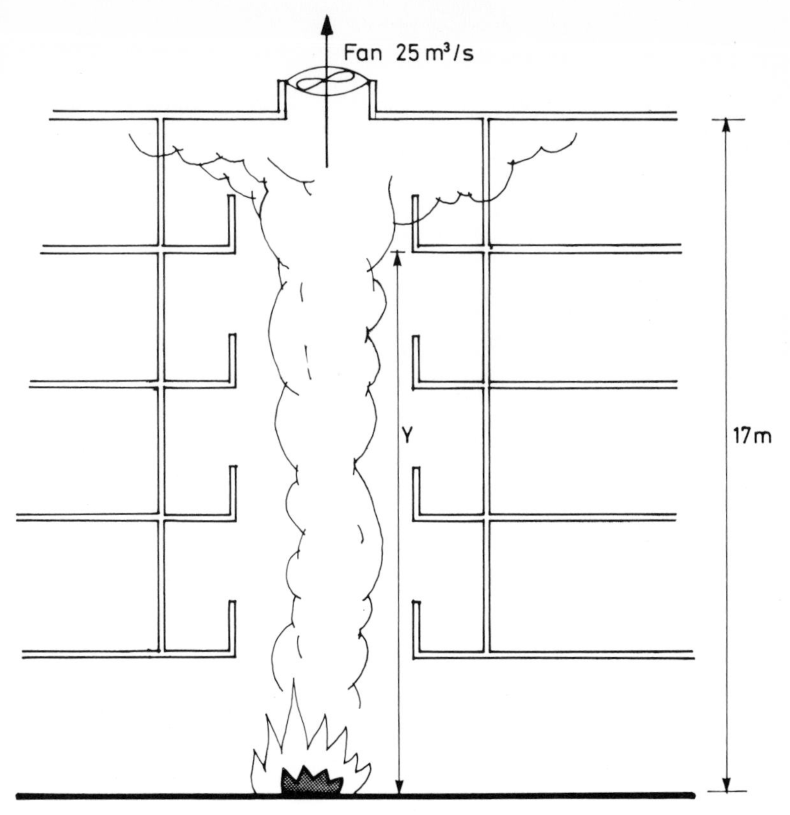

Fig. 9.34 Fire in atrium

determine the time elapse for the equilibrium situation. Consider Fig. 9.34.

Volumetric increase in smoke in building, $\dot{V}$,

$$\dot{V}=\frac{\mathrm{d}}{\mathrm{d}t}[(H-y)A]=-A\frac{\mathrm{d}y}{\mathrm{d}t} \qquad \text{see Fig. 9.35}$$

$$\dot{V}=\frac{\dot{m}}{\rho}-L$$

where L is the volume extraction rate at top of atrium,

$$\text{i.e.} \quad -A\frac{\mathrm{d}y}{\mathrm{d}t}=\frac{\dot{m}}{\rho}-L$$

$$-\frac{\mathrm{d}y}{\mathrm{d}t}=Cy^{\frac{3}{2}}-\frac{L}{A}$$

$$\text{where} \quad C = \frac{0.188 \times 4}{1.2 \times 120}$$

$$= 5.2 \times 10^{-3}$$

$$-\frac{dy}{dt} = (5.2 \times 10^{-3} y^{\frac{3}{2}}) - 0.21$$

$$\frac{dy}{dt} = 0.21 - (5.2 \times 10^{-3} y^{\frac{3}{2}})$$

Converting into incremental form

$$\frac{\Delta y}{\Delta t} = 0.21 - (5.2 \times 10^{-3} y^{\frac{3}{2}})$$

$$\Delta y = \Delta t[0.21 - 5.2 \times 10^{-3} y^{\frac{3}{2}})]$$

Starting with $y_0 = 17$ m, evaluate Δy_1 for time interval Δt then subtract Δy_1 from y_0 to get y_1, etc.

Assuming $\Delta t = 10$ secs, the height of smoke layer, y, above floor is tabulated in Table 9.5.

Thus in approximately 80 seconds the smoke layer would have reached its equilibrium state provided the fire did not grow into a bigger fire.

Other methods of removing smoke from threatened spaces include the provision of smoke shafts and chimneys Fig. 9.36.

Table 9.5

y (m)	Δy_1 (m)	t (s)
17.0		0
	−1.55	
15.45		10
	−1.06	
14.39		20
	−0.74	
13.65		30
	−0.52	
13.13		40
	−0.37	
12.74		50
	−0.26	
12.48		60
	−0.19	
12.29		70
	−0.14	
12.14		(80)

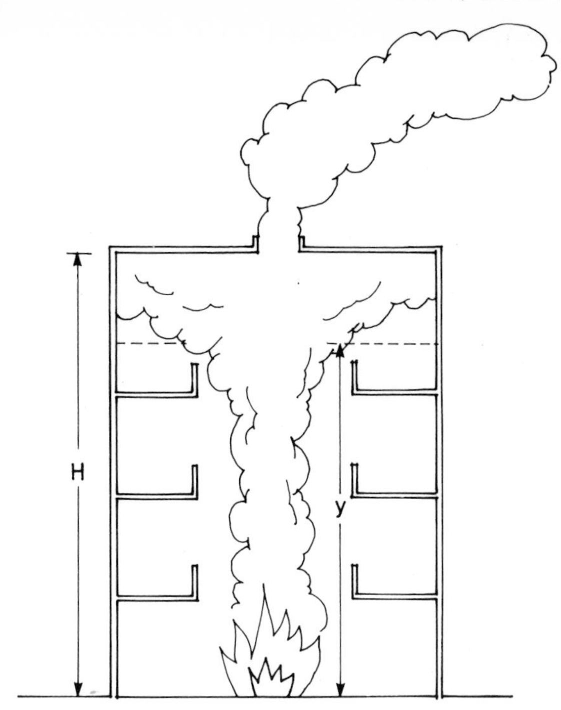

Fig. 9.35 Section of atrium for worked example

Innovations in smoke removal

Innovations in the design and construction of buildings includes the development of new technology for a specific use, or the application of existing technology to the solution of a different problem. Sometimes in the design of buildings it is necessary to incorporate explosion-relief mechanisms. These may include automatic venting or designing into the enclosing constructional envelope areas which in the event of an explosion will rupture and vent the explosion, thereby minimising damage. The student here may wish to refer to the building regulations, particularly the section dealing with structural stability, and consider the method employed to prevent progressive structural collapse of a high rise block, e.g. in the event of a gas explosion of a block of flats. Consider a single-storey building incorporating aluminium roofing. An important property of aluminium is that it melts at relatively low temperatures, i.e. around 650 °C. Thus in the event of fire, once the roof decking over the fire reaches this critical temperature it melts, a hole is formed and the fire is vented. The effect of venting is essentially three-fold:

1. Smoke-logging is prevented, thus facilitating means of escape and fire fighting
2. Convective heating of other combustibles is considerably reduced and fire spread and consequently fire damage is also considerably reduced
3. Radiative feedback from the base of the hot smoke layer is reduced if not eliminated, thereby reducing the burning rate, i.e. the fire burns as if in the open, maximum temperatures within the enclosure are not attained and a less severe fire is experienced.

Recent trials[14] have demonstrated the benefits of an aluminium roofing system compared with those of steel. Test roofs consisting of 0.7 mm thick profiled steel and aluminium sheets on steel purlins were exposed to fire and the internal temperatures recorded, Fig. 9.36. Temperatures beneath the steel sheeting rose rapidly to around 800 °C and 400 °C respectively and remained relatively high until the

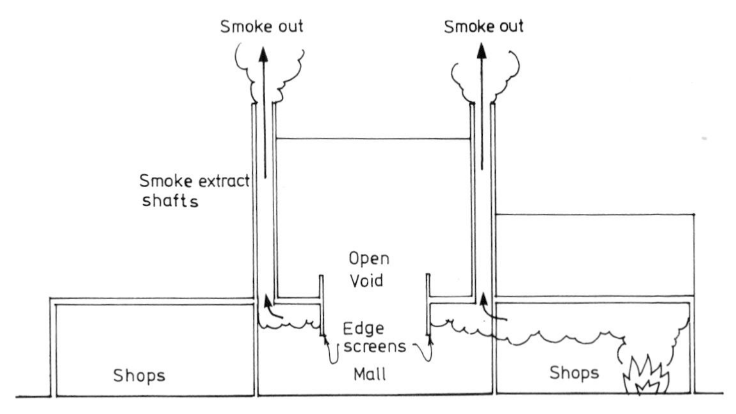

Fig. 9.36 Provision of smoke shafts

fire burned out. The temperatures beneath the aluminium roof rose to a higher peak initially, due to the different thermal inertia and emissivities of the materials. However, within five minutes of the start of the fire the aluminium melted to form a vent hole and the internal temperatures dropped rapidly. The same concept has been used in the design of schools.[15] It is possible using this method to determine how much fire damage is tolerable, i.e. how much of a building can be lost through fire without materially disturbing the activities within the building. Once the acceptable loss of part of the building is translated into volume, the building can be designed and subdivided such that the damage should never exceed predetermined levels. In Fig. 9.37, the building has been subdivided into four fire

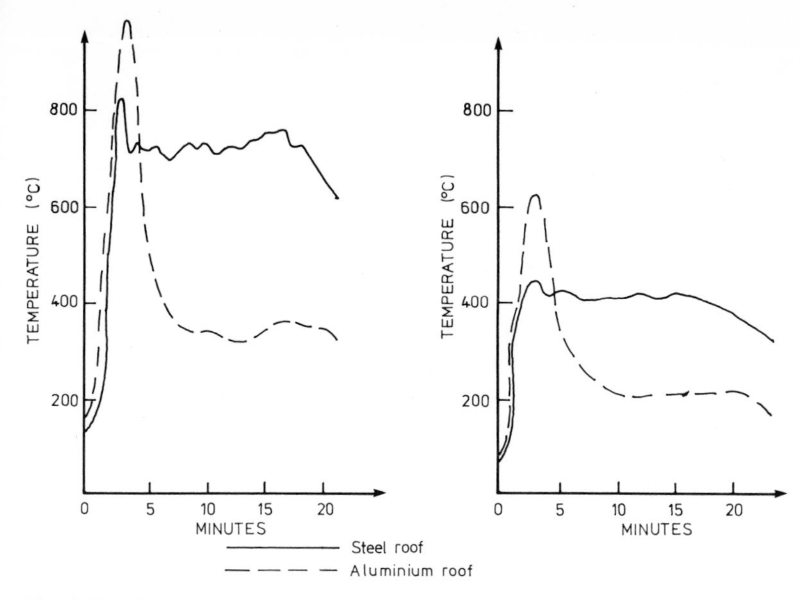

Fig. 9.37 Comparison of temperature profiles under steel and aluminium roofing (temperatures measured directly under roof 2 m and 8 m respectively)

compartments by compartment walls which will withstand a complete burnout. The roof is designed as a fire relief vent, i.e. either to melt or burn through in the early stages of the fire and provide an immediate fire vent. Thus a fire originating in compartment A is confined to compartment A and venting as a consequence of a partial failure of the roof results in acceptable predetermined fire damage.

Containment

Methods of containing smoke may range from the provision of a simple physical barrier to the inclusion of a mechanical system which prevents the ingress of smoke into the protected spaces. Figure 9.39 shows a lobby approach staircase protected against the ingress of smoke. Figure 9.40 shows screens at ceiling level along a mall creating smoke reservoirs and Fig. 9.41 illustrates how smoke can be contained using the roof profile. The performance of the latter can be enhanced by using reinforced fibreglass curtains or for large factory spaces, curtains may be used to create smoke reservoirs, Fig. 9.42. The latter methods can of course be combined with the methods discussed for removal of smoke by the inclusion of suitable extraction fans.

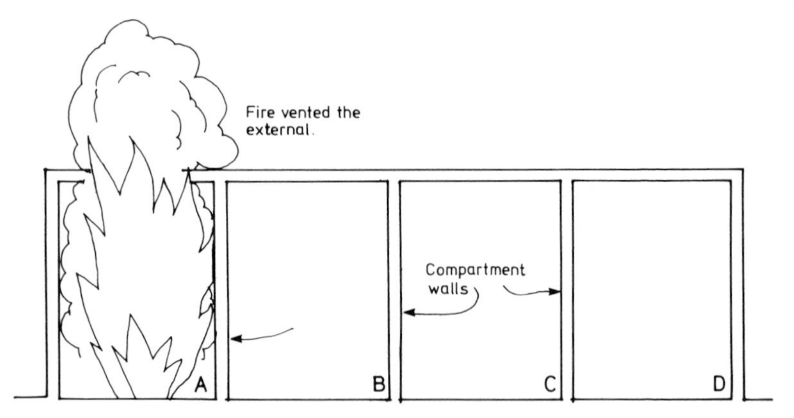

Fig. 9.38 Failure of roof at early stage of fire ensures that fire is confined to compartment A

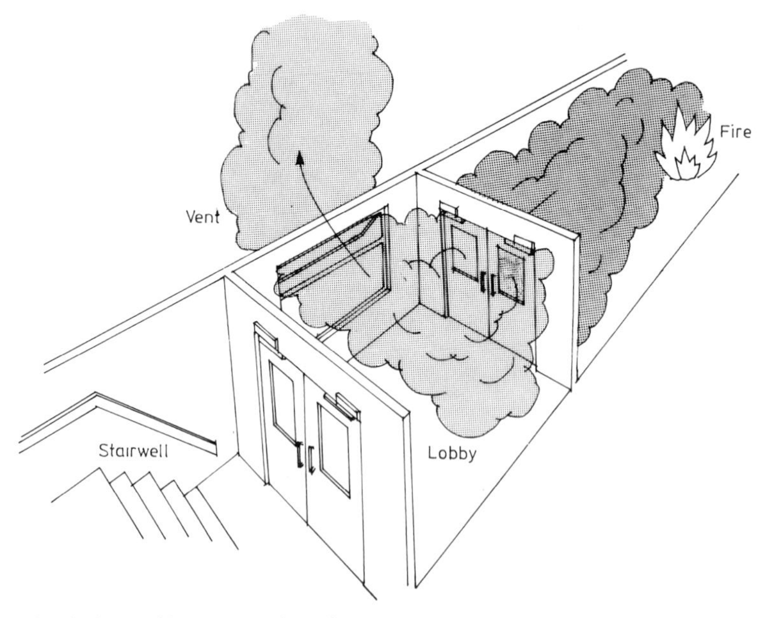

Fig. 9.39 Lobby approach staircase

Pressurisation

In Chapter 12, pressurisation as a means of smoke control was mentioned and various modes of celebration discussed. Pressurisation as a method of smoke control can be applied in various situations:

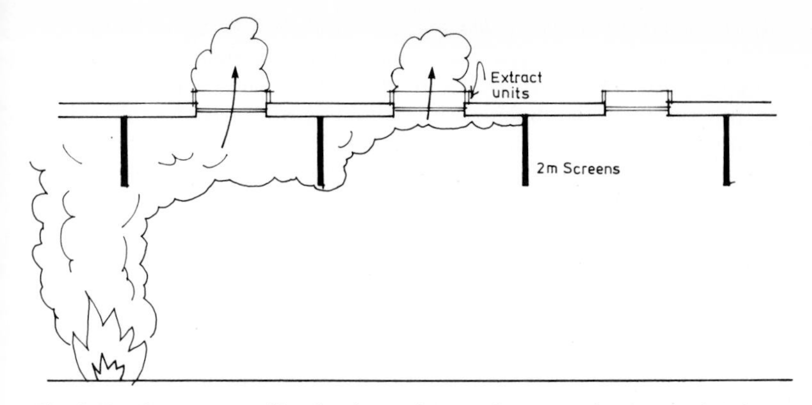

Fig. 9.40 Screens at ceiling level creating smoke reservoirs (extraction fans may be incorporated)

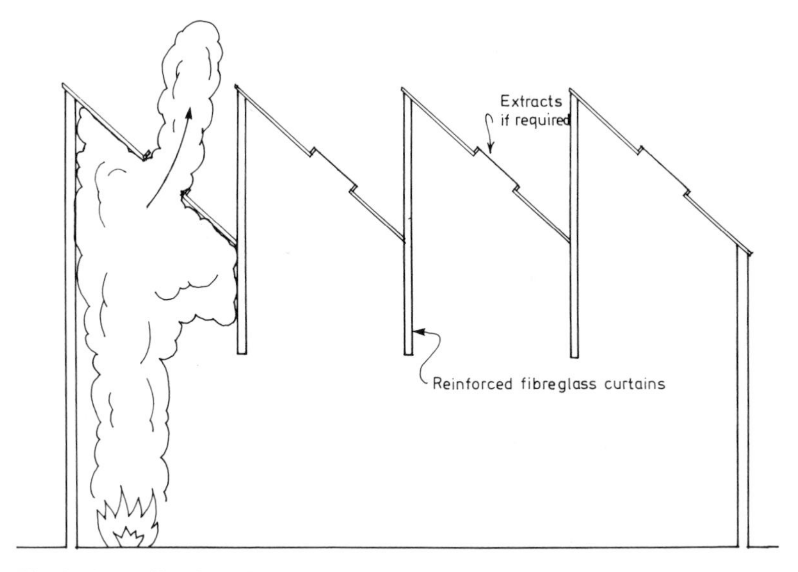

Fig. 9.41 Utilisation of roof profile for smoke control

1. Staircase only
2. Staircase and horizontal route, or
3. Lobbies and/or corridors only.

A worked example relating to the pressurisation of a staircase only will be given later. It is necessary initially to consider the separate components in the design of a pressurisation system which must be successfully integrated to provide a system which will give the required performance. These are:

The air supply to the protected spaces This must be a mechanically-operated air supply using a ducted distribution system, if necessary, to the required locations within the protected spaces. The air must be drawn in from the outside of the building and it is of vital importance that all necessary precautions are taken to ensure that smoke cannot enter the system at any point.

Air leakage out of the protected space It is perhaps unavoidable that given the function of the space to be pressurised and construction technology, that air-leakage paths from the pressurised space to adjoining spaces will exist. These leakage paths will occur around doors, windows, etc., and are essential if a pressure differential between the pressurised space and adjoining spaces is to be maintained, Fig. 9.43. However, it is the volume of air flow from the pressurised space to the adjacent space through these leakage paths which will determine the level of pressurisation achieved.

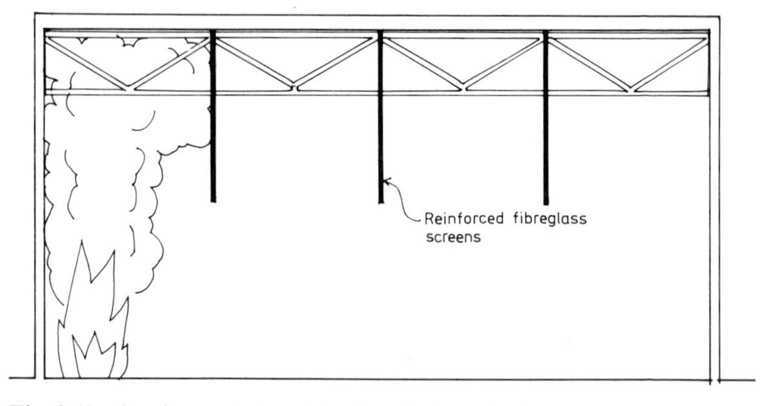

Fig. 9.42 Smoke control curtains in a factory situation

The air leakage of the external envelope As with air leakage out of the protected space, it is absolutely essential that air flowing out of the pressurised space to unpressurised spaces be able to leak out of the building to the external air. This leakage can be achieved by leakage around windows, etc., in the external walls of the building or by specially installed vents which open when the system is fully operational. If this latter leakage path is not available, the necessary pressure differential between the pressurised space and the remainder of the building will not be achieved, the whole building will be pressurised and the objective of smoke containment will be defeated.

It is essential for the student to understand that a knowledge of the air-leakage characteristics of the construction interfacing the pressurised and unpressurised parts of the building and of the external facade of the building is necessary for the design of an effective pressurisation system of smoke control. Also pressurisation as a means of smoke control cannot readily be applied to existing buildings; it is a method which can influence fundamentally the arrangement of circulation spaces in a building and consequently should be considered at a very early stage in the design process.

Pressurisation design procedure The basic steps in design procedure for pressurisation are listed below:

1. Define volume
2. Define required pressure differences. Consider: the location of the fire, volume of the fire compartment, life safety time and size of the fire
3. Estimate air-leakage characteristics of the surrounding construction
4. Modify pressure differences
5. Calculate pressure variations due to external influences, e.g. wind and temperature
6. Modify pressure differences, i.e. superimpose pressures created by the wind – laterally, and stack pressure – vertically
7. Calculate pressures induced by air-handling plant
8. Estimate combined pressure pattern
9. Combined use of air-handling plant
10. Plant design
11. Plant rooms
12. Direct layout and detection system
13. Sizing of fans and ducting
14. Select controls and prepare details.

WORKED EXAMPLE 14

Calculate:

(a) the overpressure requirements in the staircase enclosure necessary to achieve a pressure differential of 25 Pa for the conditions shown in Fig. 9.43 and comment on the results.

(b) the overpressure requirements for the conditions given in (a) when a lobby door is considered to be permanently open.

Data and relevant information:

Building	10 storeys high
Storey height	3 m
Total external facade leakage (window crack length)	400 m/floor
Crack width	0.003 m

Wind speed	$10\ \text{m s}^{-1}$	
Discharge coefficient: (C_D)	0.6	
External pressure coefficients	C_p	*Elevation*
	$+1.2$	(*C*)
	-0.2	(D)
	-0.5	(*A*)

Estimation of internal pressure coefficient

In order to take into account the external wind effect, it is necessary to estimate a suitable internal surface–pressure coefficient C_{pi}. In this case, C_{pi} relates the internal pressure to the outside pressure,

i.e. $p_i = C_{pi} \cdot p_{so}$

where $p_{so} = \frac{1}{2}\rho v^2$

$$\Delta p = (p_o - p_i)$$

$$\therefore \quad \dot{Q} = C_D A (C_{po} - C_{pi})^{\frac{1}{2}} p_{so}^{1/2} \qquad \ldots[1]$$

This equation applies for all the external surfaces.

In order to estimate p_i, an estimate of a suitable C_{pi} is required which will satisfy the physical condition that all air flowing in per second flows out. In formal terms this means that the algebraic sum of the separate flow rates per unit area over all surfaces exposed to external wind is zero.

$$\therefore \quad \sum_{r=1}^{n} \frac{q_r}{C_d A_r} = 0 \qquad \ldots[2]$$

$q_r / C_d A_r$ may be written as

$$\left(\frac{(C_{po} - C_{pi})}{|C_{po} - C_{pi}|} \right)_r \times (|C_{po} - C_{pi}|)_r^{1/2} \cdot p_{so}^{1/2} \qquad \ldots[3]$$

$\therefore$ This physical condition is written as:

$$\sum_{r=1}^{n} \frac{(C_{po} - C_{pi})_r}{(|C_{po} - C_{pi}|)_r} \cdot (|C_{po} - C_{pi}|)_r^{1/2} = 0 \qquad \ldots[4]$$

It must be noted that $(C_{po} - C_{pi})/(|C_{po} - C_{pi}|)$ is unity for each surface r and changes sign to $+1$ or -1. This allows the use of $(|C_{po} - C_{pi}|)_r^{1/2}$ to yield a value which coupled with the $+1$ or -1 can be added to other values for the exposed surfaces to give a value which will be close to zero if a good estimate of C_{pi} is chosen.

A graph of

$$\sum_{r=1}^{n} \frac{(C_{po} - C_{pi})_r}{(|C_{po} - C_{pi}|)_r} \cdot (|C_{po} - C_{pi}|)_r^{1/2}$$

against C_{pi} can be drawn from which a precise value of C_{pi} can be read off to obtain the condition given by eqn [4] above.

A graph as suggested is drawn in the worked example given.

Solution: 1

By trial and error.

Assume $C_{pi} = -0.3$

$$\frac{q_r}{C_d A_r} = \frac{C_{po} - C_{pi}}{|C_{po} - C_{pi}|} \cdot (|C_{po} - C_{pi}|)^{\frac{1}{2}} \quad \text{(A)}$$

$$= \frac{(-0.5 - (-0.3))}{|-0.5 - (-0.3)|} \cdot \sqrt{|-0.5 - (-0.3)|}$$

$$-\sqrt{0.2}$$

$$\frac{-0.2 - (-0.3)}{|-0.2 - (-0.3)|} \cdot \sqrt{0.1} \times 2 = 2\sqrt{0.1} \quad \text{(D)}$$

$$\frac{1.2 - (-0.3)}{|1.2 - (-0.3)|} \cdot \sqrt{1.5} = \sqrt{1.5} \quad \text{(C)}$$

$$\sum \frac{q_r}{C_d A} = \sqrt{1.5} + 2\sqrt{0.1} - \sqrt{0.2}$$

$$= 1.41 \quad \textit{Too High!}$$

Try $C_{pi} = -0.2$

$$\frac{-0.5 - (-0.2)}{|-0.5 - (1.2)|} \cdot \sqrt{0.3} = -\sqrt{0.3} \quad \text{Ⓐ}$$

$$\frac{-0.2 - (-0.2)}{|-0.2 - (-0.2)|} = 0.0 \quad \therefore \quad \text{ignore} \quad \text{Ⓑ}$$

$$\frac{1.2 - (-0.2)}{1.2 - (-0.2)} \cdot \sqrt{1.4} = +\sqrt{1.4} \quad \text{Ⓒ}$$

$$\sum \frac{q_r}{C_d A} = -\sqrt{0.3} + \sqrt{1.4}$$

$$= \sqrt{1.4} - \sqrt{0.3}$$

$$= 0.633 \quad \textit{Closer!} \text{ but not quite there}$$

Try $C_{pi} = -0.1$

$$\frac{-0.5 - (-0.1)}{|-0.5 - (-0.1)|} \cdot \sqrt{0.4} = -\sqrt{0.4} \quad \text{Ⓐ}$$

$$\frac{1.2 - (-0.1)}{|1.2 - (-0.1)|} \cdot \sqrt{1.3} = +\sqrt{1.3} \quad \text{Ⓑ}$$

$$\frac{-0.2 - (-0.1)}{|-0.2 - (-0.1)|} \cdot \sqrt{0.1} \times 2 = -2\sqrt{0.1} \quad \text{Ⓒ}$$

$$\sum \frac{q_r}{C_d A} = \sqrt{1.3} - \sqrt{0.4} - (\sqrt{0.1} \times 2)$$

$$= 1.14 - 0.63 - (0.32 + 2)$$

$$= -0.13 \quad \textit{Very close!}$$

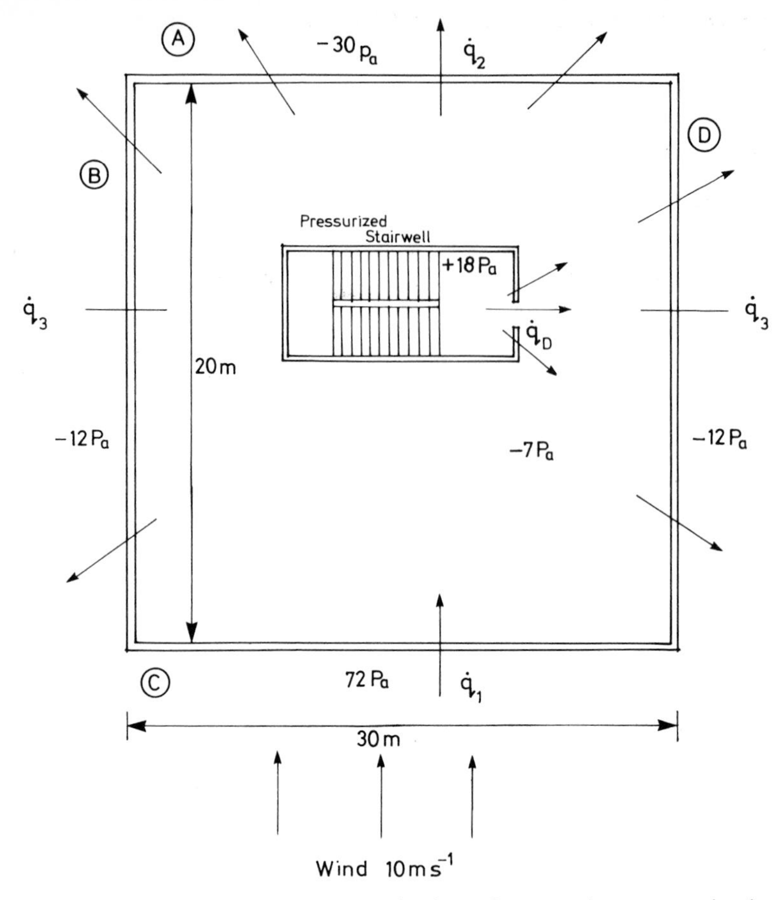

Fig. 9.43 Overpressure requirement in the staircase enclosure to maintain 25 Pa

Plot a graph of $\sum q_r/C_d A$ against C_{pi} for values obtained above (Fig. 9.44) and read off:

C_{pi} value when $\sum \dfrac{q_r}{C_d A} = 0.0$

i.e.

$$
\begin{aligned}
C_{pi} &= -0.11 \\
&= C_{pi}^{1/2} \rho V^2 \\
&= 0.11 \times 0.5 \times 1.22 \times 100 \\
&= -7 \text{ Pa (approx.)}
\end{aligned}
$$

The 7 Pa indicates an overpressure requirement of 18 Pa in the lobby, to maintain the 25 Pa pressure differential (see Fig. 9.43).

Comment: In order that the pressure differential of 25 Pa is maintained when the wind blows:

$$\dot{q}_2 + 2\dot{q}_3 \gg \dot{q}_1 + \dot{q}_D$$

where:

$\dot{q}_D$ = volume flow rate through the lobby door.

The above condition must be satisfied otherwise all of the floor spaces will become pressurised. Such an eventuality would in effect constitute a failure of the pressurisation system.

Check on leakage of external envelope

$$p_o \text{ on each external wall} = C_{pe}\tfrac{1}{2}\rho V^2$$

$$= 1.2 \times 0.5 \times 1.2 \times 100 = 72 \text{ Pa}$$

$$= -0.5 \times 0.5 \times 1.2 \times 100 = -30 \text{ Pa}$$

$$= -2.0 \times 0.5 \times 1.2 \times 100 = -12 \text{ Pa}$$

Pressure across the external walls $\Delta p_1 = 72 - (-7) = 79$ Pa

$$\Delta p_2 = -30 - (-7) = 23 \text{ Pa}$$

$$\Delta p_3 = -12 - (-7) = -5 \text{ Pa}$$

Volume flow rate (in or out) for each floor $= C_D A \sqrt{\frac{2}{\rho}\Delta P}$

ρ = density of air, 1.2 kg m^{-3}

A = area of crack

$$\therefore \quad \dot{q} = \left(C_D A \sqrt{\frac{2}{\rho}}\right)(\Delta \rho)^{\frac{1}{2}}$$

Apportion crack area in a simple proportional basis,

i.e. $\dot{q}_1 = 0.6 \times (400 \times 0.003 \times 0.3) \times \sqrt{\frac{2}{1.2}} = \sqrt{79}$

$$= 0.28 \times \sqrt{79}$$

$$= \underline{2.48} \text{ m}^3 \text{ s}^{-1}$$

Similarly:

$$\dot{q}_2 = 1.34 \text{ m}^3 \text{ s}^{-1}$$

$$\dot{q}_3 = 0.42 \text{ m}^3 \text{ s}^{-1}$$

$$\dot{q}_D = \left(C_D A_D \sqrt{\frac{2}{\rho}}\right)(\Delta P)^{\frac{1}{2}}$$

For a double leaf door 2 m high × 1.6 m wide,

Assume a leakage area $0.03\ m^2 = A_D$

$$\dot{q}_D = 0.6 \times 0.03\sqrt{\frac{2}{1.2}} \times \sqrt{25}$$

$$= 0.12\ m^3\ s^{-1}$$

Total flow into the unpressurised floor area

$$\dot{Q}_{in} = \dot{q}_1 + \dot{q}_D = 2.6\ m^3\ s^{-1}$$

$$\dot{Q}_{out} = \dot{q}_2 + 2\dot{q}_3 = 2.18\ m^3\ s^{-1}$$

Clearly $\dot{Q}_{in} > \dot{Q}_{out}$ which suggests that for the conditions specified the floor space will become pressurised. This problem may be remedied by the inclusion of automatically-operated pressure vents in the external envelope.

Solution: 2

Assume:

(i) still air conditions
(ii) effective area of open door $= 1.6\ m^2$
(iii) C_D for door $= 0.6$.

Volume flow rate through the door ($\dot{Q}_D$) is given by:

$$C_D A_D \sqrt{\frac{2}{\rho}} \times \sqrt{25}$$

Flow rate through the remaining nine doors

$$(\dot{q}_d) = 9C_D A_d \sqrt{\frac{2}{\rho}} \times \sqrt{25}$$

$\dot{Q}_D + 9\dot{q}_d =$ total flow rate from pressurised shaft

$$= \left(C_D A_D \sqrt{\frac{2}{\rho}} \times \sqrt{25}\right) + \left(9C_D A_d \sqrt{\frac{2}{\rho}} \times \sqrt{25}\right)$$

$$= (0.6 \times 1.6 \times 1.3 \times 5) + (9 \times 0.6 \times 0.028 \times 1.3 \times 5)$$

$$= 6.24 + 0.98$$

$$= 7.22\ m^3\ s^{-1}$$

Check for the effect on pressurisation of the shaft when all of the doors are closed and air is being delivered at $7.2\ m^3\ s^{-1}$.

$$\text{Flow rate out of shaft} = 10 \times C_D \times A_d \times \sqrt{\frac{2}{\rho}} \times \Delta P^{\frac{1}{2}}$$

$$= 10 \times 0.6 \times 0.028 \times 1.3 \times \Delta P^{\frac{1}{2}}$$

$$= 0.218\ \Delta P^{\frac{1}{2}}$$

Assuming no other losses

$0.218\,\Delta P^{\frac{1}{2}} = 7.22$

$$\therefore \quad \Delta P^{\frac{1}{2}} = \frac{7.22}{0.218}$$

$$\Delta P = 1{,}092 \text{ Pa*}$$

Obviously this level of pressurisation renders the staircase unusable bearing in mind that an overpressure of 60 Pa makes the opening of doors into the lobbies virtually impossible for the average person.

This problem can be overcome by the introduction of specially-designed vents which operate when the pressure in the staircase exceeds predetermined limits, e.g. 50–60 Pa. It should be noted that the influence of the wind in this example would be very marginal.

Effective areas of openings and vents In the example considered above the door openings were arranged in parallel. Situations may arise where the openings may be arranged in series or a combination of both series and parallel.

1. *Series arrangement*

In this case, Fig. 9.45, $p_1 > p_2 > p_3$

and $\quad \dot{q}_{\text{Tot}} = \dot{q}_1 = \dot{q}_2$

time $\quad \dot{q}_{\text{Tot}} = CA(\Delta p)^{\frac{1}{2}} \qquad \ldots[1]$

Then $\quad \dot{q}_1 = CA_1(\Delta p_1)^{\frac{1}{2}} \quad$ and $\quad \dot{q}_2 = CA_2(\Delta p_2)^{\frac{1}{2}}$

where $\quad \Delta p = p_1 - p_3$

$$\Delta p_1 = p_1 - p_2 = \frac{\dot{q}^2_{\text{Tot}}}{C^2 A_1^2} \qquad \ldots[2]$$

$$\Delta p_2 = p_2 - p_3 = \frac{\dot{q}^2_{\text{Tot}}}{C^2 A_1^2} \qquad \ldots[3]$$

Adding [2] and [3]

$$p_1 - p_3 = \frac{\dot{q}^2_{\text{Tot}}}{C^2}\left[\frac{1}{A_1^2} + \frac{1}{A_2^2}\right] \qquad \ldots[4]$$

also $$p_1 - p_3 = \frac{\dot{q}^2_{\text{Tot}}}{C^2}\left[\frac{1}{A^2}\right] \qquad \ldots[5]$$

Equations [4] and [5]

$$\therefore \quad \left[\frac{1}{A^2}\right] = \left[\frac{1}{A_1^2} + \frac{1}{A_2^2}\right]$$

$$A = \frac{A_1 \cdot A_2}{\sqrt{A_1^2 + A_2^2}} \qquad \ldots[6]$$

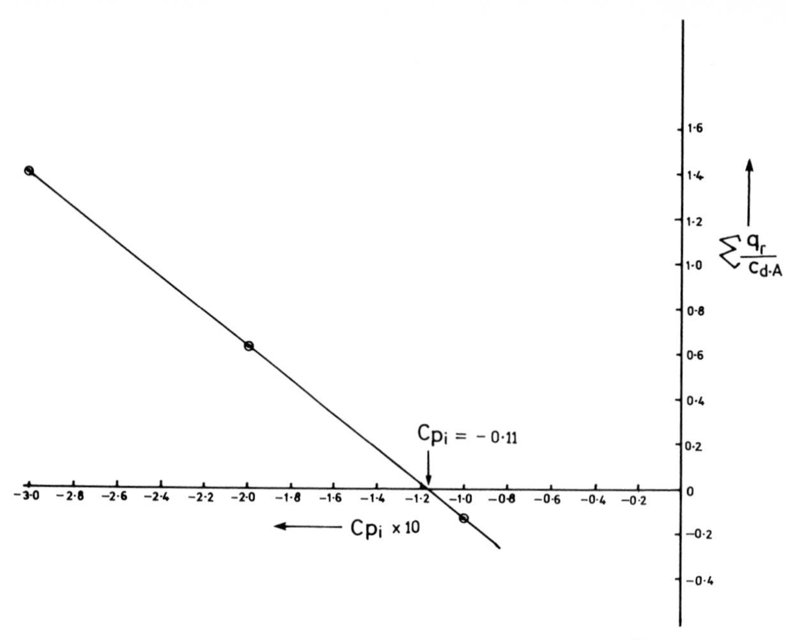

Fig. 9.44 Relationship between pressure coefficients and $\sum \frac{q_r}{C_d \cdot A}$ factor

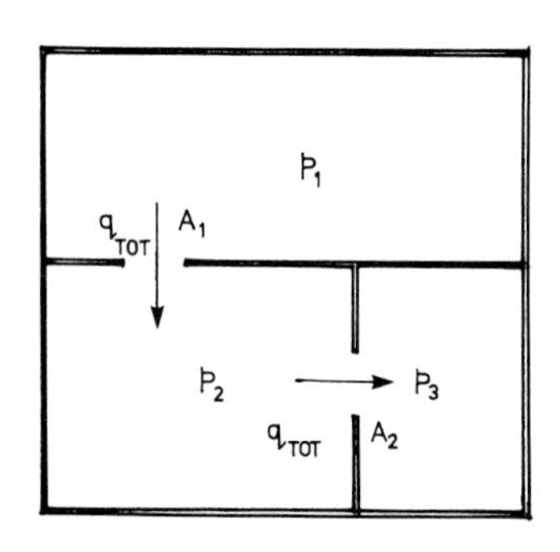

Fig. 9.45 Series arrangement

2. *Parallel arrangement*
   As shown in Fig. 9.46 $\dot{q}_{\mathrm{Tot}} = \dot{q}_1 + \dot{q}_2$

   also $\dot{q}_{\mathrm{Tot}} = CA(\Delta p)^{\frac{1}{2}}$

$$\dot{q}_1 = CA_1(\Delta p)^{\frac{1}{2}}$$

$$\dot{q}_2 = CA_2(\Delta p)^{\frac{1}{2}}$$

Then

$$CA(\Delta p)^{\frac{1}{2}} = C(A_1 + A_2)(\Delta p)^{\frac{1}{2}}$$

where

$$\Delta p = (p_1 - p_2)$$
$$A = (A_1 + A_2) \qquad \ldots [7]$$

3. *Combination of series and parallel openings*
   In this case, $p_1 > p_2 > p_3$ (Fig. 9.46)

   and $\dot{q}_{\text{Tot}} = \dot{q}_2 + \dot{q}_3$

   Referring to Fig. 9.47 using the formulae for series and parallel arrangements it can be easily shown that:

$$A = \frac{A_1 A_4}{\sqrt{A_1^2 + A_4^2}}$$

where $A_4 = A_2 + A_3$

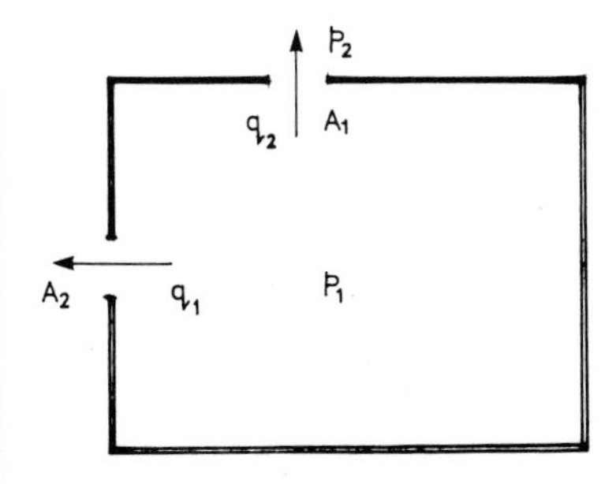

Fig. 9.46 Parallel arrangement

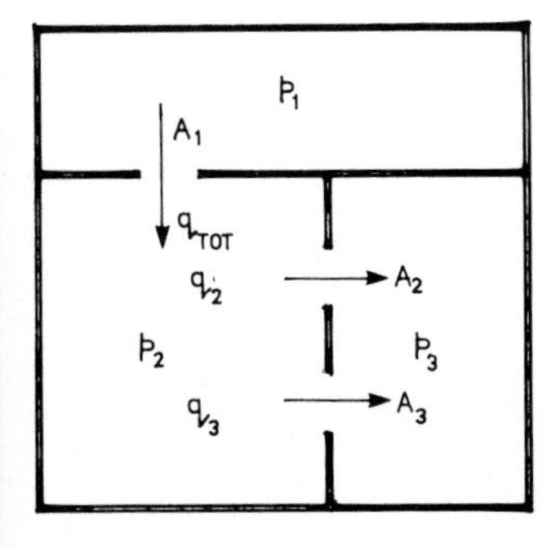

Fig. 9.47 Series and parallel arrangement

Table 9.6 Sole factors and interrelationships which influence the preliminary choice of smoke control system

Intra-space type activity	Space volume		Barrier leakage	Frequency of ignition	Criticality of space		Smoke-control system
					People	Value	
Single storey							
Store	small		low	high	low	low	Smoke detection
Store	small		low	low	low	low	–
Office	small		high	high	low	low	Self-venting
Parking garage	large		low	low	low	low	Smoke detection (natural reservoir)
Store	large		low	low	low	high	Smoke detection plus venting OR mechanical exhaust
Ward (hospital)	large		large	high	high	low	Reservoir or natural venting or mechanical systems
Inter-space type activity	**Fire**	**RS**					
Receiving space							
Store	small	small	low (high)		low		– (–)
Garage	small	large	low (high)			low	– (–)
Intensive care ward	small	small	low (high)		high	high	Mechanical system (MS)
	small	large	low (high)		high	low	Detection (mech. system)
Store/Office	large	small	low (high)		low	low	– (–)
Laundry	large	large	low (high)		low	low	– or natural vent (natural vent)
Accounts office	large	small	low (high)		high	high	Mech. system or natural vent (mech. system)
Shopping mall	large	large	high		high	low	Natural or mech. system

RS = Receiving space, Fire = Fire Space

9.7 CONCLUSIONS

Reference to Chapter 1 will indicate to some extent the magnitude of the risk to the occupants of buildings, associated with smoke.

In Chapter 6 the current suite of fire tests are discussed and future developments considered. It will be noted that there is no standard test method for assessing the propensity of building materials to produce smoke. BS 6401[17] gives a method for the measurement in the laboratory of the specific optical density of smoke generated by materials. This method, however, is intended for research and development only and not as a basis for building control purposes.

Many factors influence the preliminary selection of smoke-control systems, e.g. barrier leakage, space volume, frequency of ignition and criticality of space. Marchant[16] discussed these factors and Table 9.6 gives some indication of the relationships discussed.

BS 5588: Part 4: Smoke Control in Protected Escape Routes using Pressurisation: contains several worked examples on the pressurisation of protected escape routes and the reader may refer to these.

In recent years computer models[18] on smoke movement as design aids have become available. It is important that the user of these design tools understands the fundamentals of smoke control.

REFERENCES

1. Drysdale D D, 'Smoke production', paper given at Symposium: *Smoke Production and Control in Buildings*, University of Ulster, Sept. 1982.
2. Hinkley P L, '*Some notes on the control of smoke in enclosed shopping centres*', Fire Research Note 875, F.R.S., Boreham Wood, England.
3. Jin T, '*Visibility through fire smoke*', Building Research Institute, Tokyo, Report No. 30, March 1970 and Report No. 33, February 1971.
4. N.B.S. Test ANSI/ASTM E662-79, '*Standard Test Method for the Specific Optical Density of Smoke generated by solid materials*'.
5. Rashbash D J and Phillips L C, 'Quantification of smoke produced at fires', *Fire and Materials* **2**, 102, 1978.
6. Rabash D J and Stark G M V, '*The generation of carbon monoxide by fires in compartments*', Fire Research Note. 614, 1966.
7. Marchant E M, Principles of smoke movement and smoke control', Paper given at Symposium: *Smoke Production and Control in Buildings*, University of Ulster, Sept. 1982.
8. Butcher E G and Parnell A C, '*Smoke Control in Fire Safety Design*, SPON, London, 1979.
9. Hutchinson N B and Handegard G O P, *Building Science for a Cold Climate*, Wiley, 1983.

10. Morgan J and Marchant E W, '*Smoke Control in Buildings: design principles*', B.R.E., Digest 260, D.O.E., HMSO, London, 1982.
11. 'Some effects of natural wind on vent operation in shopping malls', *C.I.B. Symposium on the Control of Smoke Movement in Building Fires*, B.R.E., November 1975.
12. Thomas P H, *Investigations into the flow of hot gases in roof venting*', Fire Research Station, Fire Research Technical Paper No. 7, HMSO, London, 1964.
13. Spratt A and Heselden A J M, '*Efficient extraction of smoke from a thin layer under a ceiling*', Fire Research Station, Fire Research Note No. 1001, 1974.
14. McCormack S, 'Melting aluminium cools fire danger', *New Civil Engineer*, Institution of Civil Engineers, August 1985.
15. Broadsheet 11, '*Fire Precautions: an alternative approach to the design of a primary school*', Department of Education and Science, Architects and Building Group, Nottinghamshire County Council 1982.
16. Marchant E W, 'Principles of smoke movement and smoke control', paper given at Symposium: *Smoke Production and Control in Buildings*, University of Ulster, September 1982.
17. BS 6401: 1983, British Standard Method for Measurement in the Laboratory of the Specific Optical Density of Smoke Generated by Materials.
18. Klate J H and Fothergill J W, *Design of Smoke Control Systems for Buildings*, US Department of Commerce National Bureau of Standards Washington DC.

CHAPTER 10

Fire detection

10.1 INTRODUCTION

The role of a fire detector is not solely to detect a fire but to discriminate reliably between the absence and the presence of a fire. If a detector is too sensitive it may give a false alarm for a non-fire condition while a less sensitive device will not raise the alarm quickly enough to prevent possible human and material loss. Thus from a practical viewpoint the sensitivity of a detector must be optimised so that it will give an alarm for a reasonably large and potentially dangerous fire. Figure 10.1 illustrates conceptually a fire detection and warning system.

It would seem that if a minimum allowable sensitivity is required then this sensitivity would depend not only on the detector itself and its mounting situation, but also on the rate of growth and spread of the fire. The factors that control the rate of growth and the spread of fire in a building have been discussed already in some detail. It would now be a suitable point to classify detection into three distinct categories:

1. Heat detectors (point and line types)
2. Smoke detectors
3. Flame detectors.

10.2 CLASSIFICATION OF DETECTORS

This classification can be more easily understood by referring to Fig. 10.2, which shows in a clear manner this classification in better detail.

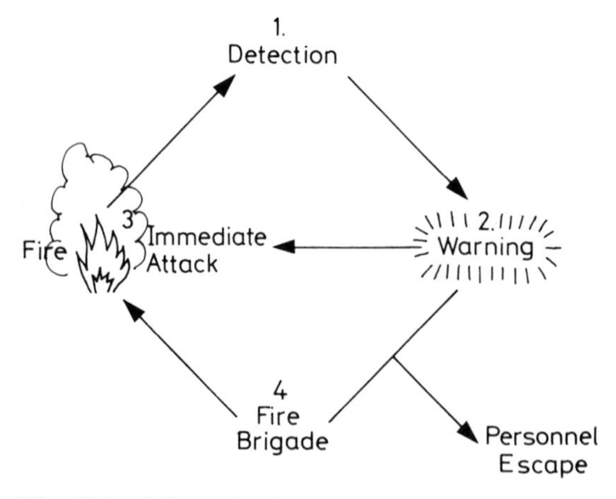

Fig. 10.1 Schematic view of detection – alarm system

10.2.1 Point detectors

Here each detector protects a small area around itself. These must operate within specified time limits for rates of temperature rise between 1 °C per minute and 30 °C per minute and when the rate of rise in temperature is less than 1 °C per minute also at fixed temperatures between 54 °C and 62 to 78 °C.

10.2.2 Line detectors

This detection system consists of sensitive elements present in a continuous line in the form of a long wire or a long tube containing a fluid. No matter which type is being used, each must be responsive to heating arising from a developing fire.

10.3 HEAT DETECTORS

There are two main types of detector:

1. Fixed temperature
2. Rate of rise of temperature.

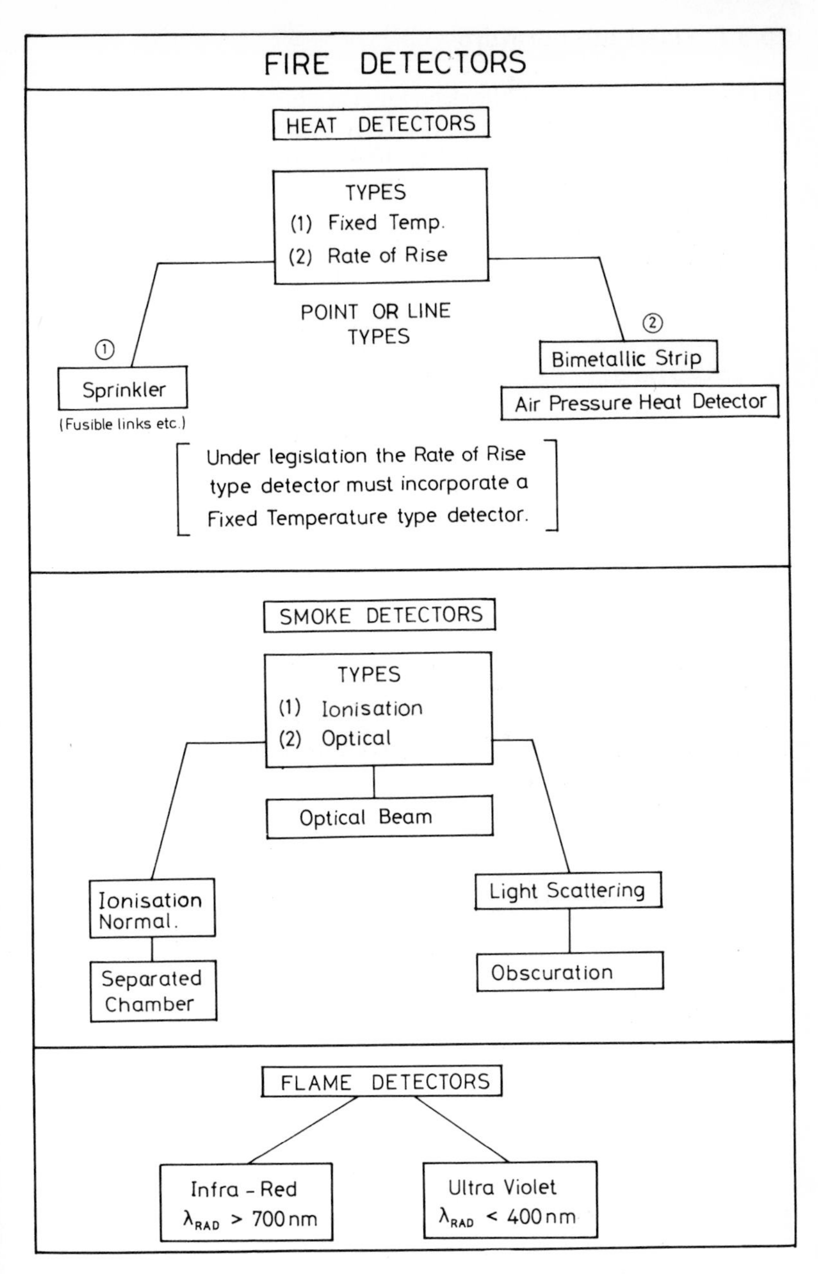

Fig. 10.2 Classification of fire detectors

10.3.1 Fixed temperature

In this case the action of the detector relies on a temperature-dependent physical property as illustrated with a bimetallic strip Fig. 10.3 and the air-pressure type, Fig. 10.4. A sprinkler also falls into this category. A sprinkler has two functions to perform in that it has

(a) to detect the fire and
(b) to provide extinction.

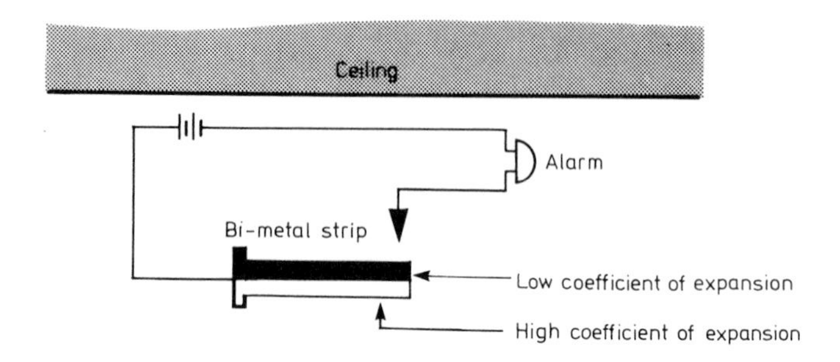

BEFORE

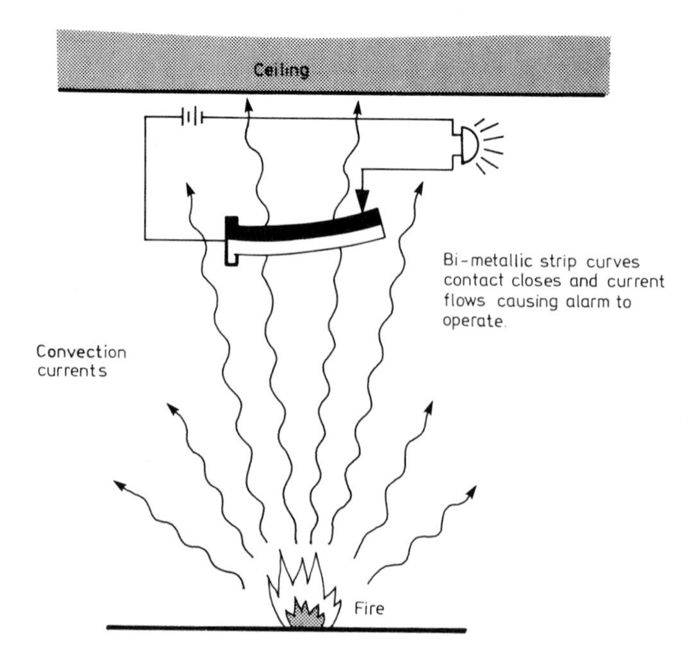

AFTER

Fig. 10.3 Bimetallic strip heat detector

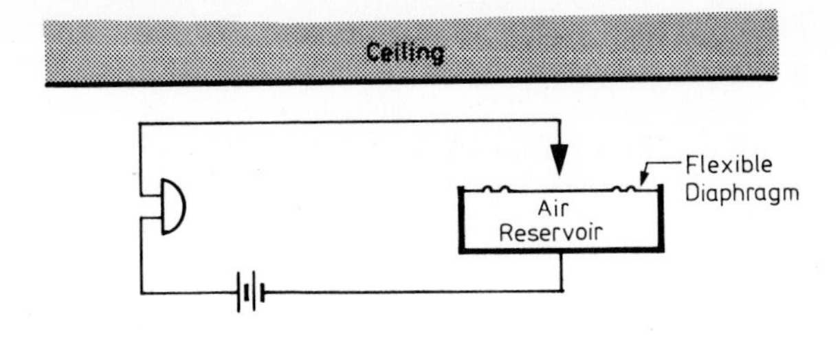

BEFORE

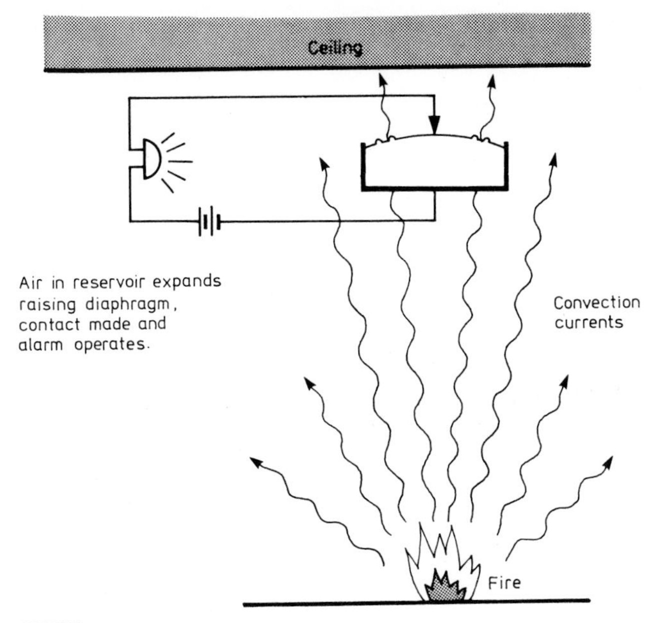

AFTER

Fig. 10.4 Air-pressure heat detector

Each function is performed separately although early detection makes fire extinction easier. The sprinkler is considered to react to the heat produced by the fire by mainly convective heat transfer and to a lesser extent radiation. Two types of sprinkler heads are illustrated in Figs 10.5 and 10.6.

10.3.2 Rate of rise detectors

This type of detector responds when the rate of rise of temperature of the air and hot gases that flow past it exceeds a minimum rate.

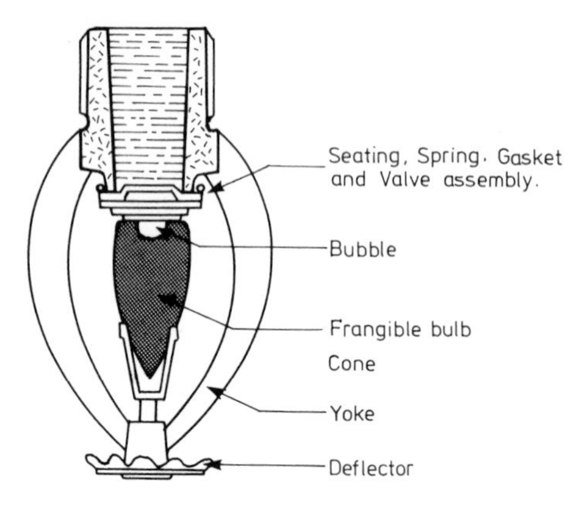

Fig. 10.5 Bulb sprinkler head

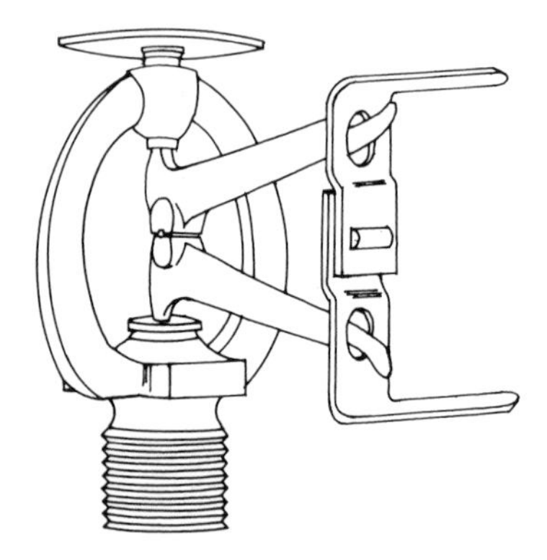

Fig. 10.6 Soldered sprinkler head

This response can be obtained by employing two opposing elements in the detector which respond at a different rate to this gas flow rate (Fig. 10.7). A suitable mechanism is the expansion of two elements of different thermal capacities say A of high capacity and B of low thermal capacity. The low-thermal-capacity element B will heat up much faster than A and eventually catch it up causing the circuit to be made. However, if the rate of rise of temperature is low enough the output from the two elements cancel each other.

Note that C is an upper stop which enables the detector to operate as a fixed temperature device even at very low rates of rise of temperature.

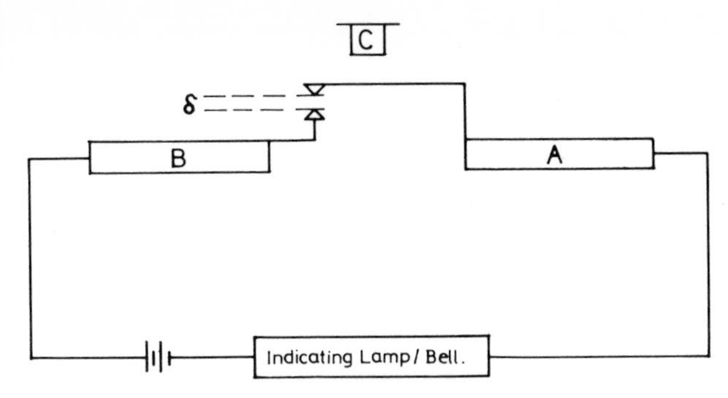

Fig. 10.7 Schematic sketch of 'rate of rise' detector

10.4 CONVECTIVE HEAT TRANSFER

From the brief comments on the fixed temperature and rate of rise of temperature detectors it will be observed that the convective heat transfer process is most influential in determining the operational characteristics of these detection systems.

As shown in Fig. 10.8 the hot products of the fire combustion process rise, pulling in or entraining air which in turn cools the combustion products by mixing. Thus the higher the location of the detector the lower the temperature of the rising plume.

Heat-sensitive detectors all have a nominal operating temperature usually some 30 °C above ambient temperature of their surroundings which ensures non-operation under normal conditions.

The following list gives the important factors that control the overall operation of the sprinkler:

1. Operating temperature of the detector
2. Thermal capacity of the parts which activate the detection system
3. Convection transfer coefficients which control the heat taken from the gaseous products to activate the detector
4. Height of ceiling on which detector is located
5. Shape of ceiling
6. Thermal properties of the ceiling
7. Distance of detector from ceiling
8. Horizontal distance of detector from fire
9. Rate of rise of air temperature surrounding the detector.

Factors 1–3 are controlled by the design of the detector, while factors 4–9 are controlled by the design of the building.

Consider the two types of sprinkler heads shown in Figs 10.5 and 10.6, one of which consists of a glass bulb containing an alcohol-based liquid while the other relies on the melting of a low-melting-

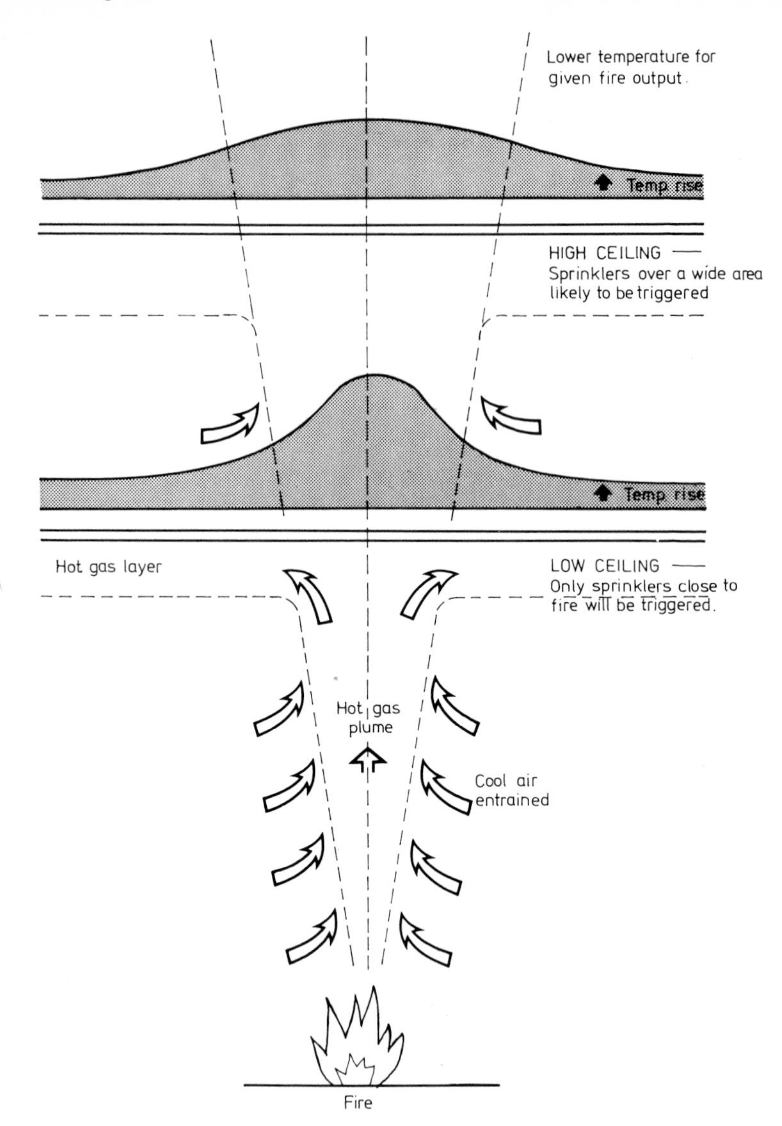

Fig. 10.8 Variation of gas temperature above fire

point solder which holds together two halves of a metal strut. In a non-fire situation the glass bulb or the metal strut holds the valve shut against the water at a high pressure.

The response time, that is the time needed for such a sprinkler to react, can now be considered in some detail using a simple model analysis.

10.5 LUMPED MODEL ANALYSIS (CONVECTIVE HEAT TRANSFER)

Here the following assumptions are made:

1. Sprinkler gains heat from surrounding gas by convection only, Fig. 10.9

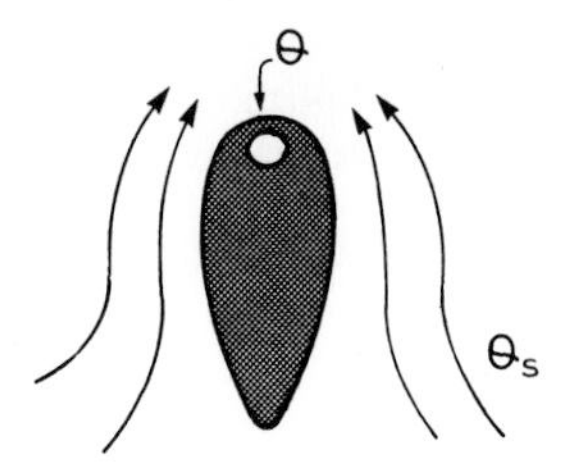

Fig. 10.9 Convective heat transfer to sprinkler head

2. No heat energy lost from the sprinkler by either convection or radiation
3. Temperature of the surrounding gas increases linearly with time,

   i.e. α = rate of change of surrounding gas temperature with time

   i.e. $\alpha = \dfrac{d\theta_s}{dt}$ (usually constant).

Using an Energy Balance Equation:

Rate of absorption of heat energy by the bulb or solder arm
= Rate of change of the lumped thermal capacity of the bulb or solder arm,

i.e. $$hA_s(\alpha t - \theta) = Mc \cdot \frac{d\theta}{dt} \qquad \ldots[1]$$

where A_s is the surface area of bulb or strut (m^2)
M is the mass of bulb or solder arms (kg)
h is the surface transfer coefficient ($W\,m^{-2}\,K^{-1}$)
c is the thermal capacity of bulb or solder arm ($J \cdot kg^{-1}\,K^{-1}$)

$$\therefore \quad (\alpha t - \theta) = \frac{Mc}{hA_s} \cdot \frac{d\theta}{dt} \qquad \ldots[2]$$

Integrating with $\theta = 0$ at $t = 0$ as boundary conditions:

If θ^1 and t^1 are the operating temperature and time, respectively, and if $hA_s t/Mc$ is large

$$\text{then} \quad \theta^1 = \alpha\left(t^1 - \frac{Mc}{hA}\right) \qquad \ldots[3]$$

$$\text{or} \quad t^1 = \frac{\theta^1}{\alpha} + \frac{Mc}{hA_s} \qquad \ldots[4]$$

where Mc/hA_s is referred to as the time constant which can have a value of $1\frac{1}{2}$ to 2 minutes for most sprinkler or solder-strut systems.

The variation for a sprinkler of t^1 with $1/\alpha$ is shown in Fig. 10.10. As can be seen from eqn [4], the operation time t^1 can be reduced by making the time constant as small as possible by making α larger or by increasing the absorption area using techniques to increase the surface using vanes or fins[2].

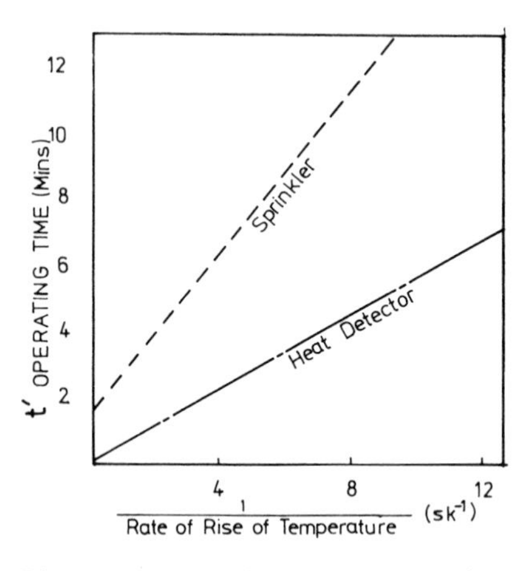

Fig. 10.10 Variation of operation time with rate of rise of temperature

Recently it was reported[1] that fast-response sprinkler heads were available of such a sensitivity that they could be utilised in domestic fires as a life safety device. The sensitivity of these heads is now expressed in terms of a 'Time Response Index' which is expressed as follows:

$$T = \frac{Mc}{hA_s} \cdot \sqrt{u}$$

i.e. $T = \tau \cdot \sqrt{u} \quad (\mathrm{m}^{\frac{1}{2}}\,\mathrm{s}^{\frac{1}{2}})$

where u is the velocity of the gas flowing over the detector head. It can be shown that the TRI remains invariant for sprinkler head over a large gas-flow-velocity range. To give the reader an idea of the new sensitivity achieved using special soldered strut-finned arrangements, the TRI is 30, compared to the usual type which has a TRI of 120.

10.6 RADIATION ABSORPTION ANALYSIS

Radiation from a fire cannot be entirely ignored, of course, since both the bulb and solder arm of the sprinkler are capable of absorbing radiation. Assuming a simple lumped model, an expression relating the operating time t^1 to the intensity of radiation I can now be developed.

Using the lumped model apprach:

Net rate of heat energy absorption

= Rate of change of thermal capacity of lumped model

i.e. $$aI = hA(\theta - \theta_s) + Mc\frac{d\theta_s}{dt}$$

(RADIATION ENERGY INPUT RATE) (RATE OF HEAT ENERGY LEAVING BODY TO SURROUNDING FLUID)

where a = absorption coefficient for the irradiated surface area,
h = heat energy transfer coefficient for surface
A = area of surface cooled by surrounding fluid.

If $\theta - \theta_s = \theta^1$ (temperature difference of surface and fluid) then

$$aI - hA\theta^1 = \frac{Mc\, d\theta^1}{dt}$$

Solving this equation using the following boundary condition:

$\theta^1 = 0$ when $t = 0$

it can be shown that the operating temperature difference θ^1_{op} is as follows:

$$\theta^1_{op} = \frac{dI}{hA}\left(1 - e^{-\frac{hA}{Mc} \cdot t^1}\right)$$

$$hA\theta^1_{op}aI = \left(1 - e^{-\frac{hA}{Mc} \cdot t^1}\right) = \left(1 - e^{-\frac{t^1}{\tau}}\right)$$

when $$\tau = \frac{Mc}{hA} = \text{time constant}$$

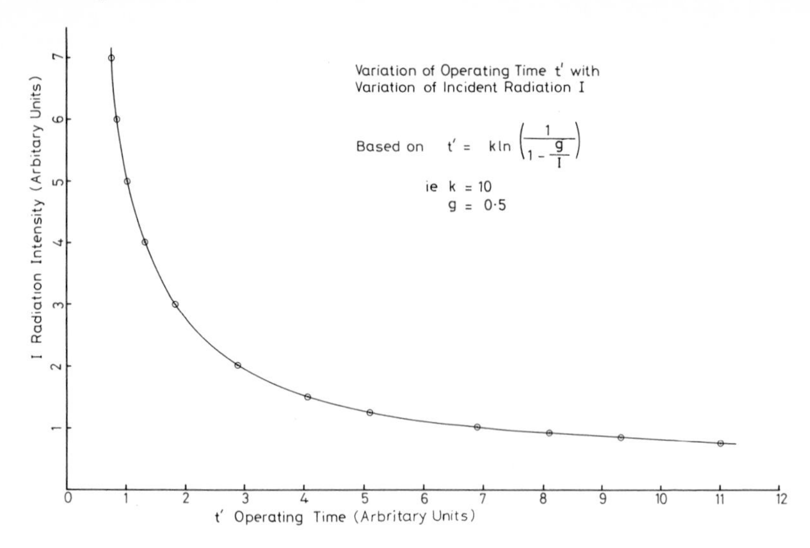

Fig. 10.11 Variation of operating limit with variation of radiation (I)

Then $$e^{-\frac{t^1}{\tau}} = 1 - \frac{hA\theta^1_{op}}{aI}$$

$$\therefore \qquad t^1 = \tau \ln \frac{1}{1 - \dfrac{hA\theta^1_{op}}{aI}}$$

The variation of t^1 with notional values of I is given in Fig. 10.11. The experimental variation of operation time t^1 with the magnitude of incident radiation intensity I tends to follow similar curves,[2] Fig. 10.12 for glass-bulb and soldered-strut sprinklers.

10.7 SMOKE DETECTORS

This type of detector must be capable of responding to smoke from smouldering and flaming combustion since the smoke from these fires is significantly different in structure and composition. Smoke from a smouldering fire tends to have much bigger particles of combustion products compared to smoke from a flaming fire where the particles are of much smaller dimensions. Thus a sensitive detector must be able to respond adequately to each type of smoke.

The various types of detector will now be considered in some detail.

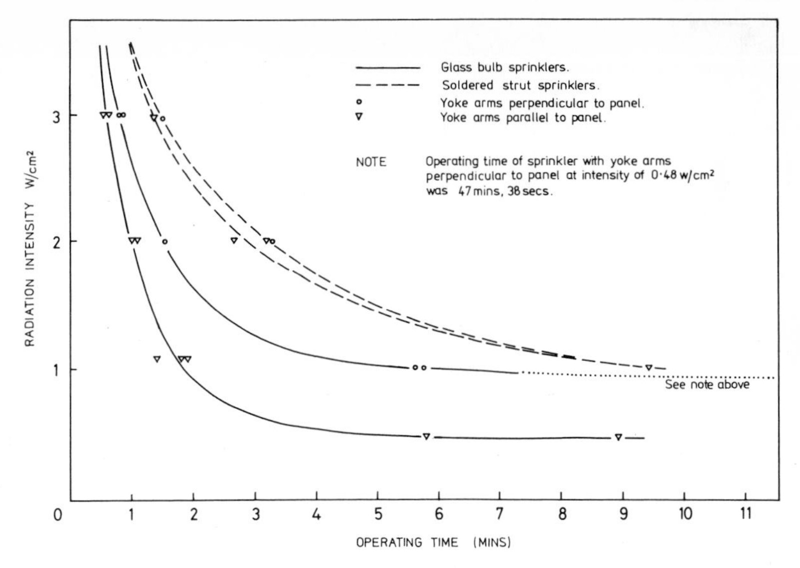

Fig. 10.12 Operating times of sprinklers at various radiant intensities

10.7.1 Ionisation detector

The standard form of ionisation detector is as shown in Fig. 10.13. Here the air is ionised by a radiation source and some of these ions interact with the incoming smoke particles to slow down and possibly recombine with ions of opposite sign while the remainder of the ions are neutralised at charged plates to produce a current. Thus the smaller the current the larger the number of smoke particles entering the device in unit time. This drop in the current sets off an alarm telling the occupants in the enclosure of the presence of smoke. It has been shown that ionisation detectors are particularly sensitive to small smoke particles. In its standard form the ionisation detector poses a problem since the pattern of ionisation within the chamber can vary from detector to detector, which causes a variance in the sensitivity of each detector.

This problem has been tackled and solved to a large extent at the B.R.E.[3] laboratories where a separated ionisation chamber has been developed and tested. Such a device is as shown in Fig. 10.14 where the smoke-laden air is ionised and then thoroughly mixed before passing into the recombination region.

This device has proved to be highly stable and sensitive, being able to detect successfully much lower smoke concentrations than the normal types are capable of detecting. This separated ionisation chamber would be especially useful in locations such as computer

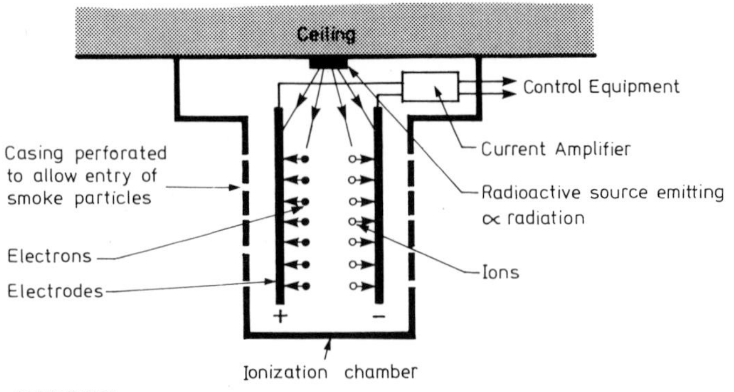

BEFORE

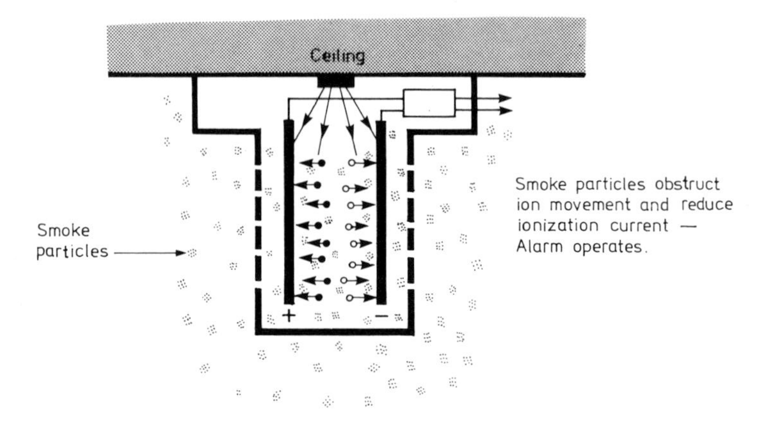

AFTER

Fig. 10.13 Ionisation chamber smoke detector

suites that are air-conditioned and no-smoking areas where even a small fire could do an enormous amount of damage to the computer hardware.

10.7.2 Optical detectors

Optical detectors function by the way in which smoke interacts with a beam of light passing through it. Two physical phenomena are important: these are **obscuration** and **scattering**.

Also for all optical-detection systems the size of the smoke particles is very important; in fact they should be comparable with the wavelength of the light passing through the smoke. In practice

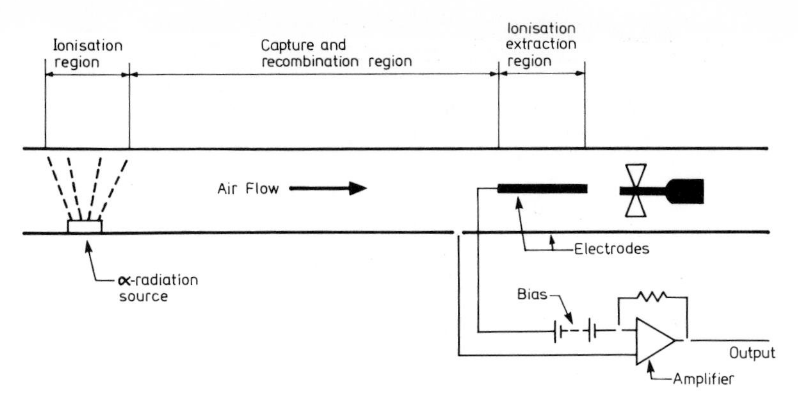

Fig. 10.14 Separated ionisation chamber

this means that only the larger particles of 500 nm or more have any appreciable effect.

Obscuration system

These are the earliest type of smoke detectors and are sensitive to the attenuation of a beam of light shining across a space, caused by the scattering and absorption of the light by smoke particles. This is described by the 'Beer–Lambert' law:

$$I = I_0 \, e^{-Kc\ell}$$

where I is the intensity at a length ℓ of a light path through smoke laden air
I_0 is the initial intensity at the light source
c = concentration of smoke particles
K = a constant.

Unlike the scattering system, to be described later, a current flows from the photocell detector all the time but is reduced by smoke which scatters some of the light away, Fig. 10.15.

Over short distances the absorption of the light by the smoke is proportional to the distance between the source and the cell. For short path lengths, say up to 1 m, the obscuration level at the correct time for giving an alarm is approximately 5 per cent, causing a change from 100 per cent to 95 per cent in the light intensity incident upon the photocell. Such a small change could be produced by other effects, such as dust on the lens or by a variation in the supply voltage to the lamp, to give rise to a false alarm. This problem can be overcome using a longer path length since the same degree of obscuration over 1 m will cause a 40 per cent obscuration

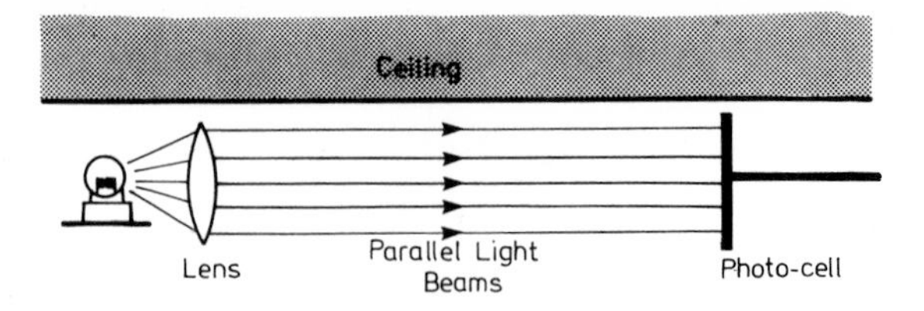

BEFORE

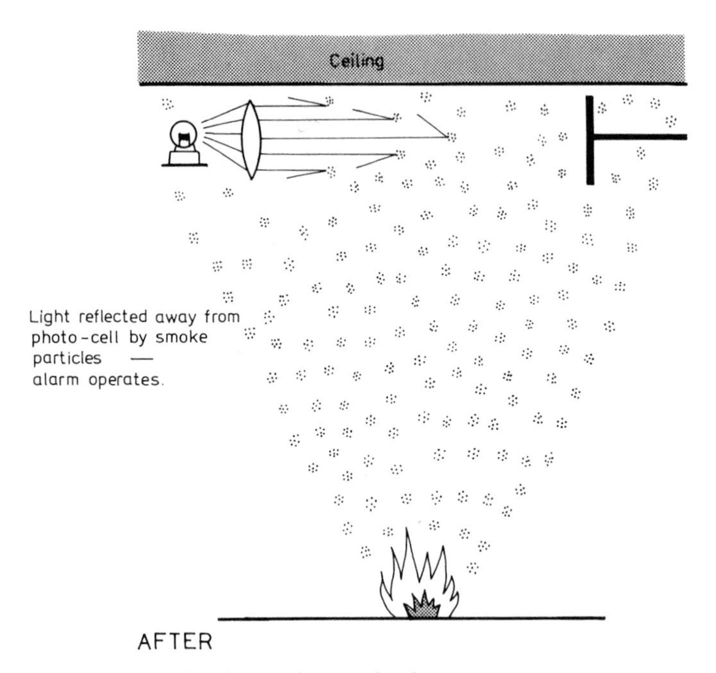

AFTER

Fig. 10.15 Light obscuration smoke detector

over 10 m. Thus if longer path lengths are employed a larger change in the photocell current can be obtained.

Scattering type

These types of detectors are shown in Fig. 10.16, and depend on the detection of scattered light from suspended smoke particles. The physical law governing the scattering of light from smoke particles is quite complex being described for very small particles by Rayleigh theory and for particles of the order of the wavelength of light by

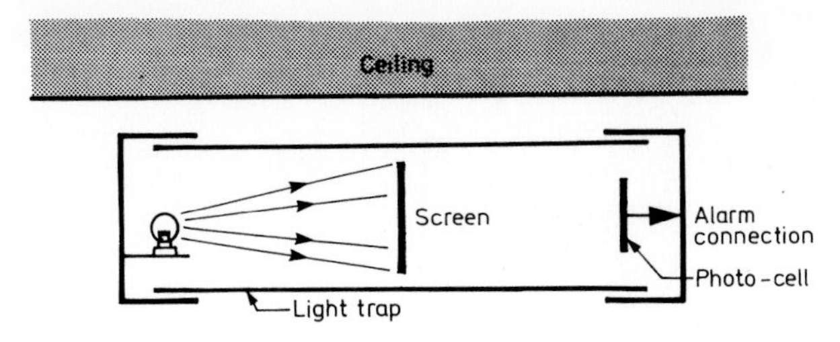

BEFORE

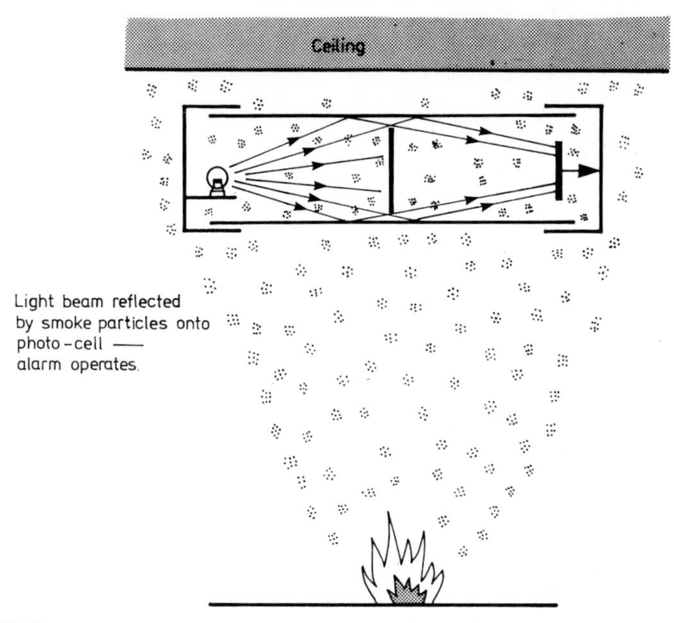

AFTER

Fig. 10.16 Light scattering smoke detection system

Mies theory. However, these laws cannot be used to describe the light scattering from smoke particles, although it has been shown that the amount of scattering produced is proportional to the wavelength.

The general construction of this type of detector, Fig. 10.17, is such that shields are included to protect the detecting element from the light paths. These labyrinths can resist the flow of smoke and have been shown to be quite insensitive to low smoke velocities. Dirt tends to block these channels and once the light source output level decreases with age the response of photo-cell decreases. Nowadays, light-emitting diodes are used to produce the light, making this device quite reliable.

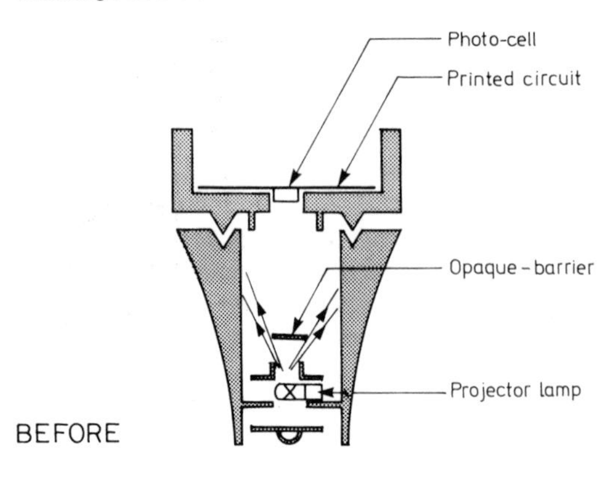

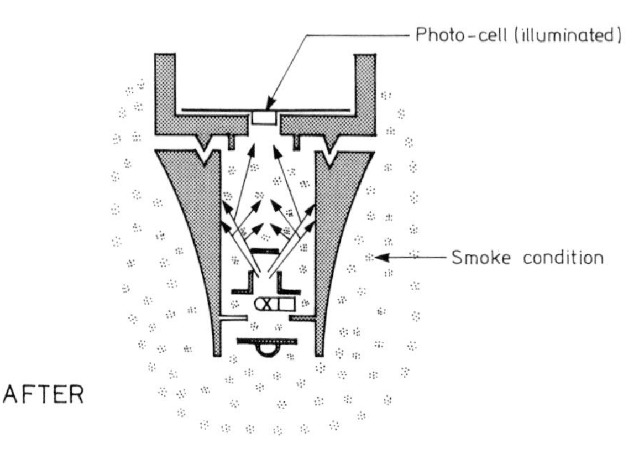

Fig. 10.17 Smoke detector

Optical beam-type smoke detector

This type of detector employs a beam of light to detect a fire at any point along its path. It consists of a pulsed light source producing a focused beam which is detected by the photo-cell (possibly after reflection). Where the pulsed beam is attenuated by smoke or irregularly deflected by air convection (excessive refraction) currents due to the presence of a fire causing the detector to respond.

10.8 FLAME DETECTORS

These detectors depend on the recognition of radiation produced in the burning zone.

10.8.1 Infra-red detector

Every fire is capable of producing radiation, particularly radiation in the infra-red region of the electromagnetic spectrum.

For typical flame combustion there is a characteristic flame flicker which is a regular changing of the flame intensity which on detection gives rise to the production of cyclic pulses. If a circuit is used that has been rendered sensitive to this flicker compared to the steady current induced by a fixed intensity of radiation, it gives an alarm signal when the flicker is received. This detector must be able to see the fire and is very useful where large areas are to be protected using several fixed heads or one or two rotating detector heads. This detector has a rapid response since it does not have to rely on smoke or heat from the fire. It can also be used in the open air, unlike the smoke detectors that need a ceiling to function effectively.

10.8.2 ultra-violet detector

Since most of the ultra-violet radiation from the sun has been absorbed by the ozone in the earth's upper atmosphere, any ultra-violet radiation produced by flaming combustion will be detected by a photo-cell sensitive to this region of the electromagnetic spectrum. Since there are few sources of ultra-violet radiation in this wavelength range there is no need to discriminate by flame flicker. However, false alarms may be caused by welding or similar activities that produce ultra-violet radiation.

10.9 CHOICE OF SYSTEM

The system eventually chosen should be based on detectors which are best suited to the prevailing conditions.

10.10 DETECTOR HEAD LOCATION

The location of detectors suitable for a particular situation depends on the likely heat-energy output from the fire and the height of the ceiling. When a fire occurs in a room of normal height, the gases and products created rise in the form of a hot plume which entrains cold air as it rises. This plume then spreads out to form a layer of hot gases ranging from 100–300 mm in thickness. The hottest part of

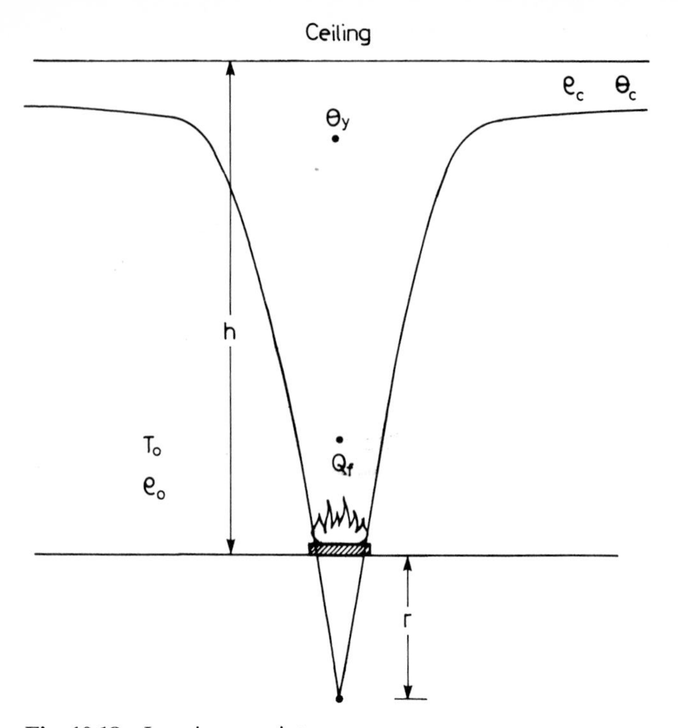

Fig. 10.18 Imaginary point source

this rising plume is directly above fire and 150 mm below the ceiling surface. If the ceiling height had been greater, say 5 m, the smoke layer at ceiling level would have been deeper and much cooler.

It can be shown by detailed analysis[4] that the heat output from a fire can be related to the temperature rise above ambient, θ_y, and the height of the ceiling, h, as shown in Fig. 10.18. So that when r is small

$$\theta_y \propto Q_f^{2/3} h^{-\frac{5}{3}}$$

$\therefore$ for a constant value of θ_y at ceiling level above a fire bed

$$\dot{Q}_f \propto h^{\frac{5}{2}}$$

This means that the minimum heat output of a fire Q_f to yield the temperature rise to cause a sprinkler or heat detector to operate will be approximately six times greater at a ceiling height of 8 m compared with one at 4 m, or we can say that the temperature rise varies with h for a controlled output fire

$$\theta_y \propto h^{-\frac{5}{3}}$$

Sometimes this has been abbreviated to

$$\theta_y \propto h^{-2}$$

It can be shown that the temperature in the layer away from the central axis of the system depends on h approximately as follows, i.e.

$$\theta_c \propto h^{-1}$$

Another physical factor that influences the operation of the detector is ceiling geometry. In the ideal case (horizontal plane), the spacing between the detectors should be such that no point on the ceiling is more than 5 m from a heat detector and 7 m from a smoke detector. However, this distance can be increased to take advantage of the channelling effects of narrow rooms, pitched and north-light roofs (Fig. 10.19), while these distances must be reduced if there are any obstructions on the ceiling, Fig. 10.20. The current code of

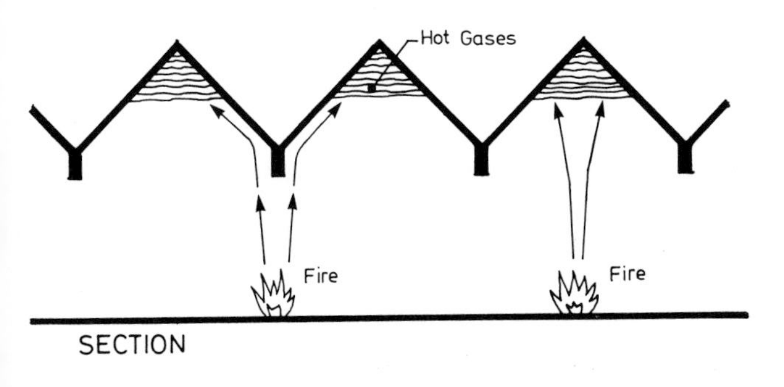

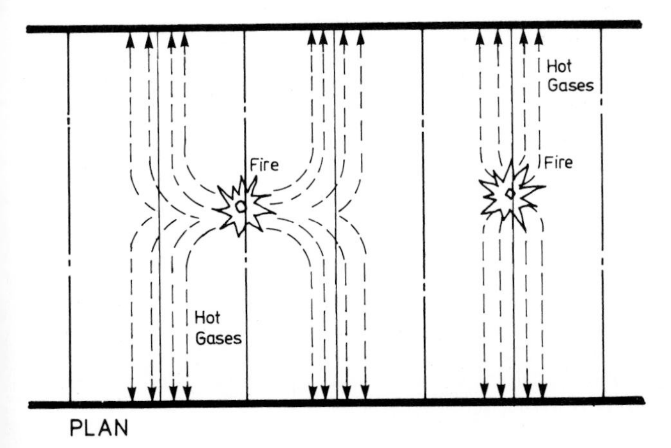

Fig. 10.19 Hot gas layers beneath profiled roofs

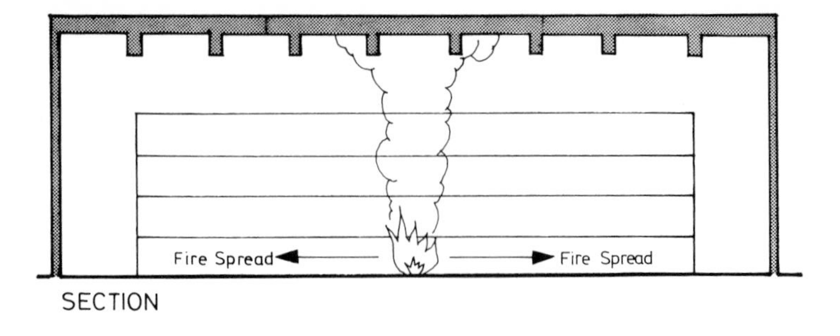

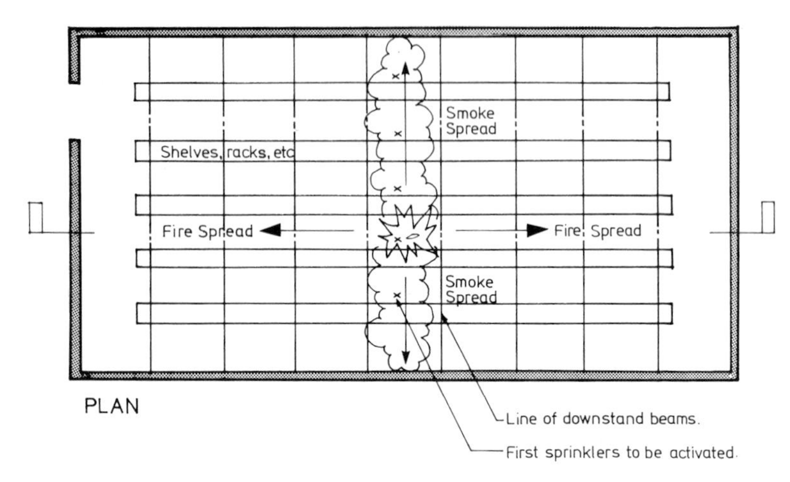

Fig. 10.20 Effect of ceiling profile on sprinkler operation

practice relating to the siting of such detector limits:

1. The distance of any point from a detector
2. The area to be covered by one detector
3. The distance between detectors and walls.

10.10.1 Heat detectors

These are suitable for small buildings since they have a greater resistance to bad environmental conditions compared to the other types. In a situation where a fire gives out heat rapidly with little smoke they may give a more rapid detection than a smoke detector.

The fixed temperature detector is not suitable for cold stores where ambient temperatures are very low. Combined fixed temperatures and rate of rise detectors should not be used in locations where rapid change in ambient temperatures are likely, for example in furnace rooms.

10.10.2 Heat detectors – location

These should be sited so that the heat-sensitive element is not less than 25 mm nor more than 150 mm below the ceiling or roof. The sensitivity of the head should be matched to the conditions prevailing and to the ceiling height.

10.10.3 Smoke detectors – Location

These should normally be located at the highest points of enclosed areas. Special problems may arise if the compartment height exceeds 11 m. The detector should also not be less than 25 mm or more than 600 mm below the roof or ceiling except where a site test rules otherwise.

This uniform distribution of smoke detectors just under the ceiling could be rendered ineffective due to smoke stratification and localisation. Thus successful location of smoke detectors requires some knowledge of factors which influence smoke production and movement within buildings.

Some recent work on the location of smoke detectors in domestic dwellings has shown that during the winter months when central heating systems are in full swing, smoke when produced is evenly dispersed within the house volume, so that air currents or open doors do not have a major effect. However, during the summer when the inside of a dwelling is usually cooler than the outside air, smoke stratification occurs at some distance below the ceiling.

This work has shown that smoke detectors, be they of the obscuration or ionisation type, are more effective as a fire detector than the heat detector. In a sheltered dwelling it would seem from experimental work[5] that an ionisation detector located in a bedroom corridor wall is almost as effective as separate detectors in each bedroom, provided there is a link between each bedroom and the corridor.

10.11 RECENT DEVELOPMENTS

It has been recently verified by an experimental investigation[6] that a light extinction or obscuration type of detector exhibits good uniform sensitivity to all types of smoke from cigarettes right through to black sooty smoke. This reaction of the detector simulates reasonably well the human visual reaction to smoke in an enclosure. Thus this type of detector could be an ideal type in the case of an emergency.[6]

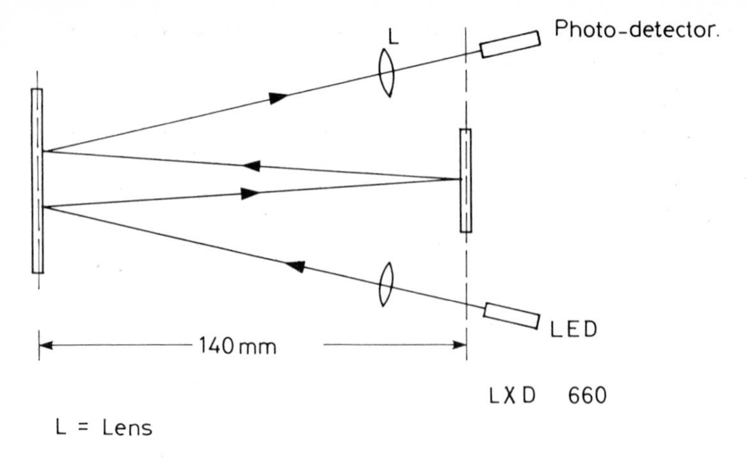

Fig. 10.21 Development in obscuration smoke detector

A sketch of such a detector is shown in Fig. 10.21 where the LED radiation-modulated pulse of wavelength 660 nm is reflected five times before being absorbed by a suitably sensitive photodiode. Thus it would seem that the prime position in the sensitivity league held by the ionisation-type detector is under threat by a detector which is sensitive to all smoke types.

10.12 CONCLUSIONS

In this Chapter the reader has been introduced to the various types of fire detectors currently in use in the domestic and industrial context. Further working details and specifications can be obtained from the manufacturers of such devices.

Intelligent alarm and detection systems are being developed now which incorporate the fire detector in a system controlled by a computer or microprocessor module. It has been reported that these decision-making systems can be very effective and are now coming within the range of the pocket of the domestic user and will thus play a big part in domestic fire safety in the future.

REFERENCES

1. Watson B, 'Fast response sprinkler heads', *Fire Surveyor*, **13**(4), August 1984.

2. Nash P and Young R A, '*The performance of the sprinkler in detecting fire*', BRE Current Paper, CP 29/75 April.
3. Burrey P E, '*Current research on fire detection at the Fire Research Station*', BRE Current Paper, CP 38/75.
4. '*Investigations into the flow of hot gases in roof venting*', Fire Research Technical Paper No. 7, HMSO reprinted 1965.
5. Kennedy R H, Riley K W P and Rogers S P, '*A study of the operation and effectiveness of fire detectors installed in the bedroom and corridors of residential institutions*', BRE Current Paper, CP 26/78 April.
6. Jun Miyama, 'Experimental research on light extinction type detectors', *Fire Safety Journal*, **6** (1983), pp. 157–64.

CHAPTER 11

Extinction of fire

11.1 INTRODUCTION

The early detection and successful extinguishment of a combustion process is highly desirable. There are many practical ways by which this can be achieved all really centred on the reduction of the flame temperature. When this happens the rate of the exothermic chemical process is reduced,[1] according to the following equation, which expresses the rate of this combustion process.

$$R = K\phi(\text{Fuel})\,\Psi(O_2)\,e^{-\frac{E}{RT}} \qquad \ldots[1]$$

where ϕ(Fuel) is a function which is dependent on the amount of fuel available, $\Psi(O_2)$ a function dependent on the oxygen present and E is the Activation Energy of the process.

From a knowledge of the elementary 'Fire Triangle', the rate of burning as defined in eqn [1] may be reduced by:

1. Reducing the oxygen
2. Reducing the temperature
3. Increasing the Activation Energy E.

By far the most efficient of these options is to reduce the temperature from 2,300 K to 1,600 K (lower flame temperature) which produces a reduction in R of some 25- to 50-fold. The reduction of the flame temperature T can be achieved by one or more of the following methods:

1. Adding dilutents to the flame (to the combustion zone)
2. Flame quenching
3. Cooling
4. Separating fuel from O_2 by a smothering process.

E, the Activation Energy, can be increased by the use of inhibitors.

These Extinction Methods will be discussed in greater detail later in this Chapter.

Another useful method to aid the understanding of the ignition and extinction processes is the 'Fire-Point' model.[2] In this theory, the ignition and consequent combustion is described by an 'Energy-Balance' equation which is based on the amount of volatiles produced and the rate at which they burn. Here the self-sustaining energy for fuel production comes from the flame itself via convection and radiation and, if applicable, from an external source of radiation. Consider the case of a diffusion flame, i.e. the burning of a candle or Bunsen burner with the oxygen inlet blocked off. The energy-balance equation is given by:

$$S = \bar{H} + R_a - R_s$$

where $\bar{H} = \dot{m}(H_f - \lambda_f)$
R_s = radiation energy loss rate per unit area from the flame
R_a = radiation fed back to the fuel (flame)
$\dot{m}$ = rate of burning of fuel per unit area
H_f = convection feedback to fuel in J/kg^{-1}
λ_f = heat energy necessary to vaporise the fuel per unit mass
S = The net rate of energy entering the fuel per unit area.

When $S > 0$ the combustion process will progress, while if $S < 0$ the process will tend to terminate.

It is obvious that if S is reduced to zero and eventually to less than zero, extinction will be considered to have occurred. This change in S may be achieved by the methods listed earlier.

11.2 TYPES OF EXTINGUISHING MECHANISMS

The first method to be discussed in detail is the addition of *dilutents* to the combustion or flame zone. The dilutent or cooling concept is used in the water and CO_2 extinguishing processes, whereas the *smothering or isolation concept* is employed in the foam process. The other processes, kalon and powder, are based on the *chemical/physical inhibitor* process which breaks the chain reactions essential to the combustion process.

11.2.1 Water

When water is sprayed onto the flame of a burning material the water evaporates and so extracts heat from the combustion zone (Fig. 11.1). If as a result of this, the flame temperature T falls below the lowest allowable adiabatic flame temperature the combustion will terminate. The amount or the rate of water needed to achieve

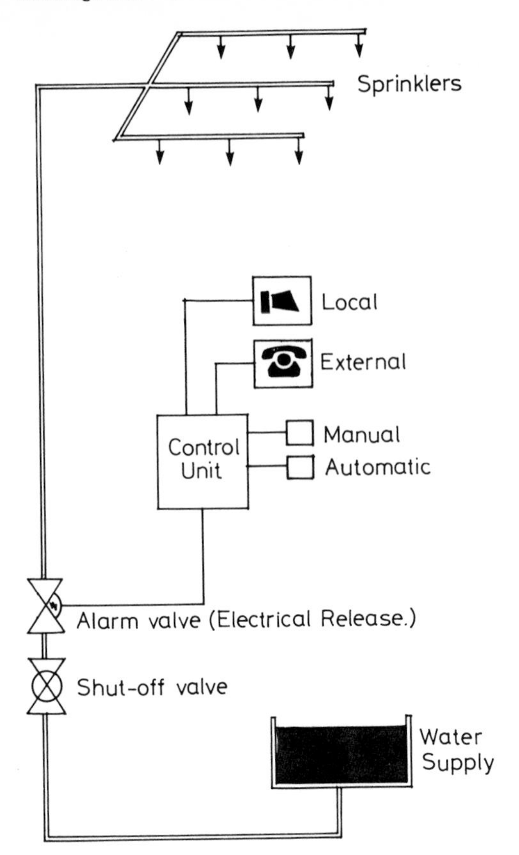

Fig. 11.1 Deluge system

this can be estimated using the 'Fire-Point' model or theory (Fig. 11.1),

i.e. $\frac{S}{L}$ = mass flow (Kg per sec) = M

L = latent heat of evaporation of water = 2,500 J kg^{-1}

$$M = \frac{\bar{H} + R_a - R_s}{L}$$

A sprinkler is usually capable of delivering at a rate of 80 g m^{-2} s^{-1} to a fire below its head and this flow rate should be sufficient to cope with most fire situations. Fuller details of sprinkler operations are given in Chapter 10.

11.2.2 CO_2 (carbon dioxide)

This process extinguishes a fire by reducing the oxygen content available so that combustion cannot be supported. For example a 30–50 per cent concentration of CO_2 can lower the oxygen content to below 10–11 per cent so that usually combustible materials cannot burn. The CO_2 gas is non-conductive and so can be used on electrical fires. CO_2 gas in small doses is non-toxic and non-corrosive. A typical CO_2 installation is shown in Fig. 11.2.

11.2.3 Foam

Here the foam moves or floats above a burning fuel or solid surface, isolating or smothering the pyrolysed vapour from the oxygen that is necessary for the combustion process.

Provided the foam has an adequate resistance to degradation it will act effectively as an extinction agent, Fig. 11.3. Foam is particularly useful in fires involving oil since if water was used as an

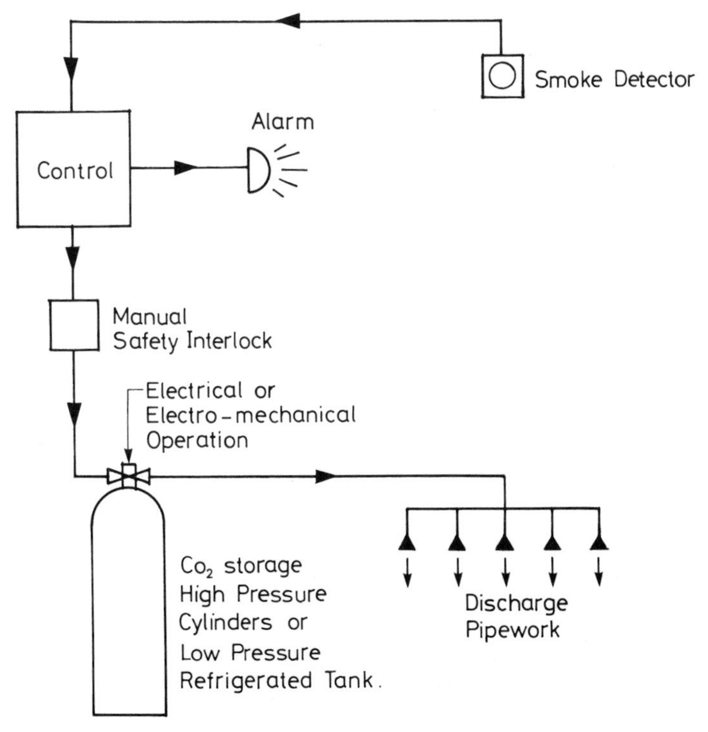

Fig. 11.2 Layout of a CO_2 sprinkler system

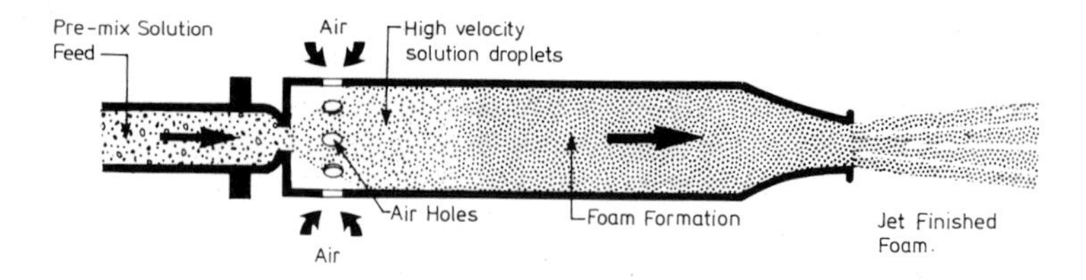

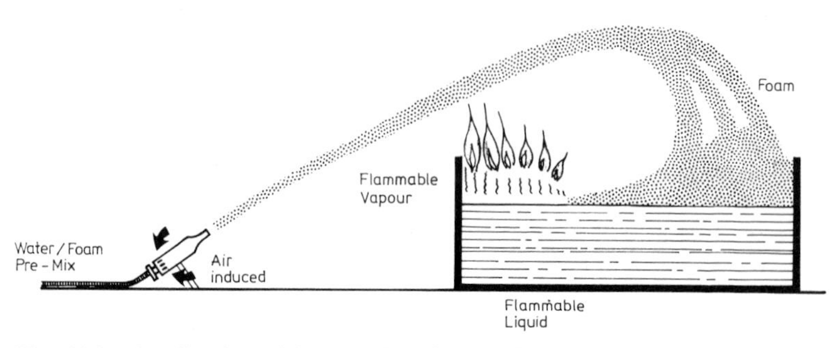

Fig. 11.3 Application of foam to burning liquid

extinguisher it would sink or possibly create steam or volatile vapour droplets which could be projected from the burning zone to help spread the fire.

11.2.4 Halon

A halon inhibits the chemical combustion process which has been shown to consist of many complex branching chain reactions which produce active hydrogen, hydroxyl, and oxygen radicals. To get some idea of how this happens consider the possible ways in which a halon, such as methyl bromide, behaves when introduced to a combustion zone.

$$CH_3Br^* \rightarrow Br^* + CH_3$$

Here the Br* radical seeks out the active H′ radicals in the combustion chain reaction $H' + Br^* \rightarrow HBr$ reducing them and thus restricting the chain reaction to some extent.

The HBr molecule can then interact with the OH* radicals as follows:

$$HBr + OH^* \rightarrow H_2O + Br^*$$

In this reaction the halon reduces the active radicals to regenerate the halon species Br*.

Thus this process causes a rapid and steady reduction of the chain carriers without the loss of the inhibiting agent.

Halon extinguishing systems are particularly useful in a situation involving electronic equipment, which is sensitive to the effects of soot and corrosion from smoke containing hydrochloric acid fumes. For example, a computer system will need an effective and fast detection and extinguishing system. Thus a halon gas of low toxicity and electrically non-conductive can extinguish a fire effectively and quickly by total flooding without wetting or leaving a residue.

A total-flooding system is one where halon gas is released and enters areas that are possibly inaccessible to water or foam. An essential requirement for this is that all doors and windows of the enclosure protected by this system must be a good fit and must remain closed during the flooding operation.

Total-flooding halon extinguishing systems exist in two forms. The first type of total-flooding system is the 'Central Bank' method that uses a network of fixed pipework in a way similar to a water sprinkler system, where the cylinders storing the halon gas are outside the area to be protected. The other system is the 'Modular System', Fig. 11.4, which by its nature eliminates the fixed pipework by relying on individual cylinders located within the area to be protected.

Usually these cylinders are linked to a highly sensitive detector system such as an ionisation smoke detector.

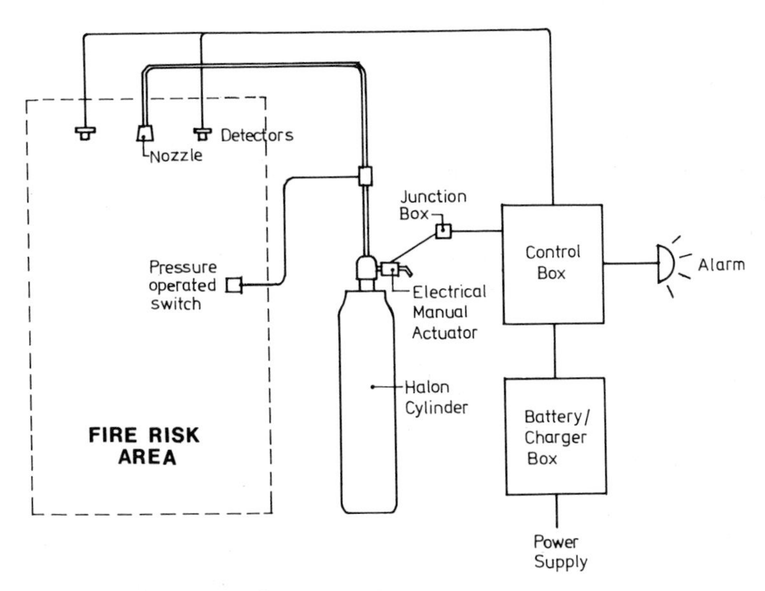

Fig. 11.4 Modular Halon system

On balance the modular system has many advantages over the fixed installation type in that:

1. It is easy to install, as cylinders can be mounted in unused wall or floor space
2. Installation causes little disturbance
3. The modular system can grow if necessary to cover any expansion
4. The system can be taken down and removed to another location
5. It is more economical to run and to maintain.

11.2.5 Powder

The exact reason why a powder can extinguish a combustion process is not known. As a result of recent research it would appear that the phenomenon of flame extinction is controlled by a gaseous reaction of the alkali metal powders and not, as previously thought, by the surface reaction process.[3] One plausible explanation centres on the capture and recombination of the chain carrier species H′ on the surface of the powder particles.

A possible process may be described as follows:

$$H' + P^* = HP$$

$$H' + HP = H_2 + P^*$$

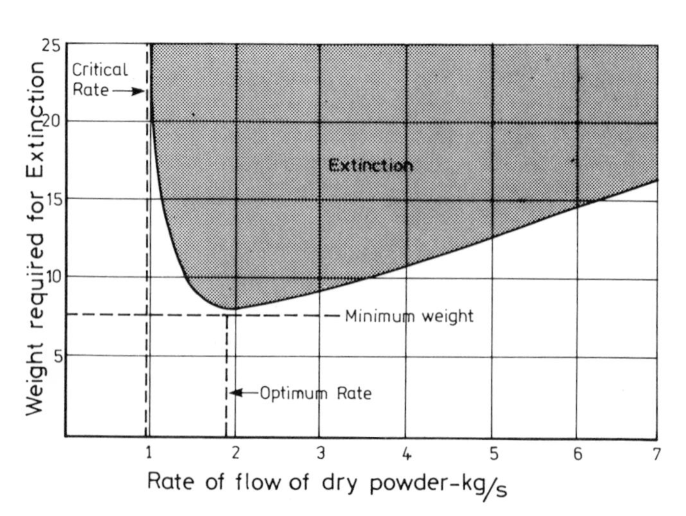

Fig. 11.5 Optimum flow rates of powder for extinction

Since this is a chemically-controlled surface effect the larger the surface area available the more effective the powder will be. The physical size of the powder particles will be as small as is possible taking into account their ballistic needs to overcome air viscosity and possible air pressure convection currents near a combustion process. As well as this surface effect, the fine wall of powder particles has also been found to offer an effective barrier to radiation from the combustion zone.

Another feature of the powder is that it can also act as a dilutent to remove some of the thermal energy from the combustion-reaction zone. Thus, taking these facts into account, the most useful particle size has been found to be of approximately 4×10^{-5} m. This size of powder particle can be successfully projected some 5 to 6 metres into a fire zone. If the optimum rate of powder is applied to the fire, Fig. 11.5, the fire may be extinguished with the least amount of powder.

Table 11.1

WHICH TO USE					
Class of Fire	Water	Foam	CO_2 Gas	Powder	Halon
A Paper Wood Textile Fabric	X	X		X	X
B Flammable Liquids		X	X	X	X
C Flammable Gases			X	X	X
Electrical Hazards			X	X	X
Vehicle Protection				X	X

Powder-type extinguishment is very effective in liquid-pool fires. The volume of powder needed to fight a fire depends obviously on the dimensions of the fuel bed, etc., so these devices can range in size from a small hand-held device to a mobile tender carrying 2 tonnes of powder suitable for typical airport fuel-spillage fires.

11.3 CHOICE OF EXTINGUISHMENT SYSTEM

In order to ensure that an effective extinguishment system is employed it is important that the extinguishing agent suits the nature and type of fire. To help with this decision process, Table 11.1 gives at a glance suggestions for suitable extinguishing agents for various fire types.

It must be noted that halon and CO_2 extinguishers should not be used where there are chemicals that can provide their own oxygen supply in the event of a fire, e.g. cellulose nitrate and gun powder are typical of such materials. When choosing a halon or CO_2 system, the toxicity of these extinguishing agents should be considered. Special care on this account must be taken where there are people in any number present when the halon or CO_2 extinguisher is activated.

To help to determine a suitably low toxic level for a halon system, Tables 11.2, 11.3 and 11.4[4] have been prepared from the most reliable information currently available.[2]

The '*F*' factor is of particular interest, Table 11.4, where

$$F = \frac{\text{concentration necessary for extinction}}{\text{concentration likely to cause narcosis}}$$

Table 11.2

ORDER OF TOXICITY
Halon 1001 Methyl Bromide
Halon 104 Carbon Tetrachloride (CTC)
25% Halon 1011 with Trichloroethane
Halon 1011 Chlorobromemethane (CBM)
Halon 2402 Dibromotetrafluroethane
Halon 1211 Bromochlorodifluoromethane (BCF)
Halon 1301 Bromotrifluoromethane (BTM)

Table 11.3

ORDER OF EXTINGUISHING EFFECTIVENESS	EXTINGUISHING CONCENTRATION PERCENTAGE VOLUME
Halon 2402 (Fluobrene)	3·5
Halon 1301 (BTM)	4·9
Halon 1211 (BCF)	5·2
25% Halon 1011 with Trichloroethane	6·35 (estimate)
Halon 1011 (CBM)	6·35
Halon 1001 (Methyl Bromide)	7·1
Halon 104 (CTC)	9·7

Table 11.4

VALUE OF THE 'F' FACTOR FOR HALONS AND CO_2	
Halon 1301 (BTM)	0·8
Halon 1211 (BCF)	4·5
Carbon Dioxide	6
Halon 2402	14
Halon 1011 (CBM)	64
25% 1011 with Trichloroethane	100
Halon 104 (CTC)	129

It is clear that Halon 1301 (BTM), Halon 1211 (BCF) and CO_2 are suitable and are thus approved by the UK Home Office for use in public places.

11.4 CONCLUSIONS

The rapid extinguishment of an unwanted fire is absolutely essential in modern society. Man discovered fire by accident and probably at the same time discovered also a method of extinguishing it.

Current knowledge of fire growth and development, and the properties of extinguishing agents such as halons, allows a method of extinction to be developed and used for particular applications.

The reader must appreciate that the method of fire control chosen must be matched to the particular application.

REFERENCES

1. Spalding D B, '*Some Fundamentals of Combustion*', Butterworth, 1955.
2. Rashbash D J, '*Relevance of Fire Point Theory to the Assessment of Fire Behaviour of Combustible Materials*'. Ed. 1975. University of Edinburgh 1975.
3. Yohru M and Takashi N, 'Extinction phenomena of premixed flames with alkali metal compounds', *Combustion and Flame* **55**(1), pp. 13–29 1984.
4. A Guide to Fire Prevention, '*The selection of gaseous fire extinguishing systems*', A current Chubb Fire Publication 1984.

CHAPTER 12

Means of escape

12.1 INTRODUCTION

At present there is no quantitative method of assessing the adequacy of any escape route provided in a building other than by empirical means. The current method of providing means of escape from buildings is by specification and rule, i.e. rules that have evolved through time and are deemed to provide a satisfactory escape route. Unfortunately the current building legislation is not based on performance standards for the design of means of escape in buildings but relies on empirical data which may or may not provide the correct solution.

Throughout building legislation relevant to means of escape, terminology such as adequate, suitable or satisfactory is used and it is obvious that differences of opinion will arise as to the 'suitability' or 'adequacy' of a particular means of escape between the designer and the controlling authority. These differences are compounded, usually because the components of escape route design have not been identified thereby rendering the concept of equivalency redundant.

The objective which demands the provision of a means of escape is that the occupants should be able to reach a place of safety, unharmed, in the event of a fire occurring. A place of safety is normally associated with an area outside the building away from the threatened space. However, a place of safety may also be:

1. A protected corridor
2. A protected staircase
3. A place of refuge within the building.

These are illustrated in Figs 12.1(a); 12.1(b); 12.1(c).

Places of refuge are necessary in very tall buildings because the evacuation of these buildings may take two hours or more.[1] Refuge floors may be provided every six or eight floors up the building,

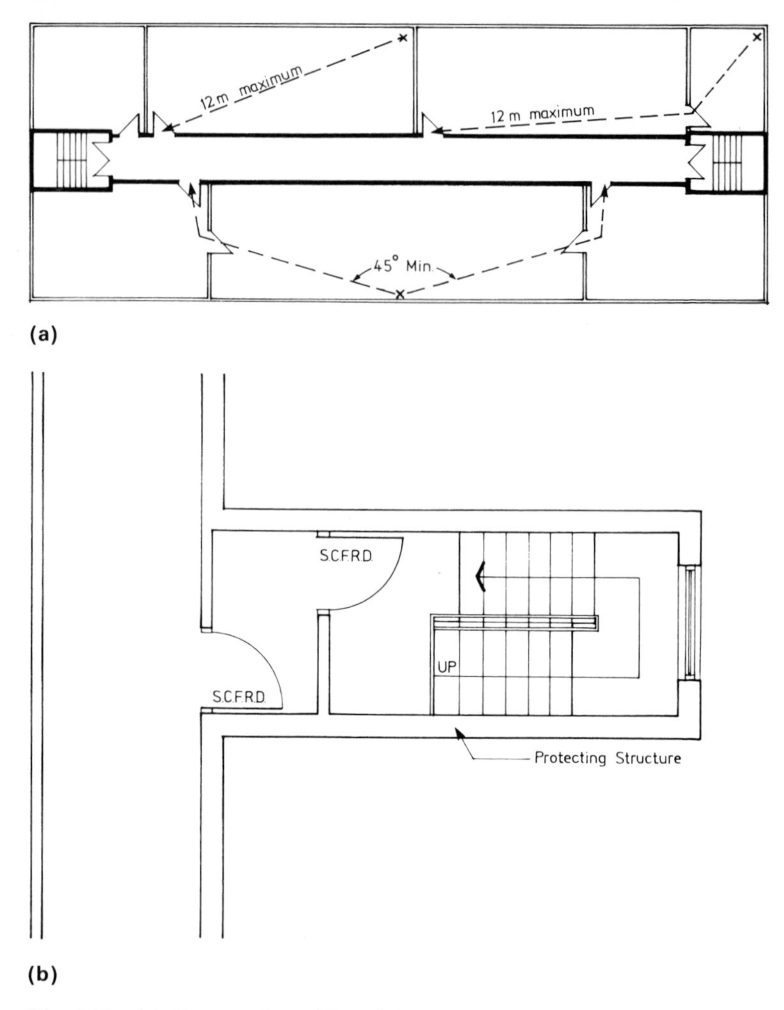

Fig. 12.1 (a), Protected corridor; (b), protected staircase; (c), provision of refuge areas

depending on the nature of the occupancy and current practice in an emergency, so that occupants of the fire floor, floor below the fire floor and floor above the fire can be evacuated to a place of safety. Thus some persons will move downwards towards lower levels whilst the occupants above the fire floor will move upwards away from the fire. The occupants of the other floors immediately threatened would be alerted and evacuation would proceed as determined by particular circumstances.

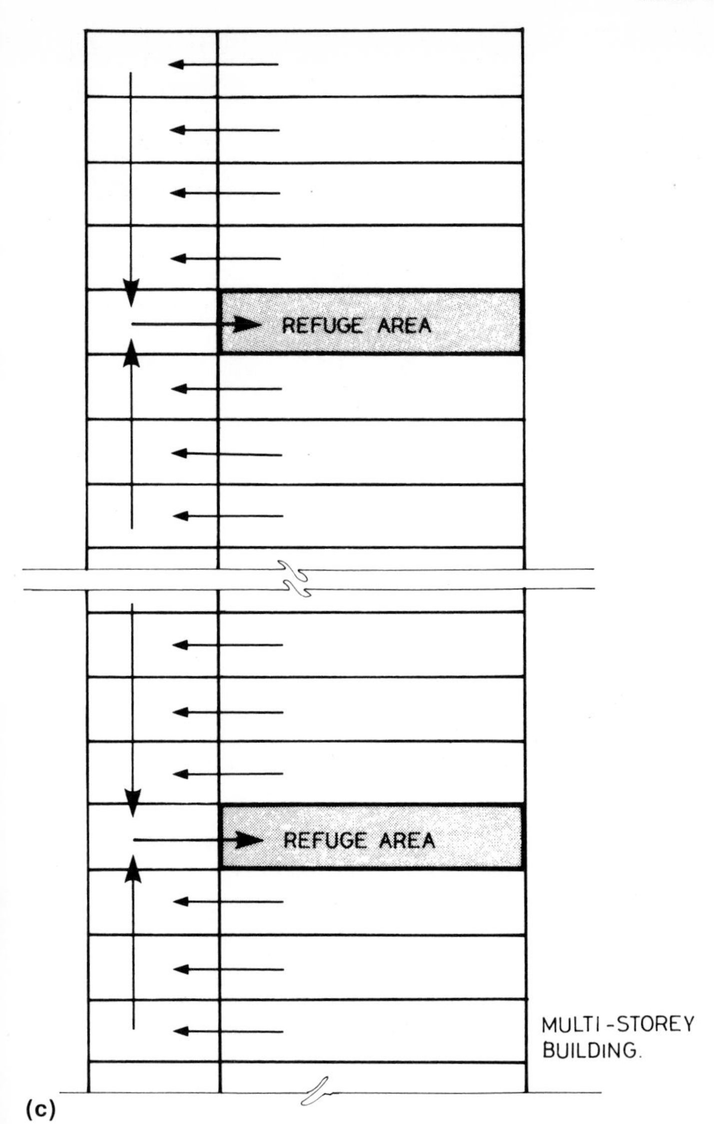

(c)

12.2 PROVISION OF ESCAPE ROUTES

In every building there should be available from each room and each storey a sufficient number of escape routes so as to ensure the safety of the occupants in the event of fire.

It is possible using various modelling techniques[2] to assess the number of escape routes available to the occupants of a building

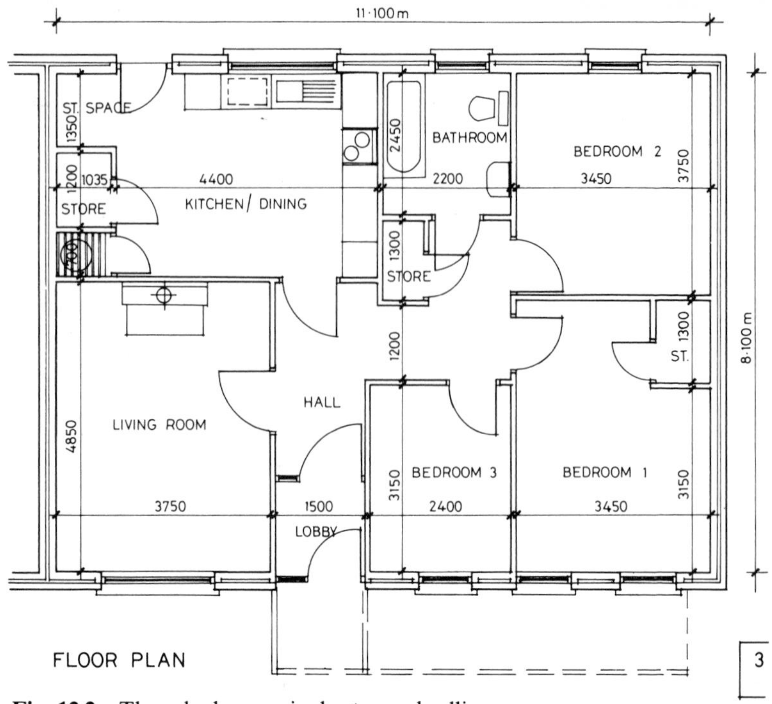

Fig. 12.2 Three bedroom single-storey dwelling

throughout the fire development sequence. Figure 12.2 shows the floor plan and Fig. 12.3 the escape analysis graph for a single-storey dwelling.

Table 12.1 shows the effect of blocked rooms on the number of escape routes available in the dwelling illustrated in Figs 12.2 and 12.3, respectively. This method[3] of escape route analysis can be applied to any building type and will assist with the difficult choices involved with escape route design. However, it should be noted that in the development of interspatial relationships within the building, the circulation of people with minimum disruption or interference to the various activities pursued in the building is essential, and consequently the planning of effective means of escape should naturally follow the flow patterns already established. In other words, natural pathways or routes developed in the design and secured by passive and active fire protection which permit the occupants to proceed to a place of safety.

Escape routes usually have three components as illustrated in Fig. 12.4. The diagram in Fig. 12.4 shows the protected horizontal routes interconnected by the protected vertical route. The internal layout of, e.g. flats and maisonettes is considered separately, i.e. the

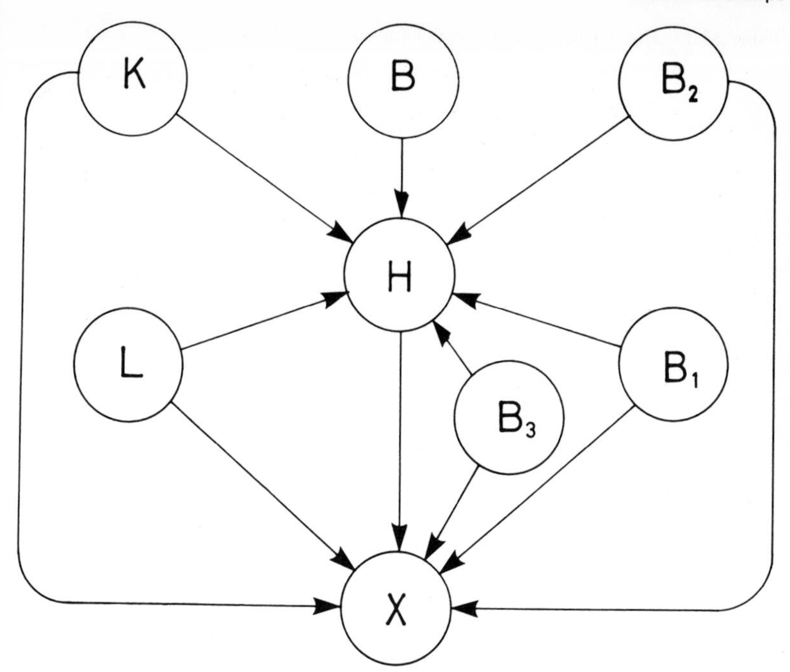

B	Bathroom
B_1	Bedroom 1
B_2	Bedroom 2
B_3	Bedroom 3
H	Hall
K	Kitchen
L	Living Room
X	Place of Safety (Ext.)

Fig. 12.3 Escape analysis graph for single storey dwelling

protected escape route illustrated in Fig. 12.4 starts at the entrance door to the flat.

12.3 COMPONENTS OF ESCAPE ROUTE DESIGN

There are at least seventeen major components in escape route design. These are:

1. Building type
2. Building contents
3. Building occupancy

Table 12.1 Escape route availability related to room of fire origin

Location	All rooms passable	Kitchen blocked	Hallway blocked
Kitchen	2	×†	1
Hall		5	×
Living room	2	2	1
Bathroom	1	1	0
Bedroom 1	2	2	1
Bedroom 2	2	2	1
Bedroom 3	2	2	1
Total	17	14	5

† '×' indicated a space that is considered impassable.

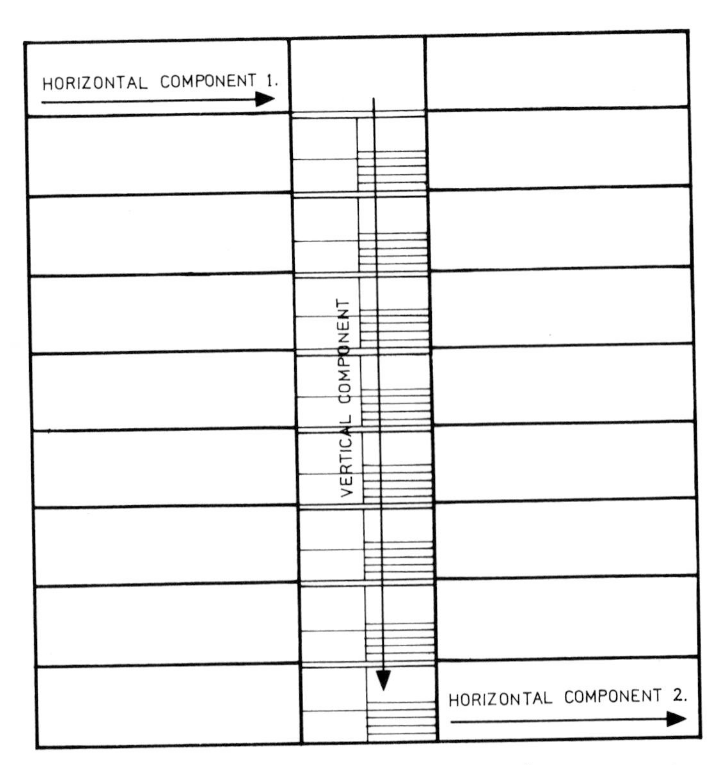

Fig. 12.4 Horizontal and vertical components of an escape route

4. Evacuation time
5. Travel distance
6. Exits
7. Escape route width
8. Enclosure of stairways
9. Lobby approach stairways
10. Doors in escape routes
11. Lighting of escape routes
12. Emergency lighting
13. Construction of and egress from windows
14. Fire detection system
15. Alarm system
16. Fire-control systems
17. Smoke-control systems.

12.3.1 Building type

Building Regulations[4] categorise buildings by notional purpose grouping which on inspection may be broadly rationalised under the following headings:

(a) residential and institutional buildings
(b) commercial and industrial buildings
(c) assembly buildings.

It is obvious that within each of the three broad classifications above, many types of building are possible and examples of the range of possible types are shown in Figs 12.5, 12.6, 12.7.

As the geometry of the building increases in complexity so also may the means of escape provided, bearing in mind that as a first basic principle escape route geometry should be as simple as

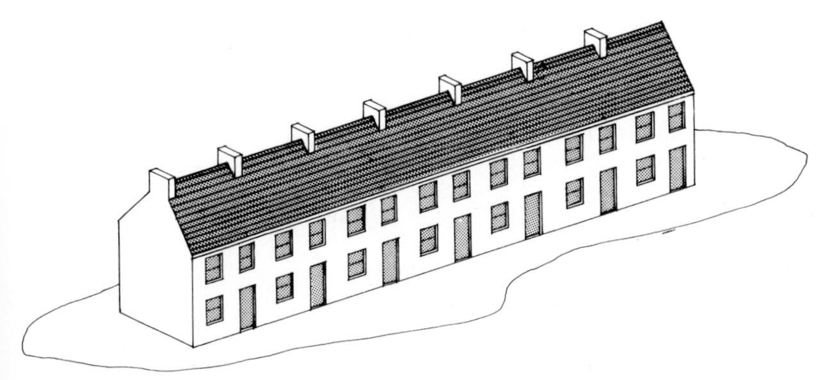

Fig. 12.5 Terrace block dwellings

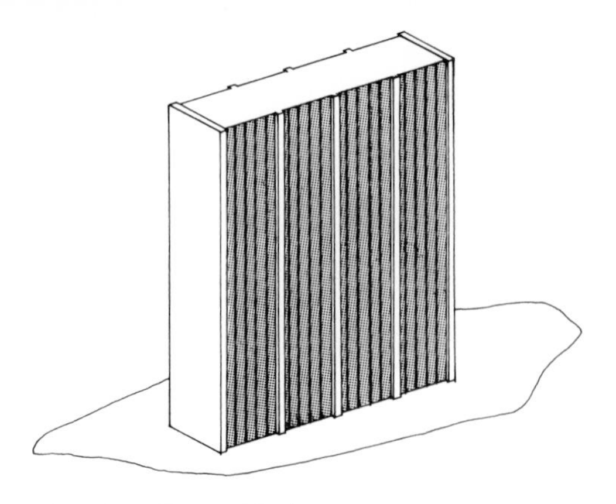

Fig. 12.6 Tower block

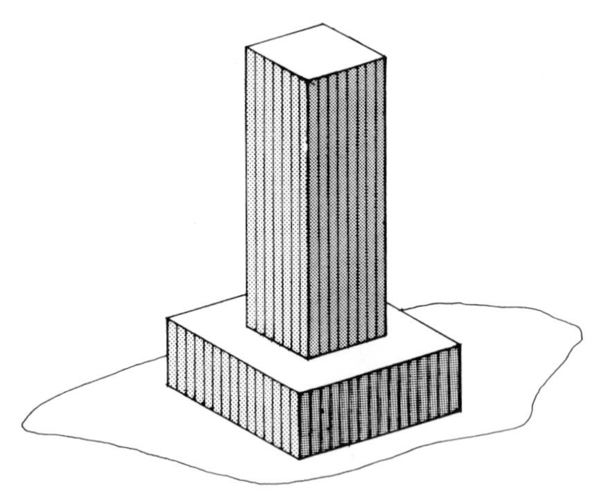

Fig. 12.7 Tower block and podium

possible. However, combining complex building geometry with multi-functional requirements and the inclusion of sleeping accommodation (Figs 12.8(a) and 12.8(b)) makes the design of a means of escape increasingly difficult, causing a change of emphasis in the other components of escape route design such as the provision of meaningful, easily understood directional signs.

12.3.2 Building contents

The inclusion and subsequent use of purpose grouping in building regulations assumes that each building within a particular purpose

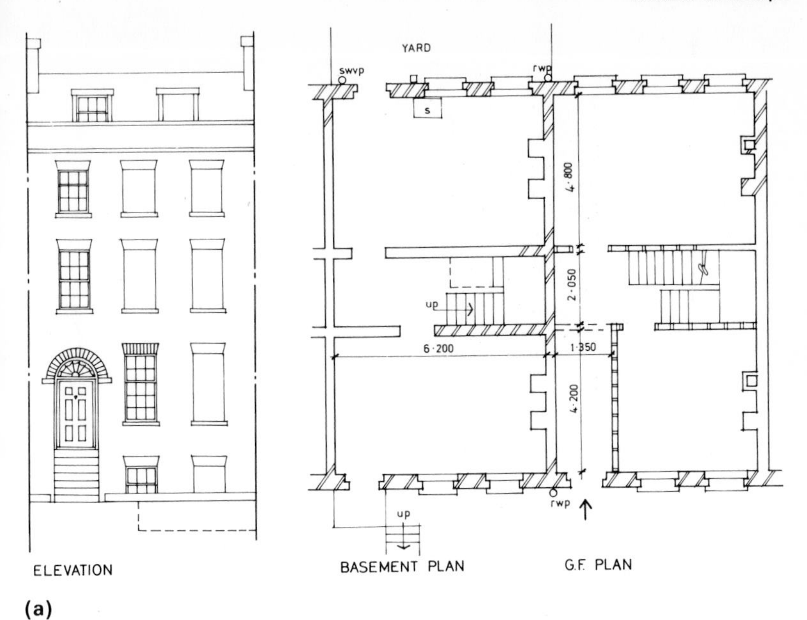

(a)

ATTIC
SOLICITOR'S OFFICE
HAIRDRESSING SALON
63 x 225 @ 350 c/c
SHOP
Studs 63x125 @ 300 c/c
NIGHT CLUB
FOOTPATH
FRONT GARDEN
YARD
2·138
2·363
2·743
3·200
2·210
SECTION
(NOT TO SCALE)

(b)

Fig. 12.8 Residential terrace building converted to commercial use

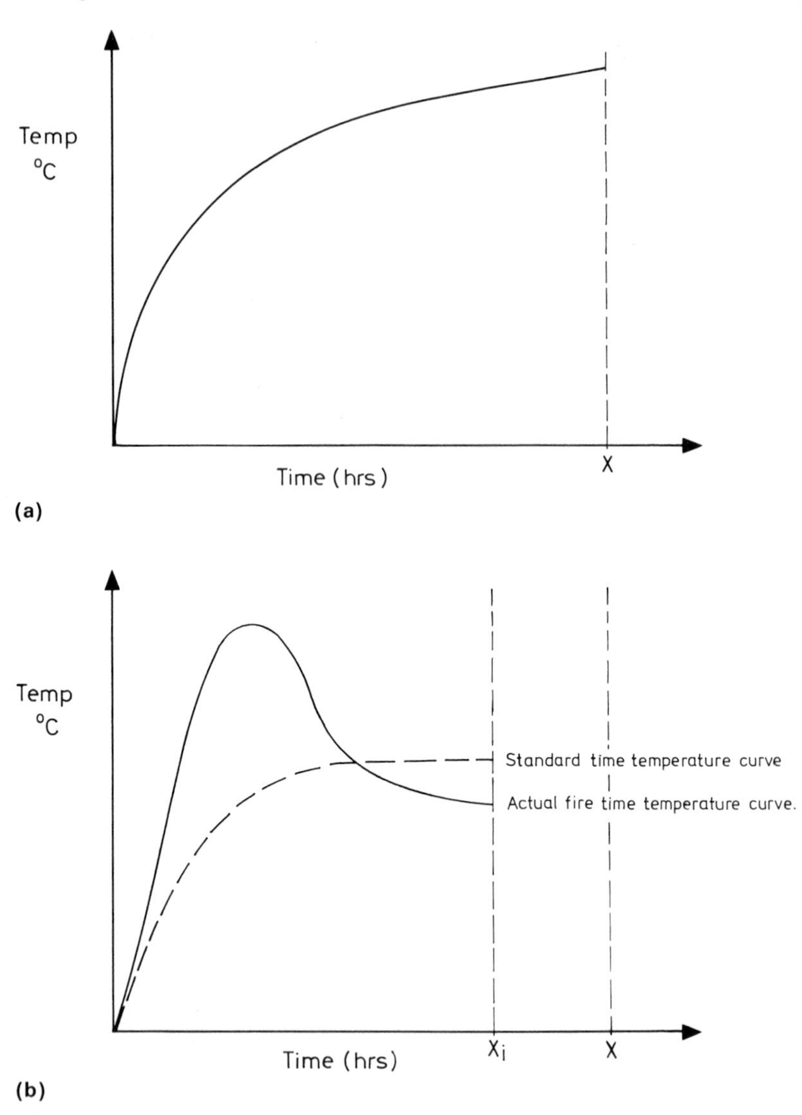

Fig. 12.9 (a), Standard time-temperature curve; (b), actual fire time-temperature sequence

group will experience the same fire exposure (Fig. 12.9(a)) and consequently that the contents of buildings within the same purpose groups are essentially the same.

In practice this assumption does not hold and the regulations assume the general case. However, in storage buildings of similar design and dimensions, non-combustible materials, for example, may be stored in one building and combustible materials such as paint and associated products may be stored in another.

In dwellings, reference to statistics[5] will show that furnishings have a considerable influence on ignition and fire growth. Modern furniture is more easily ignited by flame, burns more rapidly and produces more smoke at a faster rate of smoke production than traditional-type furniture. By changing the contents of a building, the pattern of fire growth and development can also be changed, e.g. from a potentially slow fire growth and development to a very rapid growth and development, dramatically increasing the threat to the occupants, Fig. 12.9(b). The actual fire illustrated in Fig. 12.9(b) has a duration of X_i whereas the equivalent severity under the standard time–temperature relationship occurs after a much longer duration X. Consequently the fire resistance period for constructional components derived from the standard fire-resistance test may not be appropriate for all buildings and the nature of the building contents can significantly influence the fire severity.

12.3.3 Building occupancy

The population of a building will be a major consideration in the design of a means of escape and will be considered under the following headings:

(i) Population density
(ii) Population distribution
(iii) Population mobility
(iv) Population reaction
(v) Population discipline.

Population density

In order to calculate the capacity of staircases it is necessary to know the number of persons likely to be in the building per storey or compartment. Generally speaking, load-factor data have been derived through experience and follows naturally from good design, i.e. the number of persons capable of using the building comfortably, relative to the process, activity or function.

Population density may be expressed in different ways:

(a) simply stated as the number of persons the compartment, storey or room is designed to hold, or
(b) in the case of flats or maisonettes the number of bed spaces for which the flat or maisonette is designed, or
(c) in prescribed cases the number obtained by dividing the area in m^2 of the compartment, storey or room by the occupant load factor specified.

Table 12.2

Storey/room type	Occupant + load factor
Grandstands (without fixed seating) Assembly hall (movable or no seating)	0.3–0.5
Shopping mall, dance hall, bar, dining room, coffee lounge	0.3–1.1
Offices (depending on floor area)	5–9
Shops	1.8–5
Art gallery, library, museum	

A range of occupancy load factors are shown in Table 12.2 and reference in specific cases should be made to building regulations and appropriate codes of practice.

Population distribution

It is important to know how the building population is distributed so as to avoid bottlenecks when designing means of escape. Similarly, e.g. in a hotel, parts of the building will be open to the public, these include function rooms and bars, while at the same time parts of the building will be closed to the public; these include sleeping accommodation for residents. Buildings in which people sleep, particularly dwellings, pose the greatest risk. On average 800 people per year die in fires in dwellings as opposed to 200 people per year in other buildings.

Population mobility

The occupants of a building may be very young, old, infirm, disabled, hospitalised or handicapped. If so, careful consideration must be given to designing the means of escape. Experience has shown that old people, for example, will be very reluctant to leave what they consider the safety of their own room and face the perils of an external fire escape in darkness.

It is also inconsistent to expect a physically disabled person to make his way to his workplace normally by means of a lift, and then when danger threatens require him to descend six or seven storeys via a staircase. So it may well be that although in most circumstances the use of lifts as a means of escape is discouraged, situations may arise, e.g. employment of disabled persons, where the use of specially designed and protected lifts may well prove a necessity.

The terms disabled and handicapped must not be confused, as the term handicapped is used here in the most general sense, e.g. a young mother proceeding along an enclosed shopping mall with an infant in a pram and a three-year-old child by the hand may reasonably be considered handicapped.

In buildings such as hospitals, hospices and old people's homes, many inmates will not be able to reach a place of safety by their own unaided efforts. Indeed, many will have to be transported in wheelchairs and in their beds to places of safety. Attention must also be focused on the width of the means of escape in order to allow beds to be wheeled in safety without congestion occurring.

Population reaction

This factor will depend very much on the type of building occupancy. In situations where the occupants are familiar with the building layout then orderly evacuation is a possibility, no more than that. However, in buildings to which the public have access and may be described as transient, then the popular account of human reaction to fire is that of panic and it is in this situation that the greatest dangers arise. Canter[6] suggests that the notion of panic is unhelpful and perhaps meaningless. The role-playing capability of persons and an appreciation of its significance in an emergency situation is another factor to be considered. This is particularly important for persons employed in or occupying key positions, e.g. in a communication system. Again in hotels where most of the residents are on what can only be described as a short-stay basis, there is little chance of becoming familiar with the building layout and particularly the means of escape. In this situation, if a fire should occur the residents will tend to attempt to leave the building by the way they entered it or are most familiar with, albeit that they may, in fact, be moving towards the fire.

Population discipline

As with population reaction, it is impossible to achieve this with a transient population and great reliance is placed on staff. Consequently, staff efficiency in an emergency is of paramount importance and can only be achieved by means of a thorough understanding of the risks involved and training in how to deal with emergency situations. In the Beverley Hills Supper Club fire,[7] waitresses escorted people out of the building through smoke.

However, with the best will in the world, training can be negated by a simple oversight. If, for example, reception or the switchboard is to be used to channel all communications, external and internal, should a fire occur, then such an important position must be manned at all times by trained personnel. This means even when the 'usual' person is at lunch, having coffee, on leave, or absent through illness. Only through training can any measure of population discipline be predictable.

It is also essential to the preservation of a means of escape that the indiscriminate use of decorative panels and plastics be controlled so that the fire resistance of the primary construction (elements structure) is complemented by the proper choice of materials for secondary construction (linings, etc.). In simple terms it would be completely nonsensical to require elements of structure to be fire-resisting and sometimes non-combustible while allowing the indiscriminate use of combustible wall linings. Too often the efforts of the architect can be negated by the management of buildings in use, making decisions based on expediency or convenience.

12.3.4 Evacuation time

This is the time taken for a person to go from any occupied part of the building to a place of safety. Generally speaking, evacuation times of 2–3 minutes are used in the design of the means of escape from buildings, depending on the type of construction. Obviously this time will vary quite considerably according to a person's speed of travel, e.g. very young or old people will not be able to move as quickly as the average able-bodied person.

As previously stated, a time of 2–3 minutes may be used in design even though the fire-resistance requirement of elements of structure is one hour. Consequently it must become obvious that there is no direct relationship between the two times. In fact, the 2–3 minutes evacuation criterion is derived from studies which conclude that such a time is reasonable for people to be in a stressful situation before 'panic' develops. Thus it is infinitely desirable to evacuate people before such a state of irrational behaviour occurs. Recent work has shown that in many cases people behave rationally in a crisis situation[8] and that panic is often confused with flight behaviour.

In many buildings, such times as 2–3 minutes will not be possible to achieve and each situation must be evaluated individually. Studies in Canada[1] have shown that times of $2\frac{1}{2}$ hours were required to evacuate a multi-storey building. These special circumstances may require places of safety to be provided within the building.

12.3.5 Travel distance

Travel distance is the distance to be traversed in order to reach a place of safety from which dispersal can take place (Fig. 12.10). It is not to be confused with 'direct distance' which is sometimes used (Fig. 12.11).

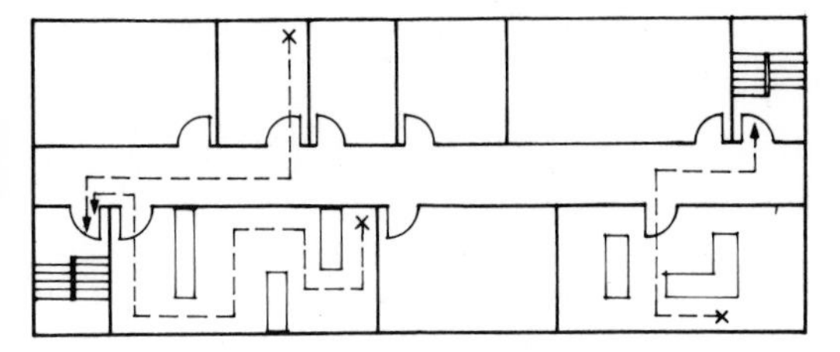

Fig. 12.10 Travel distance

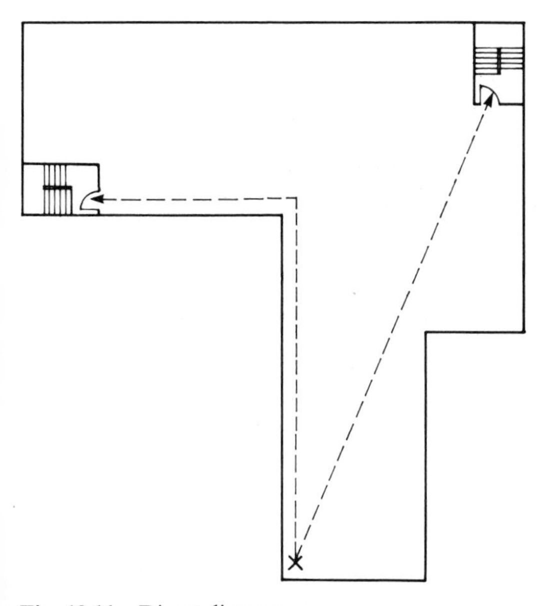

Fig. 12.11 Direct distance

The travel distance is measured around obstacles whereas the direct distance is measured over obstacles.

A range of travel distances is given in Table 12.3, varying relative to purpose grouping and particular situation.

Table 12.3

Purpose group	Travel distance related to available directions of travel	
	One direction	More than one direction
Residential buildings generally	15 m	30 m
Old people's homes	9 m	18 m
Hospital ward areas	15 m	30 and 60 m†
Institutional type buildings	15 m	30 m
Offices	18 m	45 m
Shops	15 m	30 m
Factories	18 m	45 m
Assembly buildings	15 m	30 m
Storage	18 m	45 m

† Travel distance may be increased if routed via a fire-resisting floor into a sub-compartment and in total does not exceed 60 m.

These distances have been established by experience over many years and give guidance for particular applications. However, the distance to be travelled must be related to the risk involved, i.e. the rapidity of flame and smoke spread. When a fire is in the growth stage a great deal of smoke can be produced and this smoke can move on occasions more quickly than normal walking pace. It is, therefore, essential that travel distances can be such that persons can reach a place of safety before smoke-logging of the means of escape occurs.

For buildings of regular plane shape and layout, e.g. factories, travel distances of 45 m may be sufficient, whereas in a block of flats where a dead-end situation occurs a distance of 7.5 m may be involved. By a dead-end situation is meant one where the occupants can only travel in one direction, i.e. towards the fire in order to each a place of safety.

12.3.6 Exits

Entrances, exits and circulation areas are provided in all buildings for normal use, and means of escape considerations should utilise existing arrangements wherever possible. Consequently the primary consideration should be with regard to the sufficiency of existing exits in terms of:

(i) disposition,
(ii) width, and
(iii) number,

and only if existing exits are inadequate in some respect should further design be considered.

Disposition

The positioning of exits as means of escape in case of fire is absolutely critical. In this respect the '45° rule' is introduced (Fig. 12.12).

If the angle made by lines joining the exits to any point on the floor of the storey under consideration is less than 45°, then the disposition of the exits may be considered inadequate. In Fig. 12.12, if a fire occurs at B, the other staircase is available for persons in the area. Ideally, staircases should be arranged diagonally opposite each other (Fig. 12.13). Alternatively, tower staircases may be provided.

Special situations arise where only one staircase is provided and careful consideration must be given to plan layout and methods of securing effective means of escape.

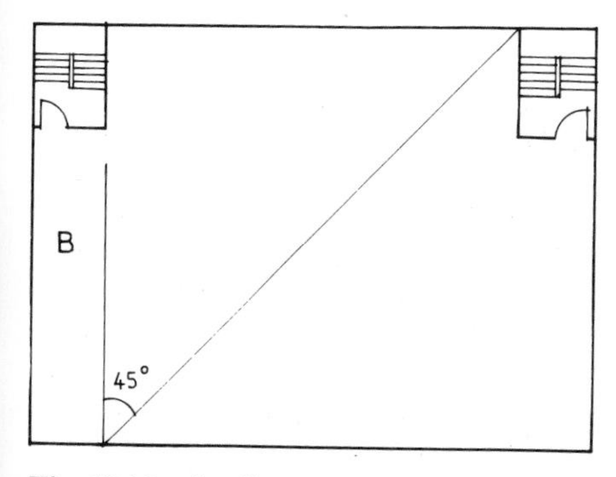

Fig. 12.12 Application of 45° rule

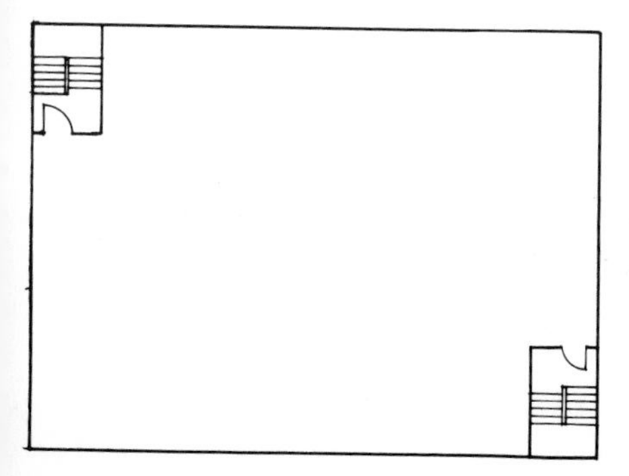

Fig. 12.13 Ideal location of staircases

Width

It is essential when designing means of escape that bottlenecks, i.e. areas where congestion will occur, be avoided. Thus corridors should not become narrower as they approach a storey exit or staircase.

In most cases the width of exits is not critical in that the number of exits provided for normal use is usually adequate to cope with the number of persons involved.

Building codes have specified a 'unit of exit width' of 533 mm, i.e. based on the assumption that 40 persons per minute in a single file will discharge through an exit 533 mm wide, or expressed as 100 persons in single file in $2\frac{1}{2}$ minutes, which relates to evacuation times previously discussed. However, a width of 533 mm is not acceptable in practice. Indeed, it is positively dangerous, and the minimum width of 0.765 m is taken to provide a discharge rate of 40 persons per minute (0.765 m equates to the standard internal single leaf door width).

WORKED EXAMPLE 1

We have thus established:

unit of exit width = 533 mm

discharge rate = 40 persons/minute

occupancy load factor = depends on situation (Table 12.2).

It is now possible to calculate the number of units of exit width and subsequently the number and width of exits required for a given number of persons, e.g.

Single-storey club

40 m × 20 m, area $= 800\ \text{m}^2$

Occupancy loading factor $= 1.1$

Number of persons $X = \dfrac{800}{1.1} = 725$

Number of units of exit width required $Y = \dfrac{X}{40 \times T}$

$$= \frac{725}{40 \times 3}$$

where T = time factor (3 minutes)
X = No. of persons
Y = No. of units of exit width required.

To determine the number of exits:

$$\frac{Y}{A} + 1 = Z \qquad \text{where } Y = \text{No. of units of exit width}$$

$$\frac{6}{A} + 1 = Z \qquad Z = \text{No. of exits}$$

$$= 2.5 \qquad A = 4 \text{ (constant)}$$

$$\therefore \quad Z \simeq 2$$

It would appear that two exits may be sufficient, incorporating an aggregate of 6 units of exit width (3 m clear width). But the complexity of the plan shape may require additional exits and it should be understood that the example given is very basic. Pauls and other researchers consider the discharge rate of 40 persons per minute to be too high and suggest a discharge rate of 30 persons per minute as a more realistic figure.

As stated previously, virtually all of data related to the design of escape routes are empirical. In very tall buildings, designers are forced to consider localised evacuation[9] of part of a building simply because total evacuation is impracticable and/or impossible.

Pauls[10] considers that to those involved with the building design and management process, evacuation is a problem which is not understood or dealt with properly and that archaic, incorrect concepts in designing and operating buildings to facilitate egress are imposed by law. To date, traditional practices have been used as a basis for the design and regulation of escape routes, e.g. the traditional approaches to stairway design are geometry based, linked to economy and standardisation of components and not just concerned with the safety of people in an emergency. Some changes have occurred in regulations controlling stairways, but these changes were only concerned with reducing accidents on stairways in normal usage.

12.3.7 Escape route width

Widths of escape routes, e.g. corridors and staircases, should be sufficient to accommodate discharge rates as indicated above. But also, particularly in the case of staircases, widths must be sufficient to accommodate these discharge flow rates when the use of one staircase is discounted. This measure provides an additional factor of safety and presupposes that in every conceivable fire scenario only one staircase is usable.

Table 12.4 gives some guidance on escape route widths.

12.3.8 Enclosure of stairways

Stairways should be constructed as protected shafts, i.e. enclosed by a protected structure which meets the requirements of building regulations regarding non-combustibility, fire resistance and surface spread of flame characteristics. Generally speaking no other part of

Table 12.4

Building type	Situation		Minimum width of escape route
Institutional	Wheelchair evacuation		1 m
	Mattress evacuation		1.225 m
	Stairway landings		3.000 m
Any other building type	Auditoria, theatres, assembly rooms with closely related seated audience		1.225 m
	Seatways		
	Gangway on one side Max. No. of seats	Gangway on two sides Max. No. of seats	
	7	14	0.300 m
	8	16	
	9	18	
	10	20	0.575 m
	14	27	
	Continental seating		0.500 m

the building should be accommodated within the stairway enclosure except, for example, sanitary accommodation where the possibility of fire occurring is extremely remote.

It is always desirable that a stairway enclosure is in part formed by an external wall. This facilitates ventilation and smoke control and access from outside by fire fighters. If, however, the stairway enclosure projects beyond the external wall of the building and is connected thereto, then either:

(a) the external wall or walls of any part of the building less than say 3 m from the stairway enclosure, or
(b) the external wall or walls of any part of the stairway enclosure within any 3 m from the building

should be imperforate and of not less than a half-hour fire resistance (Figs 12.14 and 12.15).

Basements in buildings often create difficulties regarding the provision of escape routes. In a building with one stairway serving the upper storeys as an escape route, such a stairway should not be continued into the basement storey in case persons confused in a fire situation or unfamiliar with building become hopelessly trapped.

Ideally basements should be provided with adequate ventilation and escape routes at the lower level, but this is not always possible. If the building has more than one stairway serving upper storeys as escape routes, then provided that one stairway is not continued

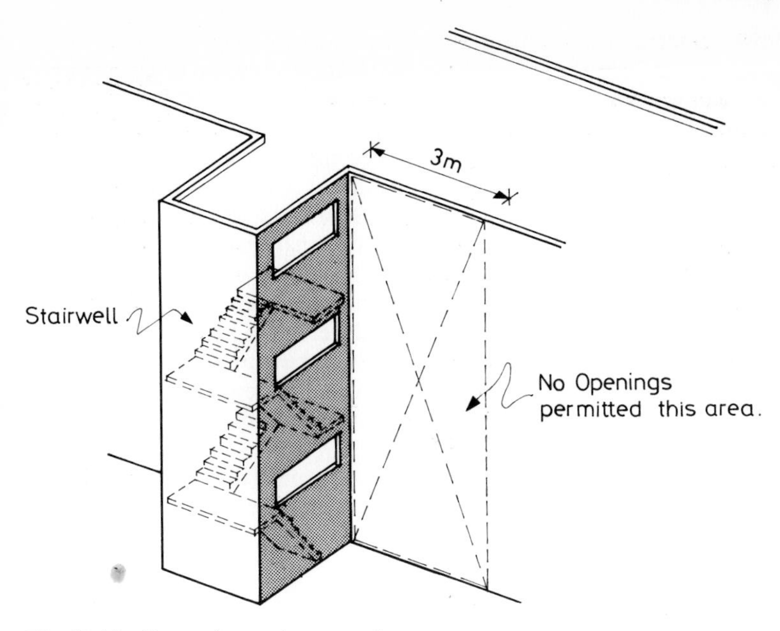

Fig. 12.14 Protecting stairway enclosures

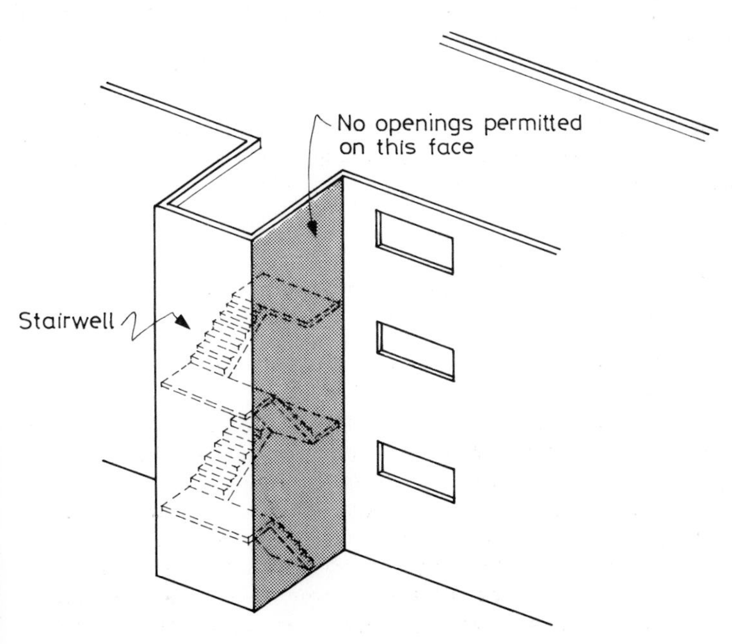

Fig. 12.15 Alternative method of protecting stairway enclosures

down to serve the basement any other stairway may be continued down to the basement. Careful consideration must be given to the planning of escape routes, especially when dealing with multiple basement construction.

The greatest personal hazard in a fire situation is that of asphyxiation due to smoke and toxic fumes. Consequently it is essential that escape routes, especially stairways, are provided with adequate means of ventilation. Where external walls enclose part of the stairway, that may be achieved by providing openable or permanent vents of say $0.5\,m^2$. If external walls are not available, vents must be provided at each storey level ducted to the external air in fire-resisting vertical ducts. With smoke outlets from stairways below ground level, care must be taken to ensure that they discharge to the external air and are sufficiently far away from any part of the escape route from the building.

Openable smoke outlets and permanent vents raise doubts as to their reliability and effectiveness. Indeed, rather than venting they may well in certain circumstances have an entirely aggravating effect on smoke movement. Pressurisation of stairways is an alternative and effective means of keeping smoke from reducing the effectiveness of the escape route. External stairways can be a very effective means of evacuating people to a place of safety.

External stairways should be limited, doors should be fire-resisting and self-closing, and the external wall of the building should be fire-resisting on either side of the stairway for a distance of 3 m. It must be ensured that persons using the stairway are sufficiently protected from any fire burning within the building. If part of the escape route to the external staircases is across a roof, then the walkway must be clearly indicated, fire-resisting, provided with handrails and roof lights, etc., sited sufficiently far from the walkway so as not to endanger persons using it.

12.3.9 Lobby approach stairways

Additional protection to stairways can be achieved by providing a lobby or ventilated lobby approach. Generally speaking, buildings over 100 m in height require a lobby at every storey level. Special situations such as basements or areas of special fire risk may require a lobby at each storey. In many other cases, depending on the height and type or class of building, protection to the stairway by means of a self-closing fire-resisting door may be sufficient.

12.3.10 Doors in escape routes

The term 'self-closing fire-resisting door' is self explanatory. Yet it is necessary to re-emphasis the significance of the words used. Self-closing does not presuppose the door being opened and then closed. The door may be held open, closing automatically in predetermined circumstances. An electromagnetic or electromechanical device activated by suitable smoke sensors may be suitable for holding open a door provided that the door will close automatically if any of the following occur:

(a) detection of smoke sufficient to activate the closing mechanism
(b) failure of the electricity supply to the device
(c) operation of the fire alarm system
(d) operation of manual override switch on the door.

The basic principle is to ensure that in a fire situation the door closes to prevent fire spread and smoke-logging of the escape route.

In Chapter 6, reference was made to BS 476: Part 8: 1972,[11] the current fire-resistance test for structural elements. It was pointed out that the insulation criterion for doors has been deleted and fire resistance for doors means satisfying the stability and integrity requirements. A great deal of confusion has arisen over the use of differing terminology regarding doors. It would be much better to forget all previous terminology regarding these doors and refer to them as self-closing fire-resisting doors related to test performance. Thus the doors shown in Table 12.5 are all fire-resisting doors to different degrees and, as such, building regulations make use of test data by specifying doors suitable for particular situations.

Further confusion exists, however, because building regulations may allow doors which satisfy BS 476: Part 1: 1953[12] (superseded by BS 476: Part 8: 1972) still to be used. The significant factor is that the Part 8 test is carried out with a positive furnace pressure as opposed to a negative furnace pressure in the Part 1 test and is consequently a more severe test. As seen in Chapter 6 and Table 12.5, doors are tested as they would be fixed and used in practice.

The performance of fire-resisting doors may be diminished considerably because of any of the following factors:

(a) small variation in door size
(b) use of hinges of incorrect mass
(c) use of hinges which extend to the outside face of the door

Table 12.5

Test material	Stability	Passage of flame	Insulation
Fitted in its frame	60 mins	60 mm	No requirement
Fitted in its frame	30 mins	20 mm	No requirement
Fitted in any frame	30 min	30 mm	No requirement

(d) locks and latches of incorrect thickness and thermal capacity
(e) bad workmanship in fitting
(f) indiscriminate use of intumescents relative to (e) above
(g) incorrect use and fitting of floor spring and other self-closing mechanisms
(h) inclusion of additional door furniture, e.g. letter plates not included in the doorset tested
(i) changed specification of the doorset facings.

Fire-resisting doorsets, i.e. (the door blank, the frame, hardware, hinges and methods of fixing) are tested as complete assemblies and variations such as those listed above from the tested assembly can seriously diminish the performance of the doorset.

Doors across escape routes generally should, and in prescribed cases of high occupant density must, open in the direction of discharge (escape), or swing in both directions and in the latter case be fitted with a small vision panel so as to avoid injury to persons who may be in the opening arc of the door.

Conflict arises between escape and security considerations, i.e. prevention of unauthorised entry from outside the building. Doors in such positions can be fitted which will open when subjected to internal pressure. Many such devices are available fitted with alarm mechanisms which reconcile the escape and security conflict. There can be no real justification in the design, construction and management of buildings today for situations arising in which such doors are effectively locked because experience has shown that even where keys are available, smoke has often obscured their location, resulting in confusion with potential catastrophic consequences.

12.3.11 Lighting of escape routes

It is generally expected that adequate artificial lighting be supplied to all parts of modern buildings. Where artificial lighting is supplied to circulation spaces designated as escape routes, it is necessary that protected circuitry be used and where a stairway forms part of the escape route the protected circuitry associated with the stairway should be separate from any circuitry which supplies lighting to any part of the same escape route.

Essentially the foregoing applies to a non-maintained system, i.e. a system where the failure of the mains supply must occur before the emergency mode is activated. Alternatively, the building designer may install a maintained lighting system. Maintained lighting systems are those which are continuously illuminated and comply with the requirements of BS 5266.[13] The advantage of using a maintained lighting system is that, should the mains supply fail, back-up safety lighting is immediately available. Also normal routine maintenance checks will identify lamps which require replacement.

12.3.12 Emergency lighting

The primary function of an emergency lighting system is to provide illumination in all or part of a building to enable the occupants to move around the building or to allow prescribed functions to be carried on within the building without interruption. Building regulations prescribe building types in which adequate emergency lighting must be provided to come into operation automatically during any interruption of the normal lighting system and Table 12.6 gives examples of situations where emergency lighting in escape routes would be normally provided.

Emergency lighting should:

(a) indicate clearly the available escape routes
(b) provide surricient illumination along escape routes to allow safe movement towards and through the exits provided
(c) be sufficient to ensure that fire alarm call points and fire fighting equipment along the escape routes can be readily located.

Table 12.6 Emergency lighting

Building type	Description of parts of a building requiring emergency lighting
Institutional	All circulation spaces, rooms having an occupant capacity of more than 60
Other residential	All circulation spaces, rooms having an occupant capacity of more than 60
Office	(a) all circulation spaces without natural lighting: and (b) in a building having a floor at a height of 18 m or more above ground level (i) all circulation spaces at every level, and (ii) any canteen or assembly room having an occupant capacity of more than 60
Shop	All sales areas and circulation spaces† Canteens and restaurants having an occupant capacity of more than 60
Factory	All circulation spaces
Assembly	All public areas and circulation spaces
Storage and general	All circulation spaces
All purpose groups	Any basement having an occupant capacity of more than 10

† Excluding all sales areas in ground floor shops not exceeding 100 m^2 in floor area.

Other factors may influence the overall effectiveness of an emergency system, e.g. the geometry of the escape route and associated spaces, the colour and distribution of lighting and the choice of finishings.

The emergency lighting power source must also have a suitable capacity for its particular function, and be independent from that of the normal lighting. The capacity or period of operation of the emergency lighting system when activated can range from half an hour to four hours depending on the particular situation.

Independent power supplies can be obtained in a number of ways:

(a) in-house generators
(b) batteries
(c) mains power from a separate grid.

Recommendations are contained in CP 1007[14] regarding the installation of emergency lighting systems in buildings which provide entertainment for the public, e.g. cinemas, ballrooms, etc. BS 5266[13] gives guidance in situations where CP 1007 does not apply.

12.3.13 Construction of and egress from windows

Over the past few years many subtle changes have occurred in the design of window assemblies. These changes have made the construction of and egress through windows an important component of escape route design. Factors such as those listed below can make windows unusable as a means of escape in an emergency and must be seriously considered.

(a) top hung opening lights (Fig. 12.16)
(b) increasing use of 4 mm and 6 mm float glass
(c) increasing use of safety glazing
(d) double glazing and triple glazing
(e) use of polycarbonate glazing.

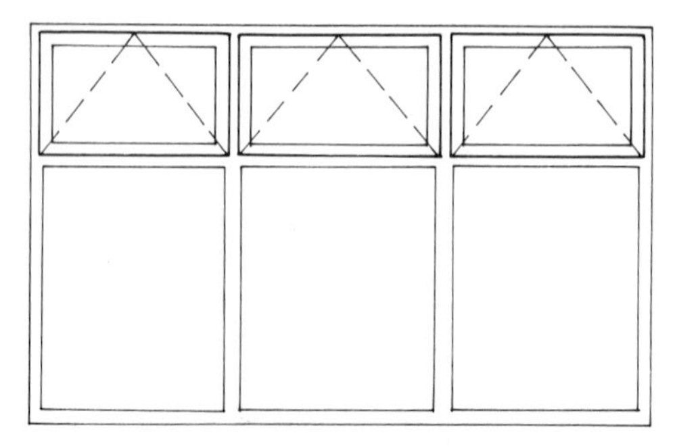

Fig. 12.16 Top hung opening sashes

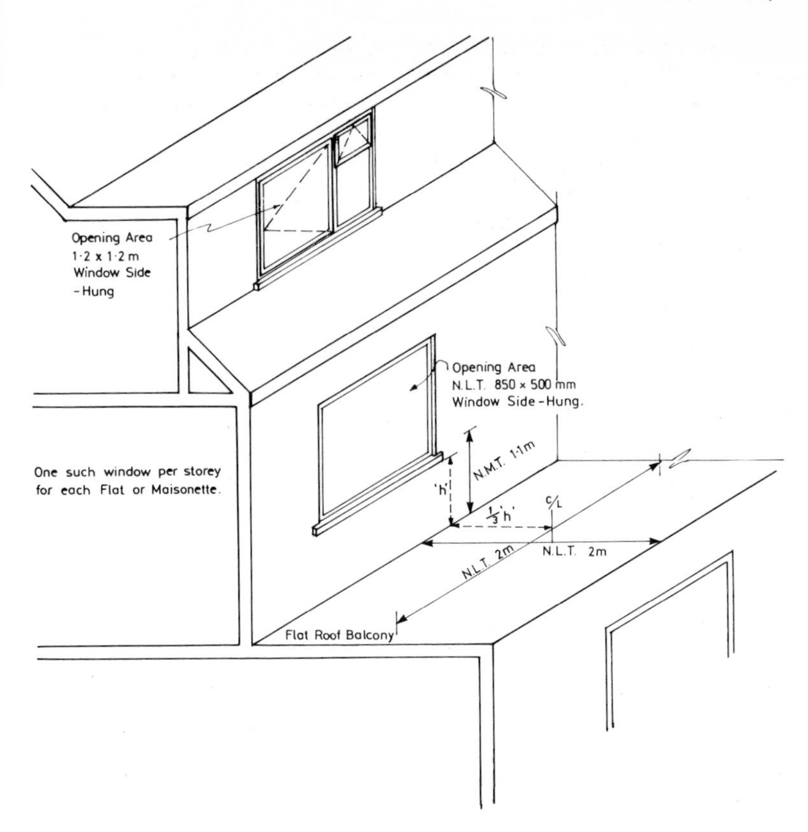

Fig. 12.17 Provision of alternative Egress routes via windows in external facade

The top hung opening lights, Fig. 12.16, are of such dimensions and so inaccessible as to make it impossible to use them as a means of escape. Items (b) and (e) above make it extremely difficult, in some cases impossible, to break through the glazing material.

Since most of the fatalities in fires occur in dwellings, and most of the current rehabilitation work is focused also around dwellings, it is towards residential properties, in general, that attention should be directed. Two-storey dwellings can easily become three storeys by having the roof space converted to provide, for example, additional sleeping accommodation. It is essential when carrying out such conversions that adequate provisions for escape and rescue in the event of fire are included.

Figure 12.17 shows the inclusion of a window as a component of an alternative escape route. The window is side hung, of suitable dimensions to allow a person to pass through, is accessible and gives access to a flat roof or balcony from which escape or rescue can be effected. Similarly with dormer windows, they must be so located as to assist with rescue, not hinder it.

It is important to be aware of other safety criteria when providing and positioning side-hung windows. It should be done in such a way that during normal usage accidents resulting from opening the window outwards are avoided.

The top window component in the provision of means of escape is only important up to about five storeys; beyond this height greater attention should be paid to the other components of escape route design.

12.3.14 Fire-detection systems

The statistics for the United Kingdom 1980[15] fire experience show that 16 per cent of the total number of fires in 1980 occurred in dwellings. The total number of fire fatalities in 1980 was 1035 and 79 per cent of these fatalities occurred in dwellings. BS 5588: Part 3: 1983[16] makes no recommendations regarding the provision of fire-detection systems. Yet it follows that if fire casualties are to be reduced significantly, methods other than increasing the fire-endurance characteristics of structural and constructional materials must be employed. It also follows that the sooner the occupants of a building are alerted to the presence of fire, the sooner effective action can be taken to secure escape from the threatened spaces.

One method of enhancing escape potential and reducing the fire casualty statistics would be the introduction of fire-detection systems, as a component of escape route design, linked to a warning system which would alert the occupants to the presence of life-threatening stimuli.

Bakauski and Zimmerman[17] list three key elements which are necessary for residential fire protection. They are:

(a) minimising fire hazards
(b) the installation of smoke detectors
(c) escape route planning.

The term planning is used here in a geometric context and although each of the three elements listed are important individually, they should be considered as integral components of escape route design contributing collectively towards the total life-safety package.

NFPA: 74[18] requires at least one smoke detector outside each sleeping area and one on each habitable storey and basement of residential buildings. Detector sensitivity and siting are very important factors which must be thoroughly researched before final decisions are taken.

12.3.15 Alarm systems

The primary function of any alarm system is to alert the occupants of a building to the presence of life-threatening agents. Consequently the alarm system must be tailored to and communicate with the building population.

People receive information continuously by hearing, seeing, touching, smelling and tasting, and the choice of a particular mode of alarm system, based on one or a combination of the above, will depend largely upon the type of building population and the nature of the information transmitted.

An aural alarm system is obviously of little use in a building occupied by deaf people; similarly a visual alarm system is no help to the blind and may do little more than confuse those people who are colour blind. Also the nature of the preferred alarm system may vary within a building, e.g. in a large hospital, where sounders may be acceptable in some areas but may be completely unacceptable in intensive-care units.

The development of a satisfactory alarm system must take into account the following factors in order to be effective:

1. The population matrix:
   (a) population type
   (b) population perception
   (c) population decision-making ability
2. The population discipline:
   (a) population action
   (b) population reaction
3. The people management systems, e.g.
   (a) prisons
   (b) hospitals
4. The nature of the information to be communicated:
   (a) warning and action required
   (b) alert, no action required
5. The method of communication:
   (a) message must be
      (i) clear
      (ii) unambiguous
      (iii) not liable to misinterpretation.

12.3.16 Fire-control systems

Fire control systems may be active, passive, or a combination of both.

Passive fire-control systems rely on the fire-endurance characteristics of structural and constructional components such as enclosing walls, partitions, doors and lobbies.

Active fire control, however, is a dynamic system comprising four elements:

(i) Detection
(ii) Warning
(iii) Calling Fire Brigade
(iv) Direct attack.

Detection

The methods of automatic detection can be considered under the following headings:

Smoke detectors:
(a) ionisation detectors
(b) optical detectors
  (i) light-scatter type
  (ii) obscuration type

Radiation detectors:
(a) infra-red detector
(b) ultra-violet detector

Heat detectors:
(a) those which measure the temperature rise
(b) those which measure the rate of temperature rise

Heat detectors generally incorporate fusible alloys or bimetallic strips and detector heads which measure both (a) and (b) above are available. Detection methods are dealt with in more detail in Chapter 10.

Warning

The nature of the warning given will depend on the type of building and building population. Detailed consideration was given to this component of escape route design earlier.

Calling Fire Brigade

When the detector senses the symptoms, e.g. smoke, heat or flame, which indicate the presence of a fire, the alarm system is activated. Coincidentally with the activation of the alarm system a call can be directed through the British Telecom cabling system and routed to the Fire Brigade Control Centre. From the mimic board display in the control room, the Control Officer can see immediately which appliances are on station and are immediately available. He can then despatch the appliances to the scene of the fire and with close communication between the fire scene and Central Control can despatch further assistance if required, whilst maintaining a reasonable general fire cover for the remainder of the area under his control.

Direct attack

The method of direct attack employed will depend on the building type, building occupancy and the nature of the hazard. Thus the methodology employed may vary in different parts of the same building. The various fixed fire-extinguishing systems are listed below:

(a) water sprinkler systems
(b) water spray systems
(c) water curtain systems
(d) water drencher systems
(e) carbon dioxide systems
(f) halon gas systems
(g) dry powder systems
(h) foam systems
  (i) ordinary-expansion foam
  (ii) medium-expansion foam
  (iii) high-expansion foam

12.3.17 Smoke-control systems

The escape route as illustrated in Fig. 12.4 is designated a place of safety. Consequently if it is to remain usable in an emergency, the escape route must be protected from fire and the products of combustion. Passive fire protection in the form of fire-resisting constructional components was discussed earlier, but passive fire protection will not prevent the movement of smoke from the fire area into the escape route and most certainly will not positively control smoke movement.

Methods of smoke control may be considered under the following headings:

(i) Natural ventilation
(ii) Mechanical ventilation
(iii) Pressurisation.

Natural ventilation

The factors which cause smoke movement within a building are:

(a) buoyancy of the rising smoke plume
(b) stack effect
(c) direction and velocity of the prevailing wind
(d) mechanical air-movement systems.

Figure 12.18 shows the rising buoyant plume. The upward buoyancy force due to the fire is of the order of 10 pascals, and in design calculations is usually ignored.

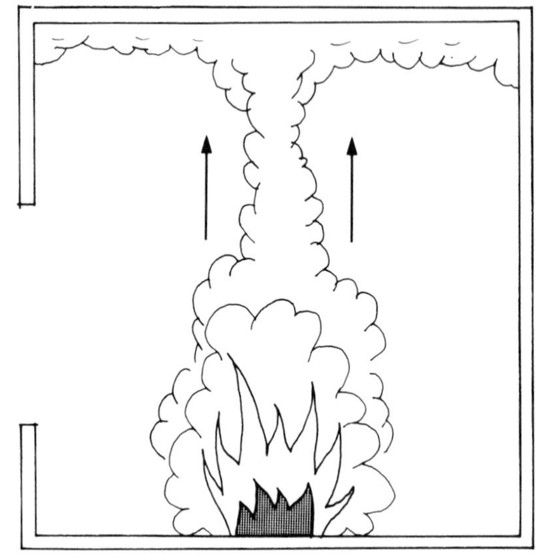

Fig. 12.18 Bouyant plume

The stack effect, Fig. 12.19, is due to the differences between the external and internal climates. These temperature differentials cause movement of the smoke either upwards or downwards until a neutral plane is reached and lateral smoke movement is induced.

The velocity and direction of the prevailing winds are most significant factors in determining smoke movement pathways. In fact the design of a successful natural cross-ventilation system, Fig. 12.20, requires a precise knowledge of the direction and wind velocity profile at the exact time of the occurrence of a fire. Clearly this knowledge will not be available to the designer and total reliance on natural ventilation includes an element of risk. CP 3[19] recognises the difficulties involved in using natural ventilation as a method of

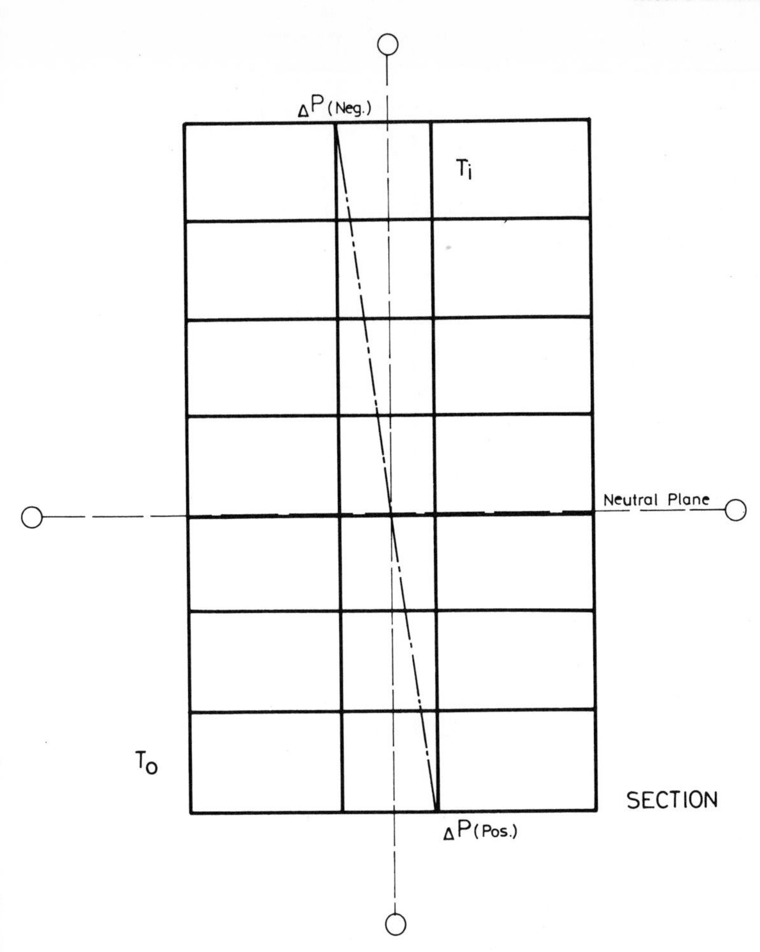

Fig. 12.19 Pressure and flow patterns caused by stack effect in the absence of wind

smoke control and accepts that there is an element of risk in its usage.

The forces caused by any mechanical air-control system will also be a factor with regard to smoke movement and will be considered later.

Consider a simple sectional elevation, Fig. 12.21. If the wind pressure is positive, the smoke will be forced back into the escape route resulting in turbulent mixing of air and smoke. Consequently the escape route may become smoke-logged much more quickly.

Figure 12.22 shows the same sectional elevation, this time with a negative wind pressure effectively sucking the smoke out. But if the

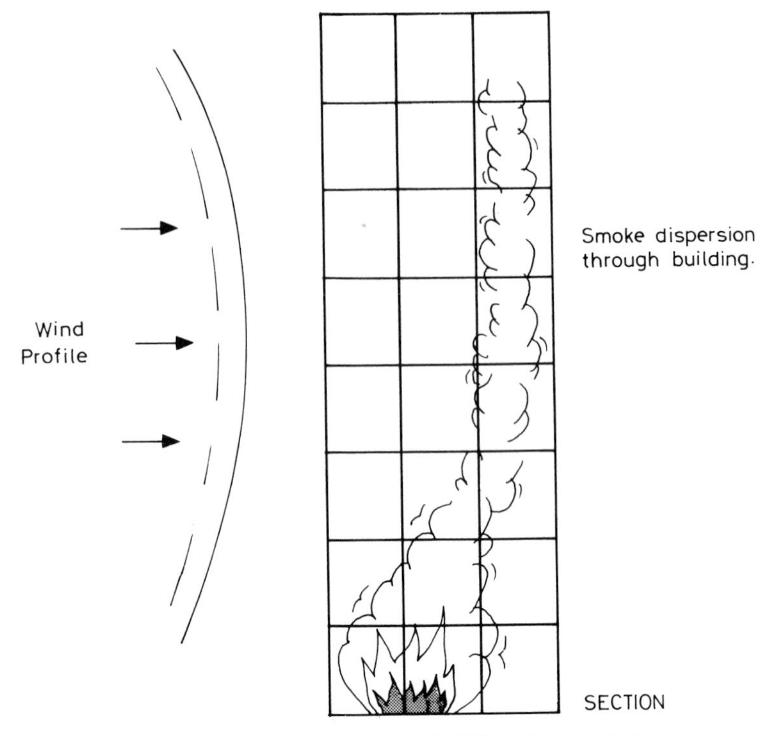

Fig. 12.20 Smoke dispersion through a building due to wind

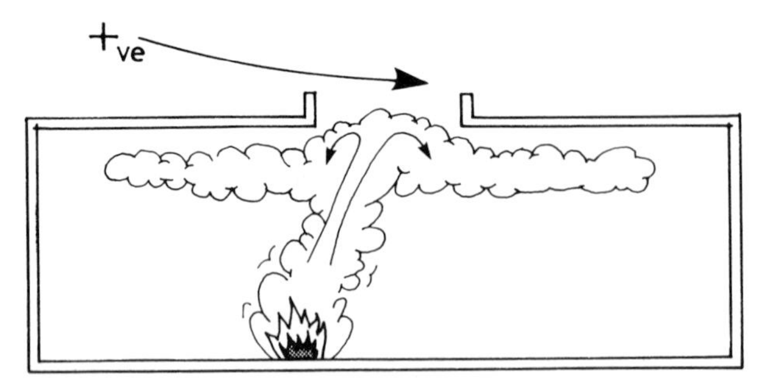

Fig. 12.21 Adverse effect of wind on smoke venting

negative pressure is too large (Fig. 12.23), fresh air is also sucked out through the smoke, causing smoke-logging in the escape route.

The use of a natural ventilation system as a means of smoke control implies, almost, control of the external environment. This cannot be so. Thus a system of smoke control is often employed in circumstances likely to be influenced by external conditions.

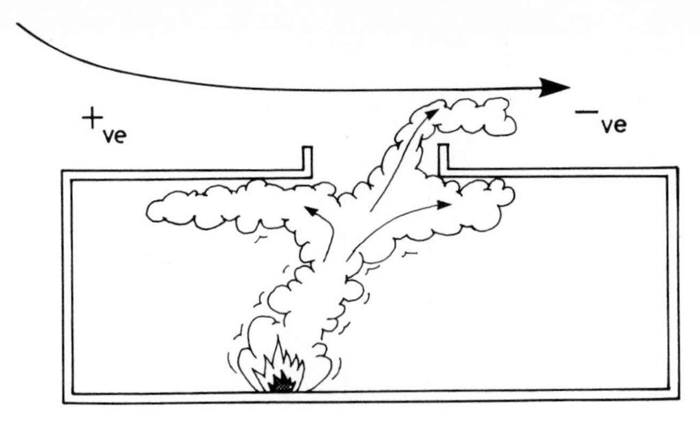

Fig. 12.22 Effective smoke extraction due to wind

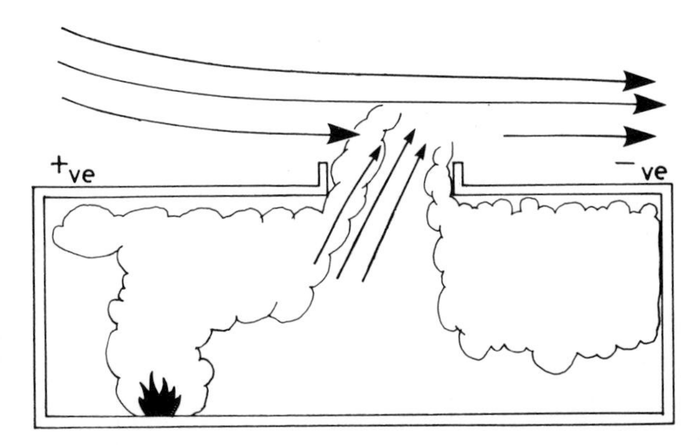

Fig. 12.23 Ineffective smoke venting where the critical wind velocity is exceeded

Figure 12.24 shows a tower block with a podium, assuming that a natural ventilation system of smoke control has been designed in full knowledge of prevailing winds, etc. Figure 12.25 shows the same building, but the surrounding area has been built up and consequently the air-movement patterns around the building using natural ventilation have all been altered, perhaps to the detriment of the building's occupants.

Figure 12.26 shows a corridor arrangement where natural ventilation may serve to spread smoke around the building rather than remove it.

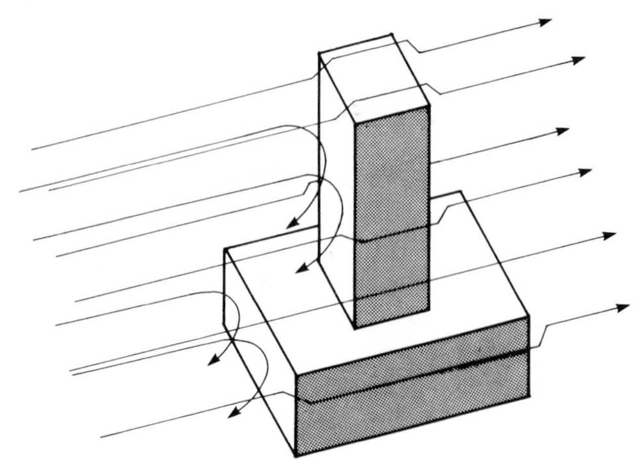

Fig. 12.24 Air flow patterns around buildings

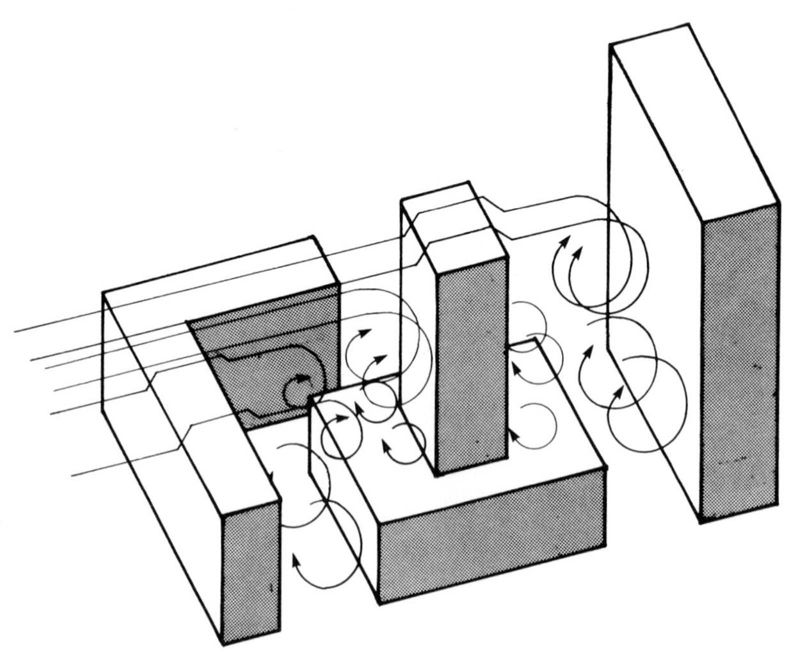

Fig. 12.25 Air flow patterns around buildings

Mechanical ventilation

From the foregoing it is clear that natural ventilation, as a smoke control system, is not entirely reliable and that the recommendations contained in CP 3 can only be taken as minimum requirements.

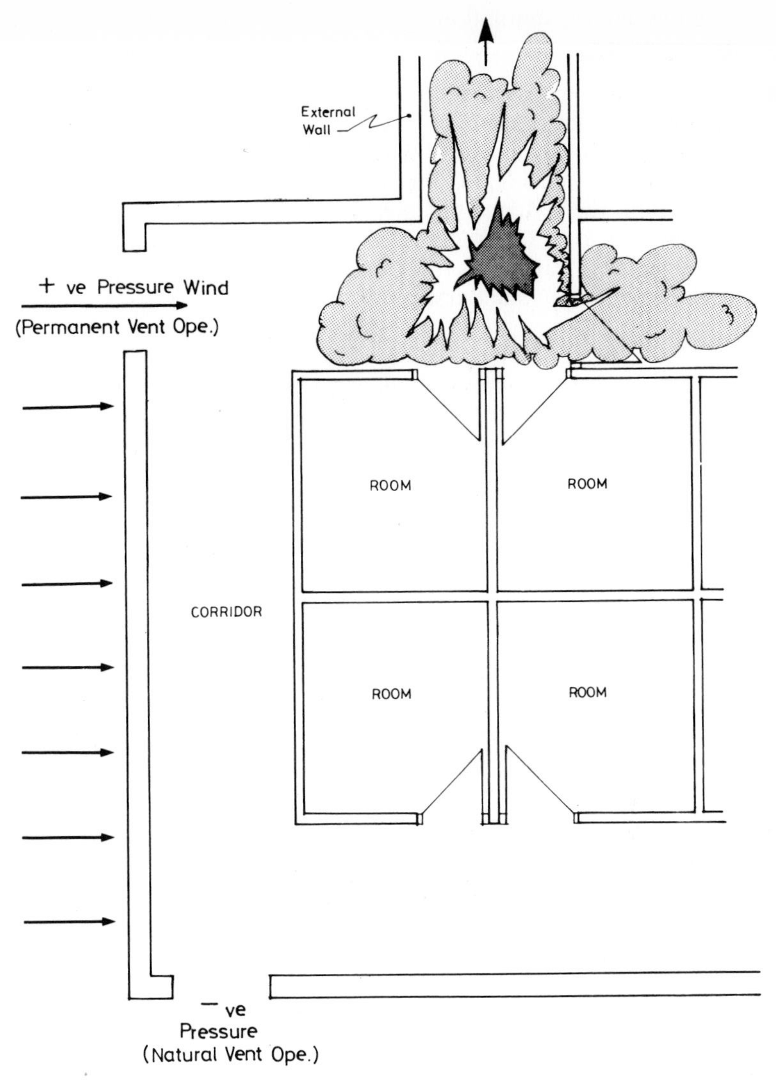

Fig. 12.26 Wind induced smoke movement

Mechanical ventilation provides an alternative to natural ventilation as a smoke-control system. The choice of mechanical ventilation may be determined by the following factors:

(a) internal climatic conditions
(b) positive smoke control possible
(c) system is controllable by fire fighters
(d) complex building geometry
(e) greater flexibility in use of floor space.

For a mechanical ventilation system of smoke control to be effective it must:

(a) prevent the flow of smoke into, and remove any smoke that has entered, the escape route
(b) be instantaneously available when required
(c) be completely reliable, i.e. the reliability of the system is a function of the system components
(d) be designed so as to remain operable during a fire situation and be of sufficient capacity so as to allow the escape to remain usable throughout the duration of the emergency.

Broadly there are three types of mechanical system:

(a) extract only: the system is designed to extract smoke from the excape route. Figure 12.27 shows that in some circumstances, extract only can induce smoke into the escape route and consequently reduce visibility
(b) extract and air inflow: this system is better than (a) above, but still smoke is induced into the protected space, Fig. 12.28
(c) Pressurisation of the escape route: here a supply of air is pumped into the escape route creating an overpressure which causes the smoke to flow away from the protected area, Fig. 12.29.

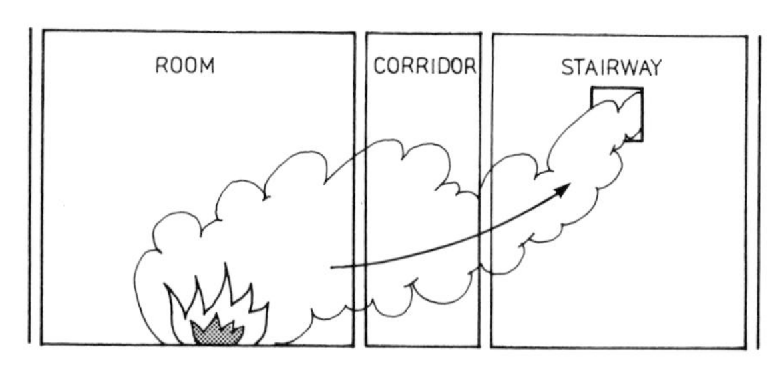

Fig. 12.27 Extract only system

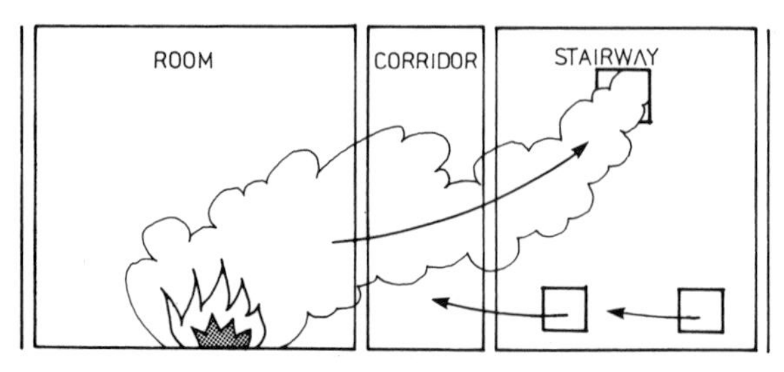

Fig. 12.28 Extract and input system

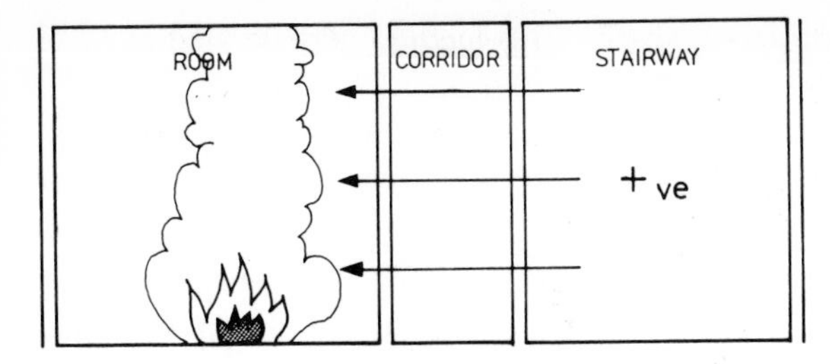

Fig. 12.29 Pressurisation of stairway system

Pressurisation

As stated above, pressurisation works by pumping air into the protected space. In this way a pressure difference across doors and divisions between the protected and adjacent spaces is created. If the maintained pressure differential is sufficiently large the pressure forcing the smoke from the fire zone towards the protected area will be insufficient to overcome the pressure differentials existing across the spatial barriers.

A pressurisation system, if it is to be effective, must be an integral part of the design and not something imposed upon the design afterwards. There are basically two types of systems:

(a) positive
(b) negative.

Figure 12.29 illustrates the positive type where clear air is forced into the protected space. This method is used for protecting staircases.

Figure 12.30 illustrates the negative type where the air in the adjacent spaces is sucked out creating a pressure differential between

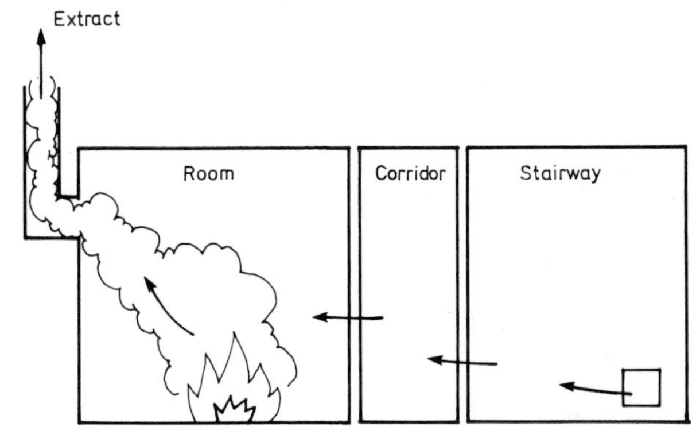

Fig. 12.30 Smoke venting from a basement

the adjacent and protected spaces. This method may be used in basement shopping complexes.

Operation of mechanical systems

Plant shut down except in an emergency In this method the plant does not come into operation until a fire situation has been detected and consequently protection is delayed.

Plant operating continuously Pressurisation is continuous using this method and consequently protection is also continuous.

Plant operating continuously at reduced capacity except in an emergency This method of operation offers some protection in the early stages of a fire and the level of protection is boosted once the fire has been detected. Obviously this is more economical than to have the plant operating at full capacity continuously.

12.4 DESIGN OF ESCAPE ROUTES

Bryan[20] listed two major principles which are used to determine the necessary exit width based upon anticipated population characteristics associated with a specific occupancy. These are:

1. Flow method, and
2. Capacity method.

12.4.1 The flow method

This method employs the theory of evacuation of a building within a predetermined maximum length of time. Van Bogaert[21] illustrated the use of this method, Fig. 12.31(a)–(d).

The Figures represent a four-storey building with two staircases A and B. Staircase B rapidly becomes untenable and all the occupants must evacuate the premises via staircase A.

The total time T_t is the sum of the times required to complete phases

(A and B) : C : and D

(T_1) : $(T_2)+(T_3)$: (T_4)

where $T_1 \ldots T_4$ is the time required to complete each phase,

i.e. $T_t = T_1 + T_2 + T_3 + T_4$

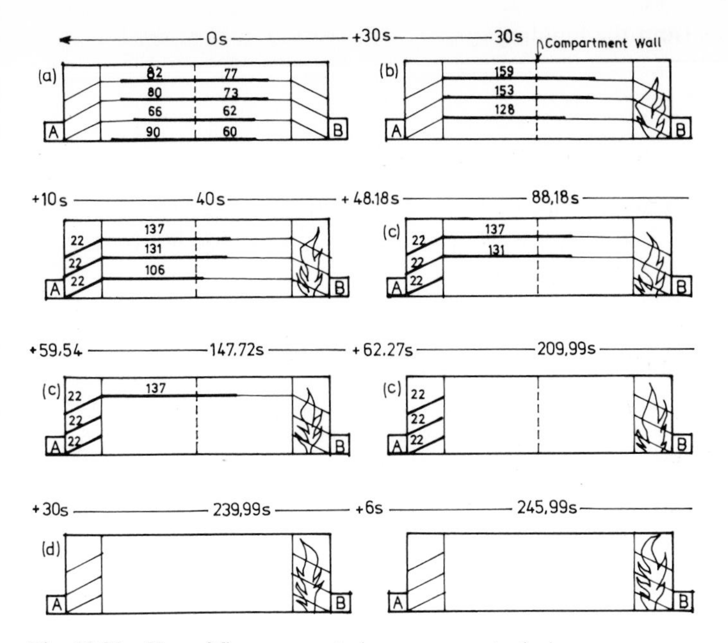

Fig. 12.31 Use of flow concepts in escape route design

If the travel distance X^1 on the upper floors can be taken as 45 m and the walking speed (horizontal) as 1.5 m s^{-1}:

$$T_1 = \frac{X^1}{V_h} = \frac{45}{1.5} = 30 \text{ secs}$$

Taking the occupant capacity of the staircase between each floor level (Y) as 22 and the discharge rate down the stairs of 2.2 (persons/sec)

$$T_2 = \frac{Y}{V_d} = \frac{22}{2.2} = 10 \text{ secs}$$

Completion of phase C requires the movement of the total population from the upper floors.

Thus:

$$T_3 = \frac{P}{D_d \times U} = \frac{440}{1.1 \times 2} = 200 \text{ secs}$$

where:

P = total remaining population on the upper floors, including the staircases
D_d = discharge rate (persons per unit width per sec)
U = number of units of exit width.

Phase D is the movement of people from the foot of the staircase to the exits from the building:

$$T_4 = \frac{N_e}{V_h} = \frac{9}{1.5} = 6 \text{ secs}$$

where N_e = the distance to the exits (m).

Another phase must also be considered and its corresponding time T_5 which is the time necessary to evacuate the ground floor.

$$T_5 = \frac{X}{V_h} + \frac{P_1}{D_h \times U} = \frac{16.5}{1.5} + \frac{150}{1.5 \times 9}$$
$$= 22.11 \text{ secs}$$

where X = travel distance on the ground floor (in this case 16.5 M because more exits are available on the ground floor)
P_1 = population of the ground floor.

T_5 is only of importance in this case when $T_5 > T_1 + T_2$, i.e. if the time to evacuate the ground floor is greater than the time to complete phase A and B and the first part of phase C, Figs 12.31(a) and 12.31(c).

If $T_5 > T_1 + T_2$, then

$$T_t = T_1 + T_2 + T_3 + T_4 + T_5 - (T_1 + T_2)$$
$$= T_3 + T_4 + T_5$$

If $T_5 < T_1 + T_2$, then

$$T_t = T_1 + T_2 + T_3 + T_4$$

In this example, $T_1 + T_2 = 40$ secs

$$T_5 = 22.11 \text{ secs}$$
$$T_5 < T_1 + T_2$$

then the total time to evacuate the building

$$T_t = T_1 + T_2 + T_3 + T_4$$
$$= 246 \text{ secs}$$

12.4.2 The capacity method

This method is based upon the theory that enough stairways are provided in a building to house adequately all the occupants within the stairway enclosure in the event of a fire, without requiring any movement out of the stairways. Provided that the stairway enclosure provides an effective barrier against fire, evacuation of the building could proceed at a pace related to the physical ability of the occupants.

The standing capacity on stairways and landings has been estimated as 3.5 persons per square metre.[1,22] The horizontal staircase area, per unit width, excluding landings, is $d \cos \theta$

where:

d = distance along the line of slope (typically 6 m per storey height)
θ = the slope.

The length of the landing is likely to be about four times the staircase width. Taking θ as 30°, from these figures:

$$S = (bd \cos \theta + 4b^2)\rho$$

$$= 18b + 14b^2$$

where S = the number of people that can be housed per storey
b = staircase width (m)
ρ = density of people per unit area.

For prolonged stays where it is prudent to avoid undue stress and possible irrational behaviour of some people a lower density is desirable. A value of 1.5 persons per square metre has been recommended.[23]

Earlier in this Chapter, the horizontal and vertical components of escape routes were illustrated, Fig. 12.4. Table 12.7 shows the measured flow rates of people[23] and the average flow rates for level passageways, upward and downward motion on stairs, i.e. 1.5, 1.1 and 1.15 persons per metre per second.

Figure 12.32 illustrates the variation of velocity and flow with crowd density for corridors of 1.1 m to 3 m width.[22] It can be seen that flow rate is a maximum when the crowd density falls between one and five persons per square metre.

Table 12.7 Measured rates of flow of people

Source	Rate of flow, persons per second per metre width		
	level passage	stairs up	stairs down
British data	0.8–1.4	1.0	0.9
French data	1.1–2.7	1.4	0.9–1.6
American data	1.4	0.9–1.7	1.0–1.5
Cinema construction	0.9–2.1	–	–
North American Transit Authorities	1.5	0.9	1.1
Paris metro	1.7	1.0	1.3
London Transport Board	1.5	1.1	1.1
Average	1.5	1.1	1.15

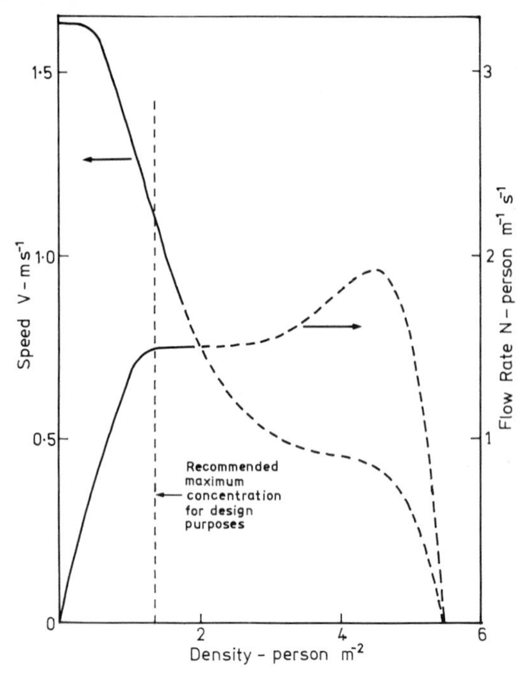

Fig. 12.32 Effect of population density on velocity and flow rate for level passages

Togawa[24] derived the following formula experimentally:

$$V = V_0 \rho^{-0.8}$$

where:

V = crowd walking velocity
V_0 = constant ($1.3\ \mathrm{m\,s^{-1}}$)
ρ = density in persons per square metre.

The flow rate, N is given by:

$$N = V \cdot \rho$$
$$= V_0 \rho^{0.2}$$

Taking: (i) $\rho = 2$ persons per square metre
$N = 1.5$ persons per metre per second
(ii) $\rho = 5$ persons per square metre
$N = 1.8$ persons per metre per second.

These values of N correlate well with Fig. 12.32.

Table 12.8 Movement on staircases

Source	Direction	Density (people per square metre)	Mean velocity V' (along line of slope) ($m\,s^{-1}$)	Rate of flow, N' ($= V'$) (persons per metre per second)
London Transport Board	up	2.05	0.51	1.05
	down	1.94	0.59	1.15
Togawa	up	2.6	0.5	1.3
Galbreath	up	0.7	0.8	0.55
		1.1	0.8	0.9
		1.8	0.7	1.25
		2.2	0.6	1.3
		2.6	0.5	1.3
		3.4	0.4	1.35
		4.1	0.3	1.25
		4.4	0.25	1.1
		5.4	0	0

Table 12.8 summarises some of the data relating to crowd movement on stairways. The time required for a given number of people to pass through an exit can be calculated from the formula:[23,25]

$$T = Q/Nb$$

where:

Q = total number of people
N = flow rate
b = exit width in metres.
N is usually taken as 1.7 persons per metre per second.

Melinek and Booth[26] concluded that:

(a) the maximum population M which can be evacuated to a staircase, assuming a permitted evacuation time of $2\frac{1}{2}$ minutes, is given approximately by $M = 200b + (18b + 14b^2)(\eta - 1)$

where b = the staircase width in metres
η = the number of storeys served by the staircase.

(b) the minimum total evaucation time T_e for a multi-storey building is given by:

$$T_e = \left(\sum_{i=r}^{\eta} Q_i \right) \bigg/ (N^1 b_{r-1}) + rt_s$$

where:

r = the floor number (1 to η) which gives the maximum value of T_e
Q_i = population of floor i
b_{r-1} = width of the staircase between floors $r-1$ and r
N^1 = flow rate of people per unit width down the stairs
t_s = the time taken for a member of an unimpeded crowd to descend one storey.

If the population Q and the staircase width, b, are the same for each floor then $T_e = T_1$ or T_η whichever is the larger

where $T_1 = \eta Q/(N^1 b) + t_s$ and
$T_\eta = Q/(N^1 b) + \eta t_s$

Typical values of N^1 and t_s are given as 1.1 persons per metre per second and 16 seconds respectively.

Much of what might be described as escape route design data is contained in building legislation and associated approved documents. The legislative picture is a very confused one[27] and terms such as reasonable and adequate are still much overworked in prescribing the requirements for the provision of means of escape from buildings.

12.5 COMPUTER MODELLING FOR EMERGENCY EVACUATION OF BUILDINGS

Berlin *et al.*[29] describe a methodology for estimating the necessary time for the evacuation of residents from group homes. Figures 12.33 and 12.34 illustrate the resident and staff factors, respectively, considered in modelling emergency evacuation from group homes.

The method consists of constructing a network which precisely describes the building egress system. Building floor plans are used as the basis for constructing the network which in essence is a collection of nodes and links. Nodes represent the building spaces and the links represent the paths to travel between paired nodes.

A variable time simulation is incorporated to model evacuation based on the network of the building. The simulation model provides for:

1. Different types of resident disabilities and staff capabilities to be considered
2. Number of occupants to be varied at any specific location

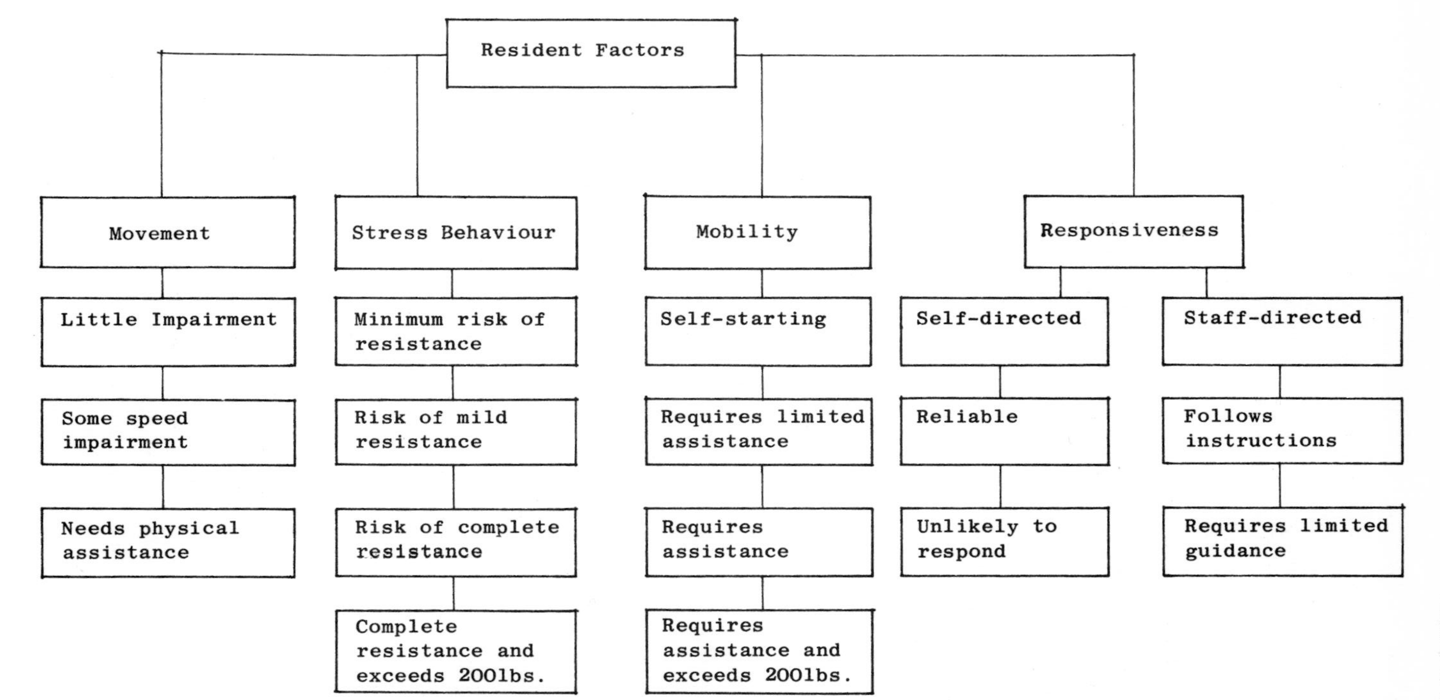

Fig. 12.33 Resident factors considered in modelling emergency evacuation from group homes

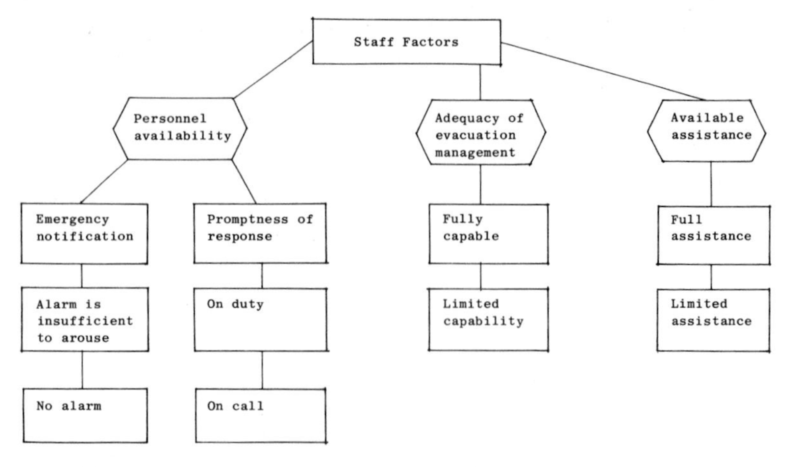

Fig. 12.34 Start factors considered in modelling emergency evacuation from group homes

3. Alternative rescue policies to be considered
4. Alternative egress policies to be considered.

The model in effect is a useful aid to management when deciding upon rescue and egress policies for a particular building complex.

Stahl[30] outlines the B FIRES computer program which was designed to simulate emergency egress behaviour of building occupants during fires. This computer program simulates the perceptual and behavioural responses of building occupants involved in fire emergencies. The structure of the program is such that a wide variety of emergency scenarios can be simulated by treating human behaviour at a very fundamental level. The basic unit of occupant behaviour generated by B FIRES is the 'individual momentary response' to the environment at a discrete point in time t. A building fire event is considered as a chain of discrete 'time frames' $(t_1, t_2, \ldots, t_n)$ and for each such frame a behavioural response for every occupant in this simulated building is generated.

A complete 'picture' of the building fire event is obtained by replaying each time frame sequentially, i.e. the simultaneous egress performance of all occupants in response to an increasing fire threat.

The responsiveness of B FIRES is based upon an information processing explanation of human behaviour which suggests that building occupants act in accordance with their perceptions of a constantly changing environment. The environment undergoes change between two time frames, t_i and t_{i+1}, i.e. people have changed their respective locations, smoke has spread into different spaces and other physical changes to the building may have occurred. The behavioural response of a simulated occupant at the time t_i is determined by information gathered by the occupant which describes the state of the environment at the time t_i. The occupant then interprets this information within the contextural framework of emergency egress objectives which influence the individual's overall behaviour. The interpretation of the information received at time t_i is accomplished by comparing current and previous distances between the occupant, the fire threat and the exit goal and by comparing knowledge about the threat and exit locations possessed by the occupant with amounts possessed by other simulated individuals nearby. The simulated occupant then is required to evaluate alternative responses and select an action as the response for time t_i.

A model to enable the time available for occupants of a building to move to a place of safety in the event of an unwanted fire has been described by Cooper and Stroup.[31] In this model t_{HAZ} represents the time when hazardous conditions start to prevail, i.e. the occupants of a building would be safe if they can egress from a threatened space before t_{HAZ} occurs. By similar reckoning, t_{DET}, the time to detection of the fire, represents the earliest time by which the occupants could reasonably be expected to initiate egress activity.

The Available Safe Egress Time (ASET) is defined simply as:

$$\mathrm{ASET} = t_{\mathrm{HAZ}} - t_{\mathrm{DET}}$$

However, in this model, if the building design is to be considered safe, the ASET from each threatened space within the building must be longer than the actual time required for people successfully to evacuate these spaces. Thus the Designed Safe Egress concept states that a building is safe if:

$$\mathrm{ASET} > \mathrm{RSET}$$

for all threatened spaces, where RSET is the Required Safe Egress Time.

The computation of ASET requires:

1. Identification of the burning characteristics of the combustible contents
2. Physical description of the building spaces
3. Enclosure fire model to simulate analytically the dynamic environment which evolves in each building space
4. Identification of criteria for fire detection and onset of hazard
5. Application of (3) and (4), estimation of t_{DET} and t_{HAZ} and computation of ASET where $\mathrm{ASET} = t_{\mathrm{HAZ}} - t_{\mathrm{DET}}$.

A computer program to determine optimal building evacuation plans, called EVACENT +, is described by Kisko and Francis.[32] This model, like others, requires the development of a network for the building, the network consisting of nodes and arcs. The nodes represent the building spaces, such as rooms and circulation spaces,

Table 12.9 Components of an escape route system

Component		Time				
		T_p	T_a	T_{rs}	T_s	T_f
Detection system:	Gas	1				1
	Smoke	2				2
	Heat	3				3
	Human	1	2			4
Alarm system:	Audible	1	2			
	Visual	1	2			
	Tactile	2	3			
	Olfactory	2	3			
People factors:	Sensory:					
	Hearing	1	2	3	3	
	Vision	1	1	2	2	
	Touch	4	4	4	3	
	Smell	2				
	Psychological function	1	2			4
	Physical function			1	1	
	Physiological function		2	2	2	
	Social habits	2	2	2	3	2
	Number		3	1	1	4
	Type of group		3	1	1	
	Distribution of people	4	4	1	2	
	Density			1	1	
	Mobility			1	1	
	Type of escape possible			1	1	
	Training	3	2	1	1	2
	Management		2	2	2	2
	Time of day	2	2	3	3	2
	Season, year			3	3	4
Building factors:	Size of room		3	2		3
	Shape of room	3	3	2		
	Smoke load	2	2	2	3	1
	Fire load	2	2	2	3	1
	Sources of ignition	2	2	2	3	1
	Exit location		3	1		
	Exit size			1		
	Escape route size			2	1	
	Direction of finding				1	
	Stair geometry				1	
	Lifts				2	
	Refuges				2	
	Maintenance			3	2	2
	Secondary power supply			3	2	

Table 12.9—*continued*

Component		Time				
		T_p	T_a	T_{rs}	T_s	T_f
Fire-Control Systems:	Fire resistance			3	2	3
	Active suppression			2	3	2
	Smoke control			3	2	1
	Emergency lighting			3	2	
	Communication systems	3	2	2	3	
Fire Brigade:	Attendance time			3	2	3
	Access to building			2	1	
	Care of escapers			3		
	Fire fighting					2

and the arcs represent the connections between the building spaces. For each node it is necessary to specify the number of people in each building space, and for each arc a traversal time must be specified. The computer program calculates the optimum evacuation time and allows several patterns to be identified.

12.6 CONCLUSIONS

Many factors must be taken into consideration in the design of escape routes from buildings. Much of the empirical data in use assume a homogeneous population. In the past such an assumption would have been reasonable in so far as the requirements of building legislation could be met. With the introduction of statutory requirements for access of disabled persons to buildings, the human variable in the equation assumes a new dimension. Access to buildings is not the same as egress in an emergency, and the latter must be carefully considered at the design stage.[28] It is also opportune at this juncture to introduce a further component of escape route provision, i.e. management of premises in use. For example, given the provision of access for disabled persons to public assembly buildings, the means of egress in an emergency, the layout or plan of the premises, and the provision for movement of people in and between different volumes must be considered against the management backcloth which will control the premises in use.

Escape route design is much more difficult than has been

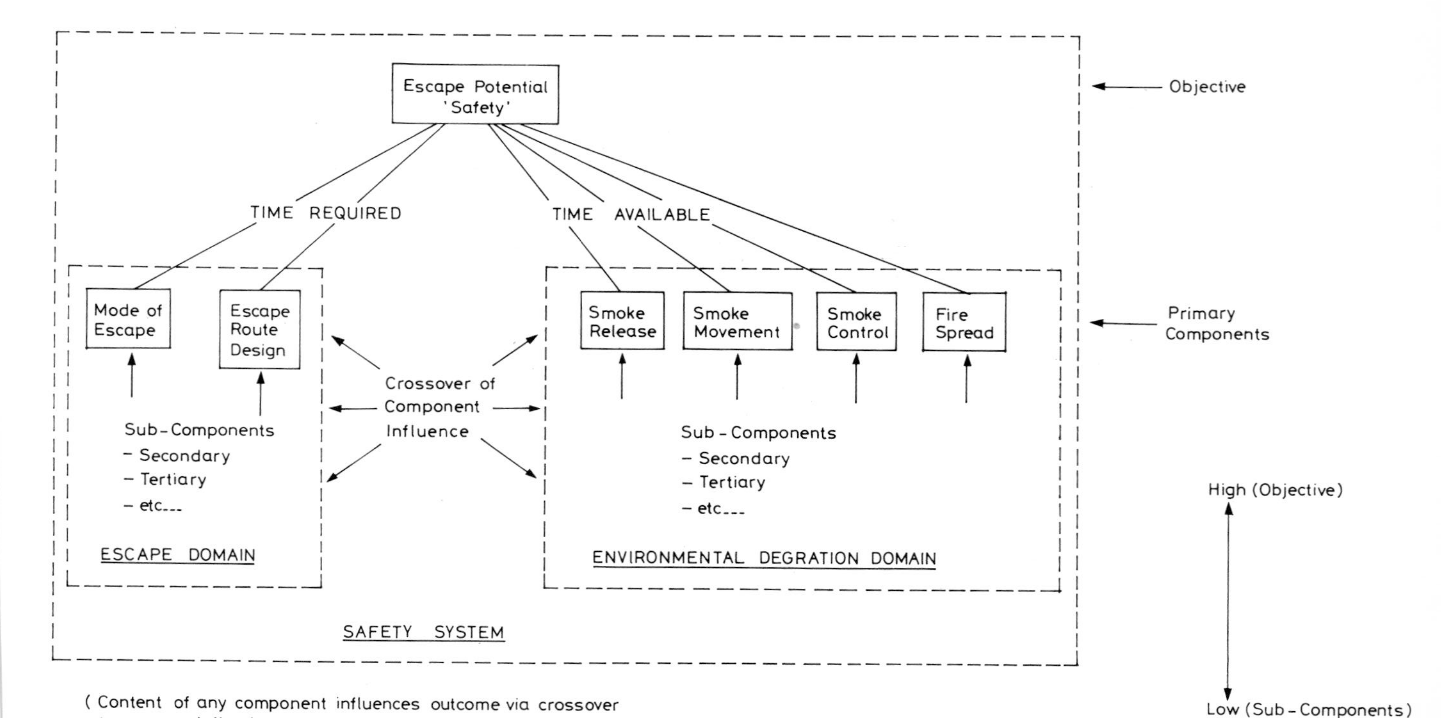

Fig. 12.35 Primary components that determine escape potential

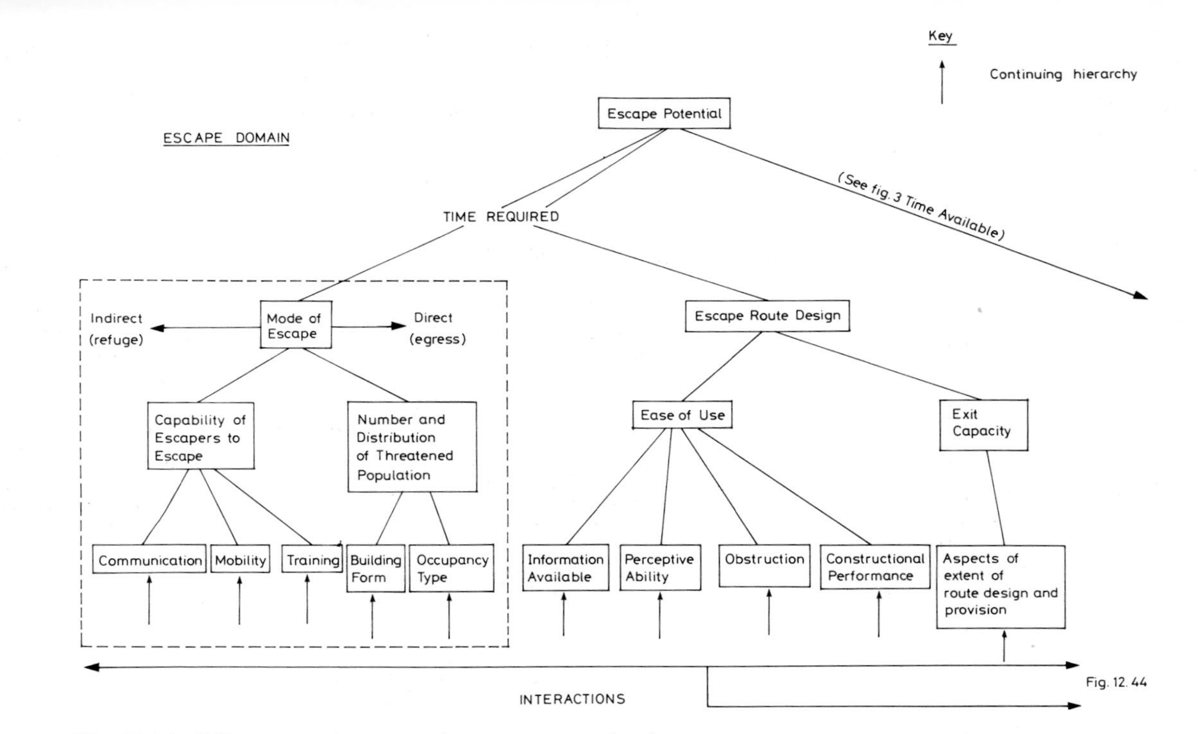

Fig. 12.36 T(Req) overview to tertiary component level

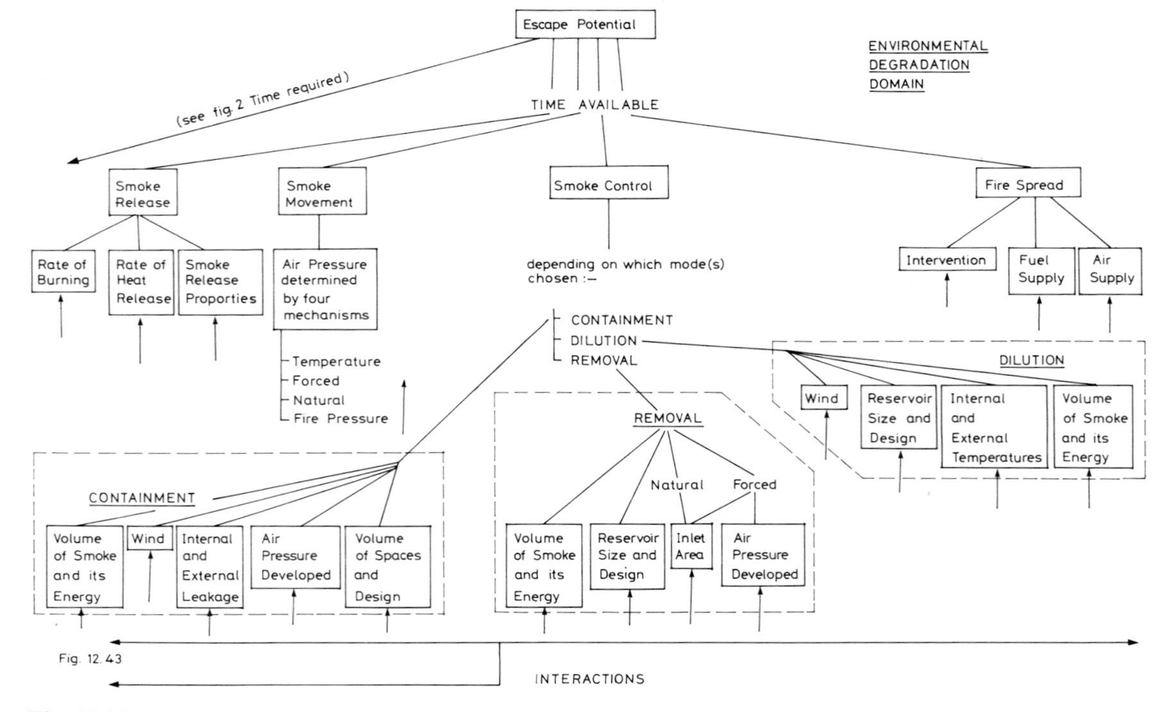

Fig. 12.37 T(AVAIL) overview to tertiary component level

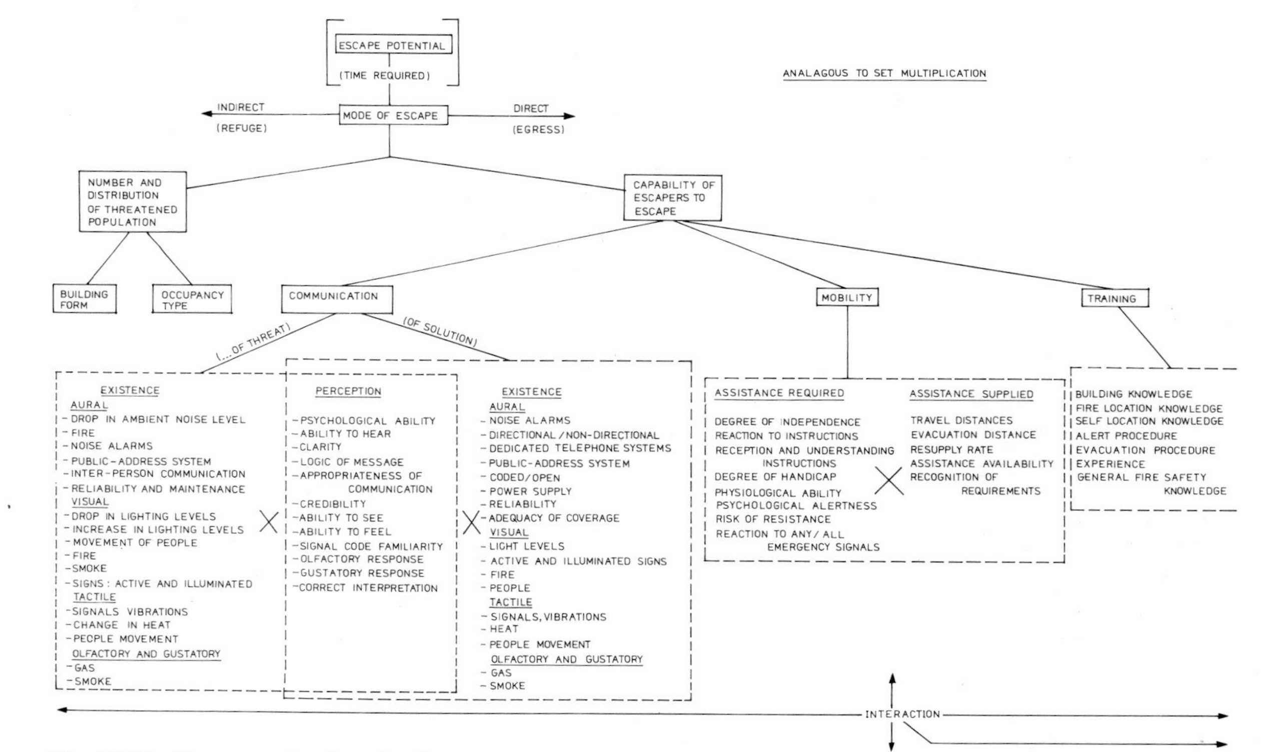

Fig. 12.38 Components of mode of escape

illustrated in this Chapter, requiring a deeper understanding of:

1. Fire science and technology
2. Behaviour of people in an emergency
3. Building design, and
4. Interactions related to (1), (2) and (3) above.

Marchant[2] has identified some fifty one components of escape route design, Table 12.9,

where:

T_p is the elapsed time from ignition to the beginning of safety action
T_a is the time from perception to the beginning of safety action
T_{rs} is the elapsed time from initiation of safety action to reaching a place of relative safety
T_s is the elapsed time from reaching a place of relative safety to reaching a refuge or open air
T_f is the elapsed time from ignition for fire to develop untenable environmental conditions.

The relative importance of each component is ranked between 1 and 5, 1 being the strongest influence.

The relationship between the components to escape route design is given by:

$$\frac{T_p + T_a + T_{rs}}{T_f} < 1$$

If the concept of equivalency is accepted the possible interactions provide a very flexible approach to effective escape route design.

Hinks,[33] Figs 12.35, 12.36, 12.37, 12.38 illustrates the many factors and complex interactions which must be considered in any attempt to determine 'escape potential' from a building. He concludes that assessment of escape potential in buildings requires a full complement of factors, and a correct model to evaluate the level of importance, interactions and relative values, including changes in relative values of components.

At present it is not possible to assess the escape potential of buildings, and while research continues towards this long-term objective, the empirical data available plus an enlightened attitude to escape route design and management should prove adequate.

REFERENCES

1. Galbraith M, '*Time of evacuation by stairs in high buildings*', Fire Research Note No. 8, National Research Council of Canada, Division of Building Research, 1969.

2. Marchant E W, 'Modelling fire safety and risk', *Fires and Human Behaviour*, Canter, D (Ed.), John Wiley & Sons, 1980.
3. Berlin G W, 'The use of directed escape routes for assessing escape potential', Vol. 14, No. 2, pp. 126–35, *Fire Technology*, 1978.
4. Building Regulations (England and Wales), HMSO, 1976.
5. *United Kingdom Fire Statistics*, Home Office, London.
6. Canter D, 'Fires and human behaviour – an introduction', *Fires and Human Behaviour*, D Canter (Ed.), John Wiley and Sons, 1980.
7. Best P L, '*Reconstruction of a tragedy: the Beverley Hills Supper Club fire, Southgate, Kentucky, May 28th*', Boston National Fire Protection Association, 1977.
8. Wood P G, '*The behaviour of people in fires*', Fire Research Note, 953, British Research Establishment, Fire Research Station, Borehamwood, UK, 1972.
9. 'Fire safety in high rise buildings', Papers presented at the *International Fire Protection Conference*, Amsterdam, April/May 1975. Fire Prevention Association.
10. Pauls J L, '*Building evacuation*'. Research Findings and Recommendations.
11. BS 476: Part 8: 1972: Test methods and criteria for the fire resistance of elements of building construction, British Standards Institution, London.
12. BS 476: Part 1: 1953: Test methods and criteria for the fire resistance of elements of building construction, British Standards Institution, London.
13. BS 5266: Part 1: 1975: Code of practice for the emergency lighting of premises, British Standards Institution, London.
14. CP 1007: 1955: Maintained lighting for cinemas, British Standards Institution, London.
15. *United Kingdom Fire Statistics*, 1980, Home Office, London.
16. BS 5588: Part 3: 1983: Fire precautions in the design and construction of buildings, British Standards Institution, London.
17. Bukauski and Zimmerman, 'Household Warning Systems', *National Fire Prevention Handbook*, Fifteenth Edition, NFPA, 1981.
18 NFPA: 74, *Standard for the Installation, Maintenance and Use of Household Fire Warning Equipment.*
19. British Standard Code of Practice, CP 3: Part 1: 1971: Code of basic data for the design of buildings: Chapter IV, Precautions Against Fire, British Standards Institution, London.
20. Bryan J L, 'Concepts of Egress Design', *National Fire Protection Handbook*, Fifteenth Edition, NFPA, 1981.
21. Van Bogaert A F, *Prospective Dans La Construction Scolave*', pp. 100–109, Vander, Louvain, Belgique, 1976.
22. Galbreath M, '*Time of vacation by stairs in high-rise buildings*', Fire Research Note. No. 8, National Research Council of Canada, Division of Building Research, Ottawa, 1969.
23. Freun J J, '*Pedestrian planning and design*', Metropolitan Association of Urban Designers and Environmental Planners Inc., New York, 1971.
24. '*Second report of the operational research team on the capacity of footways*', London Transport Board Research Report. No. 95, London, 1958.

25. Togawa K, '*Study of fire escapes based on the observation of multitude currents*', Report No. 14, Japanese Building Research Institute, Tokyo, 1955.
26. Melinek S J and Booth S, '*An analysis of evacuation times and the movement of crowds in buildings*', British Research Establishment Current Paper, CP96/75, BRE, 1975.
27. Butcher E G and Parnell A C, '*Designing for Fire Safety*', John Wiley & Sons, 1983.
28. Shields T J and Silcock G W H, 'Access to buildings for the disabled – is it enough?', *Journal of the Institution of Building Control Officers*, 1984.
29. Berlin G N, Dutt A and Gupta S M, 'Modelling emergency evacuation from Group Homes', *Fire Technology*, pp. 38–48, Feb., 1982.
30. Stahl F I, 'B Fires – 11', A behaviour-based computer simulation of emergency egress during fires', *Fire Technology*, pp. 49–65, Feb., 1982.
31. Cooper L Y and Stroup D W, 'ASET – a computer program for calculating available safe egress time', *Fire Safety Journal*, **9**(1), 20–46, May 1985.
32. Kisko T M and Francis R L, 'A computer program to determine optimal building evacuation plans', *Fire Safety Journal*, **9**, 211–31 (2), July 1985.
33. Hinks A J, 'Towards the development of a heuristic structure for escape potential', *Fire* (The Journal of the Fire Protection Profession), Vol. 77, No. 984, pp. 21–4, October 1985.

CHAPTER 13

Fire safety evaluation

13.1 DEVELOPMENT OF FIRE SAFETY MODELS

The concept of Building Fire Safety Evaulation has been around for quite some time, but the development of useful, reliable fire safety evaluation techniques is proving elusive and is still very much in the embryonic state.

Since the twelfth century in this country, reliance for fire safety has been enshrined in prescriptive methods of building control. The early ordinances[1] were consolidated following disastrous conflagrations, eventually taking the form of current building regulations and codes of practice. Because of the *ad hoc* development of building legislation, many buildings may be over protected or under protected with respect to fire safety.

To determine the level of safety required demands, as a prerequisite, a rational approach to decision making which in turn requires, as a minimum, a model or method of analysis to define and structure the problem under consideration. It would appear that some systems rely on a simple but naïve accounting technique. The costs are entered in one ledger and the benefits in another. Thus by the application of rudimentary cost–benefit analysis, optimum solutions are distilled. Idealistically, cost–benefit analysis offers a precise aggregated sum of respective needs. But even a simple fire scenario is not unidimensional; rather it is dependent upon a more complex structuring of influence and interactions.

The most readily recognisable model is that of 'ends-versus-means' in which an objective is targeted and the most cost-effective means for achieving it is then sought, with the final determination being subject to internal and external constraints. This approach, i.e. single objective model, only makes sense where a single objective dominates over a range of criteria.

A fire safety system contains more than one objective and although compatible within the totality of corporate policy, there may be occasions where individual tactics employed to attain objectives may not be complementary. Consequently, with multiple objectives to consider, an objective matrix is required which will facilitate considerations of trade-off and redundancy between the various objectives. Analysis of the fire safety system will reveal complex interactions between the constituents of the system, e.g. between policy (corporate strategy), objectives, tactics and components, indicating the development of several matrices in order to arrive at some meaningful conclusion.

Simply stated, if the objectives and constraints upon the system have been unambiguously and adequately specified, the choice of tactics for the attainment of the objectives may be treated as a technical matter. But even technical matters involve, to some degree, a choice which may be influenced by considerations external to the system. So even in the most idealistic setting, the model is described according to the analyst's judgement and cannot represent the complexity of reality. It does, however, render the problem more manageable by identifying those factors judged relevant to the decision-making process and enables these to be categorised:

1. Objectives
2. Constraints
3. Tactics.

For the individual analyst the problem is one of perception, i.e. his perception of reality. He makes judgements which guide his view on how best to proceed, but these judgements will vary with the evolution of ideas, experience, exposure and position of the analyst. Through the development of ideas, issues may change character, as when the number of fires occurring in buildings ceases to be viewed as a case for increasing the fire resistance of structural and constructional elements and becomes viewed as a case for controlling occupant density or fire-load density, providing additional means of escape or providing automatic detection, alarm and suppression systems.

Even a cursory examination of the prescriptive legislation currently in existence reveals a lack of expressly-stated policy and objectivity. Consequently it is extremely difficult to attempt to quantify or evaluate fire safety in buildings.

At present there is no existing universally-accepted methodology for the evaluation, analysis or design of fire safety in buildings. Rasbash[2] outlined twenty steps necessary to the attainment of a balanced set of fire precautions. These steps are:

Step	Criteria
1. Define the hazard	hazard area occupancy risk
2. Define the objectives	optimum balance of cost of precautions and residual risk reduce injury, minimise damage
3. Identify and quantify materials that burn	cellulosics, non-cellulosics, distribution
4. Identify and quantify sources of ignition	ignition sources
5. Identify and quantify conditions for fire spread	fuel package, interspace and interspatial relationships
6. Identify and quantify agents that cause fire	failure: human, mechanical, electrical
7. Estimate the probability of a fire being caused	analysis: event trees, decision trees, statistical method, stochastic modelling
8. Survey available means of controlling fire	detection, control, extinction
9. Estimate the course of fire behaviour	systems analysis
10. Identify harmful agents produced by fires	heat, pressure, smoke, toxic gases, corrosive agents
11. Estimate the production of harmful agents by fire	temporal–quantity relationships
12. Survey methods of protection against harmful agents	isolation, segregation, compartmentation, escape route design, smoke control
13. Estimate direct hurt and damage	fault-tree analysis
14. Identify possible harm to processes at risk	key result area analysis
15. Survey methods to protect processes at risk	Step 12 above, contingency planning
16. Estimate total expectation of loss and harm by fire	probabilistic basis of expectation of loss and damage
17. Postulate changes in fire hazard situation	review through step by step procedure
18. Estimate the effects of changes	calculation of cost-effectiveness
19. Define acceptable methods of achieving objectives	formulate methods to attain objectives
20. Formulate and express requirements	increase range of options open to designers

Figure 13.1 shows the interactions within the twenty-step process.

It is however very clear that the development of any system entails, at some stage, preferences, the concept of preferences being linked inseparately to the logical conditions of rational choice. The important factor which pervades system analysis is that the analyst must connect the identified alternatives with each other; they must be ranked transitively (e.g. if A is preferred to B and B to C then it must follow that A is preferred to C), and any state which is close to one alternative (say A) must be regarded as standing in the same relation as does A to the other alternatives.

The process of analysis may be separated into a sequence of discrete steps or phases. For example:

1. 'Identify the problem and understand it'.
   Unless the problem is clearly defined it is impossible to proceed in a rational fashion any further in the sequence of analysis.
2. 'Define and clarify the objectives'.
   Given an understanding of the problem, the next step is to determine what eventual outcomes are most highly desired.
3. 'Consider alternatives for the attainment of the objectives'.
   Once agreement on the desired objectives has been achieved, various alternative courses of action must be identified or devised which will have some probability of leading to the attainment of these objectives.
4. Analyse the anticipated consequences of each alternative'.
   After listing potential courses of action each alternative must then be evaluated critically in terms of its end results and their desirability.
5. 'Select a course of action'.
   Select the most desirable alternative and implement the programme. For Step 5 to be truly effective it must be preceded by the previous four analytical stages.
6. 'Monitor progress towards the desired objectives'.
   It is an essential ingredient of any system that a review component be included to monitor actual progress against planned progress.

Figure 13.2 illustrates diagrammatically the planning process for model development.

Marchant[3] identified three approaches to fire safety evaluation:

(a) the qualitative approach
(b) the quantitative approach
(c) the rationalised systematic approach.

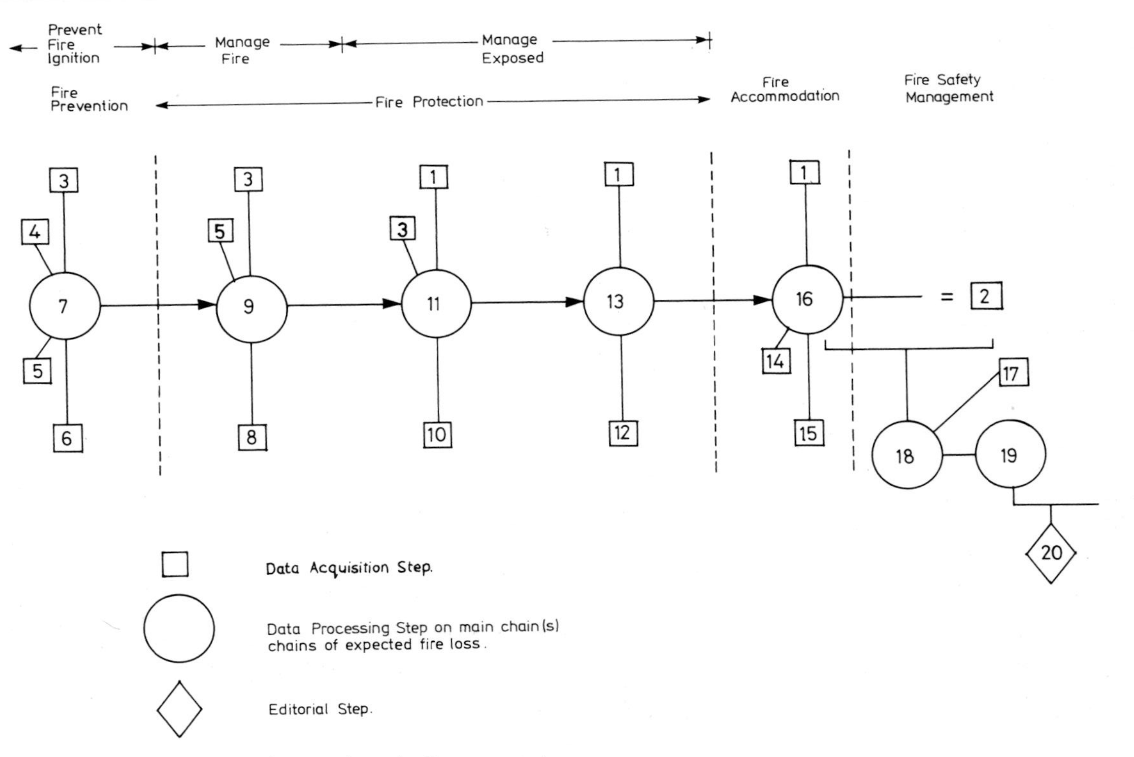

Fig. 13.1 Component interactions in fire precautions

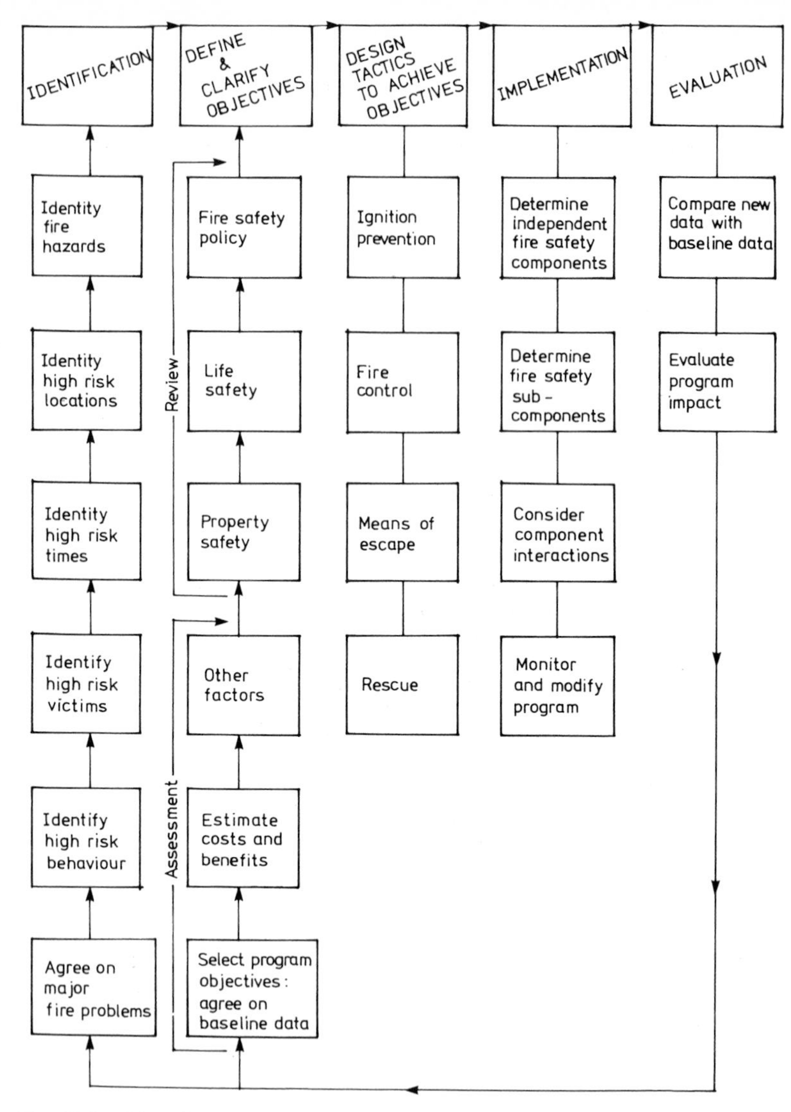

Fig. 13.2 Planning process for model development

13.2 THE QUALITATIVE APPROACH

Difficulty in measuring the physical and environmental quality of building units and the inclusive built environment remains the most niggling problem in evaluating the various attributes of our town and city developments.

Measuring the performance of buildings, people and processes in fire safety engineering terms is just as difficult and demands as a prerequisite an analysis of risk and the determination of an acceptable level of risk. Risk perceptions will depend upon the types of risk, scope of risk, the effect of risk and will be influenced by the following considerations:

(a) technological
(b) economic
(c) legal
(d) managerial
(e) political
(f) social
(g) cultural
(h) environmental.

The qualitative approach is quite simply based upon expert/professional judgement. Much of the current building legislation related to fire safety in this country is based upon a corporate qualitative assessment of risk and the measures necessary to mitigate such risk. This approach would be more acceptable if only individual buildings were considered. However, the methodology employed in the building regulations, for example, is to categorise building by occupancy type and then apply specific provisions within each occupancy classification. The underpinning assumption is that occupancy classification defines the totality of the building function, ignoring the reality of complex multi-function, and hence multi-occupancy, buildings.

Another sweeping assumption is of a static rather than dynamic nature of occupancies within the same classification. Simply stated the traditional approach, as incorporated into building legislation, assumes that all the factors which affect fire development are exactly represented in every building within an occupancy classification and that these factors do not change during the life of the building. Clearly this cannot be so and reality is not adequately represented. Hence the growing awareness that such an approach may leave some buildings over-protected and others under-protected.

13.3 THE QUANTITATIVE APPROACH

The quantitative approach to fire safety engineering requires:

(a) an analysis of risk
(b) the determination of an acceptable level of risk
(c) adoption of measures to ensure that the prescribed levels of risk are not exceeded.

Thus it would appear that a deterministic methodology should be readily available. However, the deterministic approach presumes an ability to determine the precise behaviour of any fire at any time in the future, given complete knowledge of the building, the contents and the people. Thus to develop a deterministic building-fire-safety model requires full and common knowledge to produce:

(a) a building-fire-dynamics model
(b) a smoke and toxic gas production and movement model
(c) a people behaviour model

and the integration of all three to produce a total building-fire-safety model.[4] Deterministic modelling therefore requires complete physiochemical and thermodynamic knowledge of the fire system including people behaviour and building geometry. The deterministic approach is the most idealistic approach and while remaining a laudable objective in the development of fire safety engineering, is beyond immediate capabilities.

Other less precise quantitative systems have been produced based on:

(a) event trees
(b) decision trees
(c) stochastic modelling.

13.3.1 Event trees

Event trees are logic diagrams which illustrate in sequential order the factors involved in ignition, fire development, fire growth and fire control. Given an initial critical event or final event, probabilities can be determined for intermediate events which lead to a calculation of the probability of the total fire system. If fire safety objectives can be quantitatively expressed then it is usually necessary to create a model utilising probability theory to estimate the likelihood of attaining the stated objectives.

Event trees may be in the form of:

(i) fault trees, or
(ii) success trees.

Fault trees

Fault trees may be defined as logic diagrams which illustrate the various ways in which a system can fail, given a final event. Figure 13.3 shows the fault-tree analysis for a fire at a Horton Sphere, Fig. 13.4, following a small leak.[5]

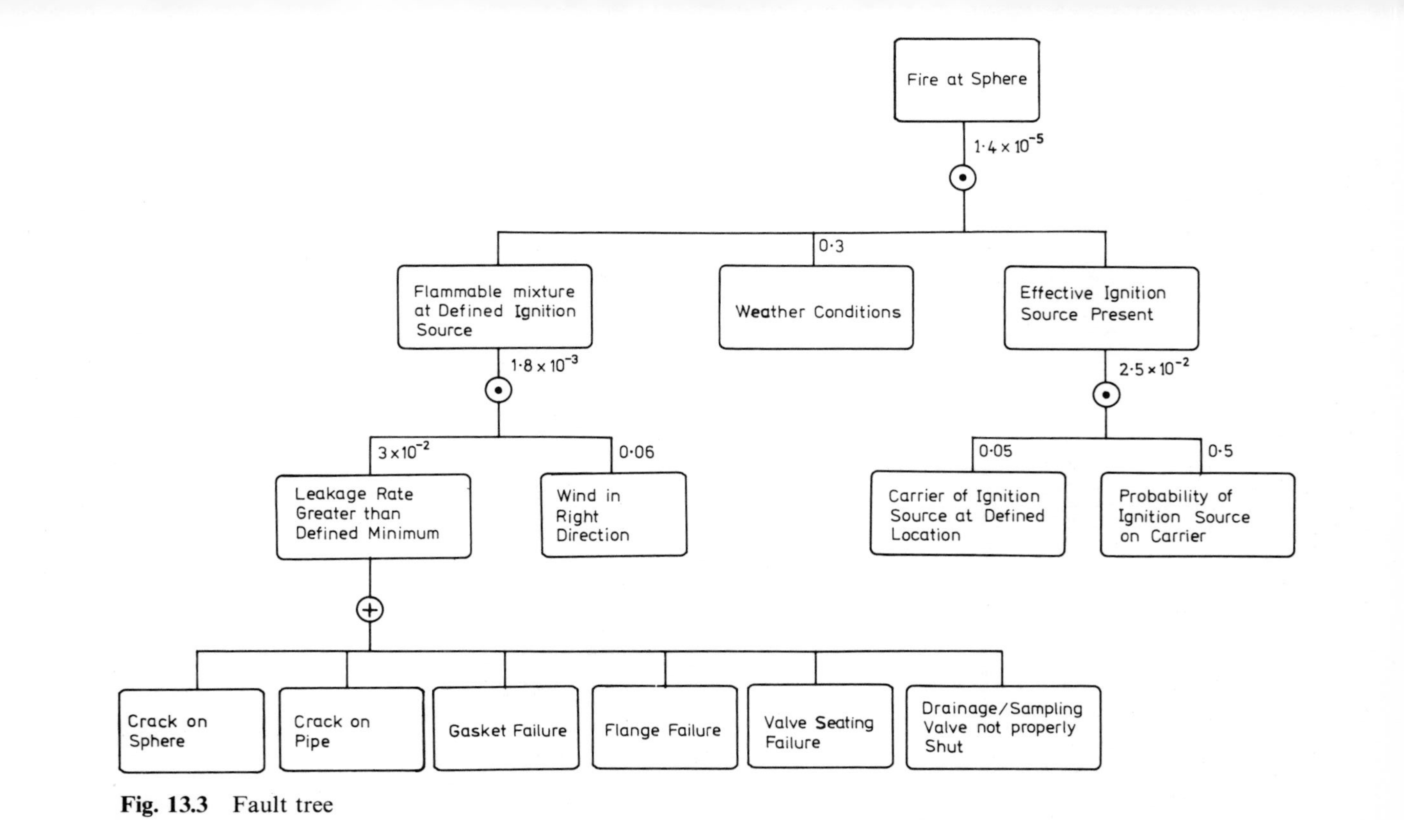

Fig. 13.3 Fault tree

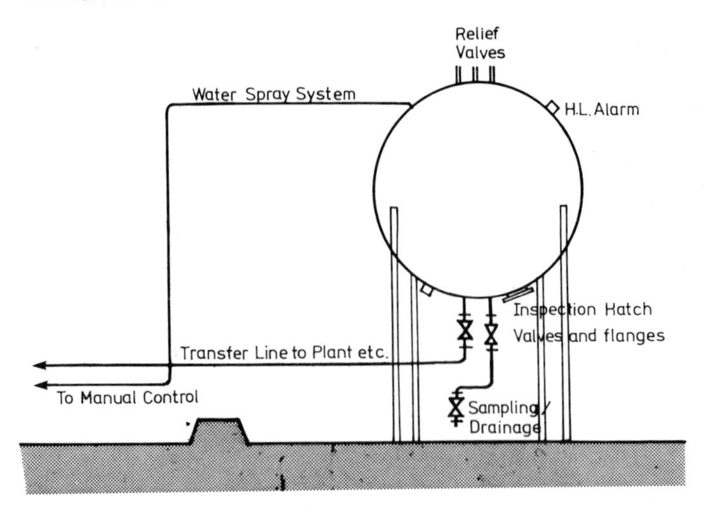

Fig. 13.4 Layout of Horton Sphere

The construction of the fault tree requires the specification of an unwanted event, known as the top event, and the objective of the analysis procedure is to determine the various ways in which the top event can occur. Given the final unwanted event as a starting point, it is necessary to work backwards through the tree to intermediate events and finally to a set of basic events. The events in the tree are connected at node points termed logic gates which show what combination of particular events could cause the next particular top event. These logic gates are AND gates, i.e. all the component events must be presented and OR gates, i.e. any one of the component events need only be present to cause the unwanted specified event.

AND gate $\odot$

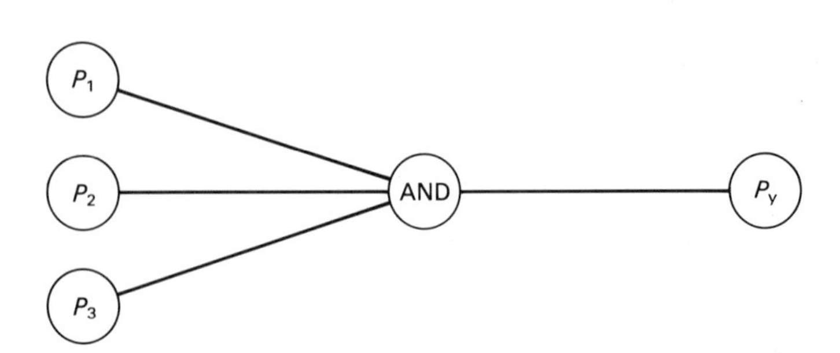

P_1, P_2 and P_3 represent respective probabilities. The events are independent but not mutually exclusive.

Then $P_y = P_1 \cdot P_2 \cdot P_3$

OR gate $\oplus$

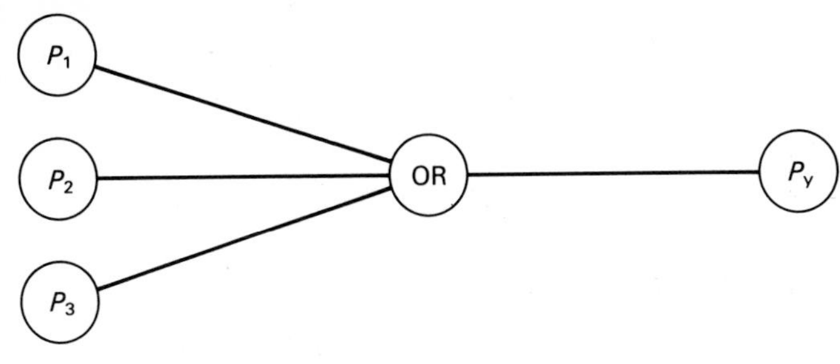

$P_y = 1 - \bar{P}_1 \cdot \bar{P}_2 \cdot \bar{P}_3$

where $\bar{P}$ = complement of P, i.e. $\bar{P}_1$ is the probability of not being in condition (1). Because the probabilities at OR gates are often very small

$P_y \simeq P_1 + P_2 + P_3 \ldots.$

Success trees

Success trees may be described as logic diagrams used to illustrate the alternative ways in which a specified objective may be attained. In fact the success tree is the inverse of the fault tree, i.e. the former requires the achievement of a top event whilst the latter requires the avoidance of a top event. A variation of the success tree is the fire safety concepts tree[6] which adopts a systems approach to fire safety, the elements being concepts rather than events and thus they are qualitative rather than quantitative methods. Figure 13.5 illustrates the conceptual approach mentioned above.

13.3.2 Decision trees

These can be described simply as logic diagrams used to represent the outcome of decisions taken at various progressive levels. Decision trees are a management tool used to assess the consequences of decisions with reference to a particular problem. The approach involves linking a number of event branches which, when fully developed, resemble a tree. The process commences with a primary decision that has at least two alternatives to be evaluated. The probability of each outcome must be ascertained as well as its monetary value.

Figure 13.6 illustrates the decision-making process in flow chart form. As an illustration, consider an organisation with a large stock of buildings. Assume, at the beginning of the financial year,

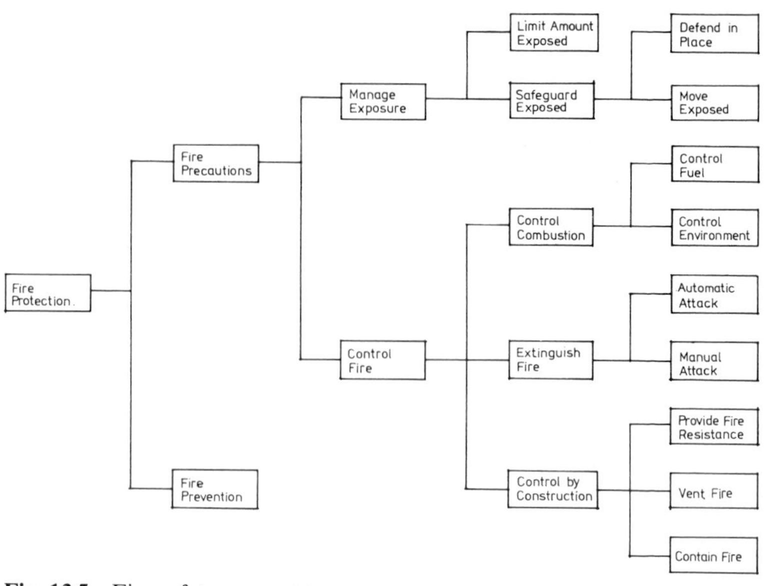

Fig. 13.5 Fire safety concept tree

management has two alternatives – using the existing stock without modification or upgrading the existing building stock to an acceptable level of fire safety. Obviously increasing the levels of fire protection requires some additional investment. In the event of fire the net cost of the fire, i.e. the value of potential losses less the cost of additional protection, are illustrated in Fig. 13.7. Multiplying the net costs by the respective probabilities and summing them gives a net potential cost of upgrading of £1,500 and £19,000 for not upgrading. Figure 13.7 could be extended over a time range in years to illustrate the impact of decisions over a prescribed period. Sphilbert[7] developed a decision tree, Fig. 13.8, which incorporated the following considerations:

(a) no sprinklers
(b) sprinklers incorporated
(c) deductibles in insurance.

It should be noted that the decision-making process involves recognition of a problem, identification of alternative courses of action, evaluation of potential outcomes, and finally a choice. Information is the raw material for the decision-making process and consequently any decision tree is only as good as the information which spawned it.

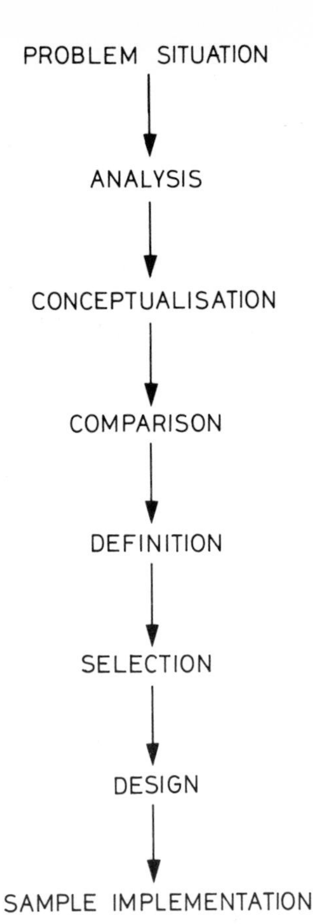

Fig. 13.6 Flow chart for decision making

13.3.3 Stochastic model approach

Unlike the deterministic process, this type of model develops in time and space in accordance with the laws of probability. There are many stochastic processes within the various branches of science and technology.[8] The probabilities of each of the stages of the development of a stochastic process are independent of each other.

To give the reader a simple example of such a process, consider the 'Random Walk' process which is demonstrated graphically in Fig. 13.9, where the position of a particle X_τ is shown at various times τ.

The relationship $X_\tau = X_{\tau-1} + Z_\tau$ relates the position at time τ to that at $\tau - 1$ where 1 is the time step.

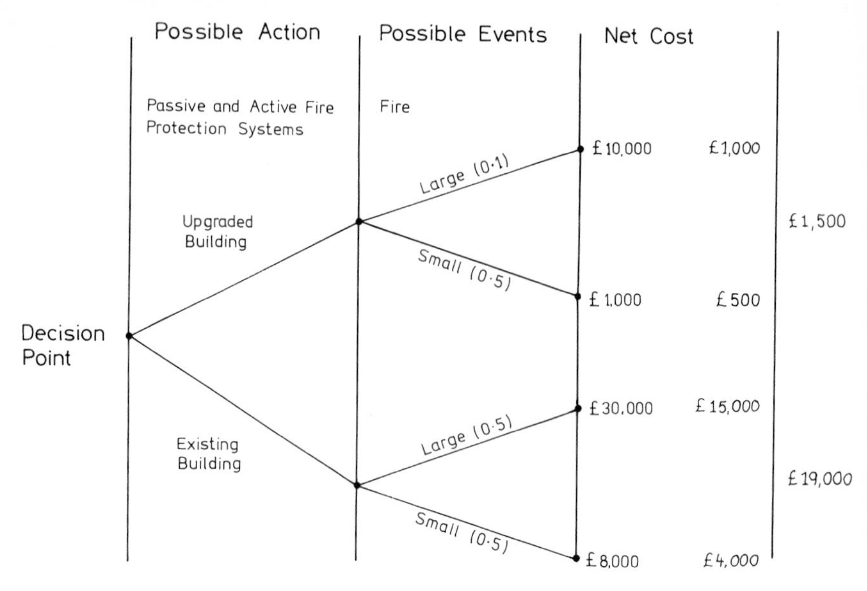

Fig. 13.7 Branching decision tree model for fire safety

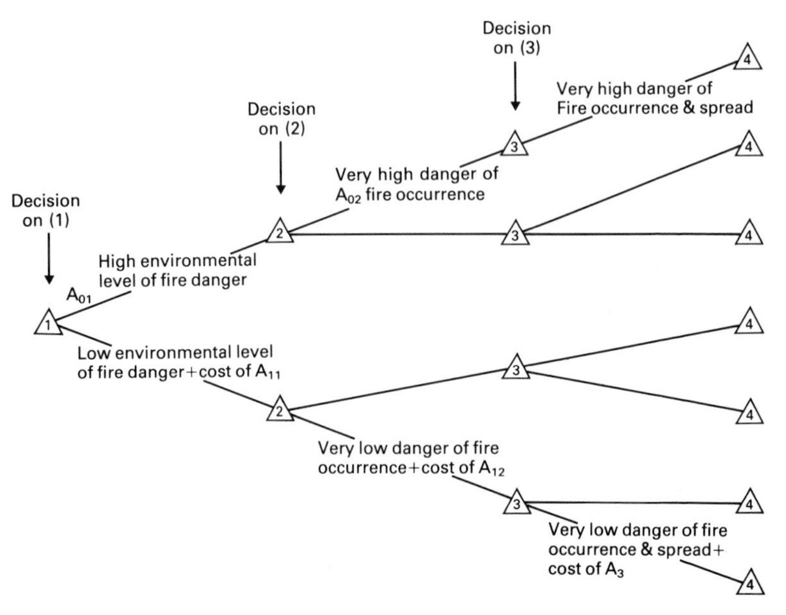

Fig. 13.8 Branching decision tree model for fire safety

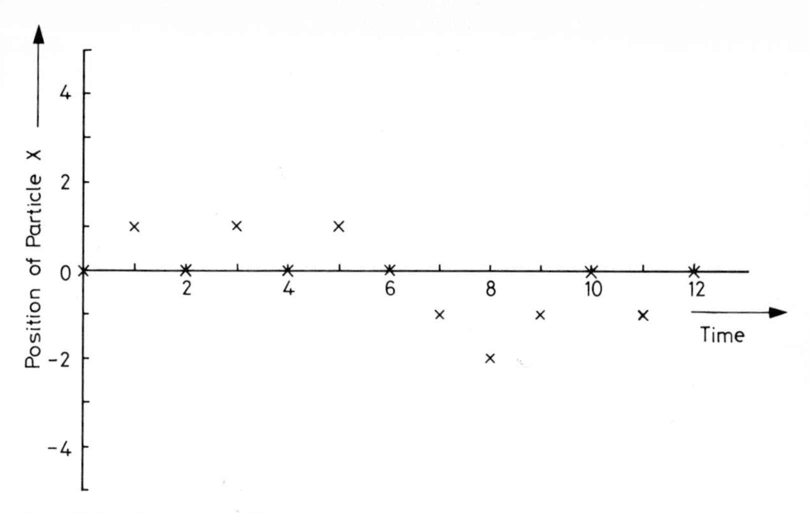

Fig. 13.9 Random walk process

Also $X_\tau = \sum_{i=1}^{\tau} Z_i$

Hence the location of particle at time τ is the sum of the separate steps

$Z_\tau = \pm 1$ or 0

taken up to time τ.

It is usual to assign probabilities for these steps,

i.e. $\text{prob}\,(Z_\tau = +1) \quad = \rho$

$\quad\quad prob\,(Z_\tau = -1) \quad = \alpha$

so $\quad \text{prob}\,(Z_\tau = 0) \quad = 1 - \rho - \alpha$

13.3.4 Poisson processes

These are processes that range from random radiation decay through to the spread of fire within a building.

Consider the case of fire spread within a building where the events occur singly and at random with the passage of time. It is possible to express the probability of an event occurring to move from one state to another in the time interval

i.e. $\text{prob}\,(P(t, t + \Delta t) = 1) = \rho \cdot \Delta t$

and the probability of an event not occurring,

i.e. $\text{prob}\,(P(t, t + \Delta t) = 0) = 1 - \rho \cdot \Delta t$

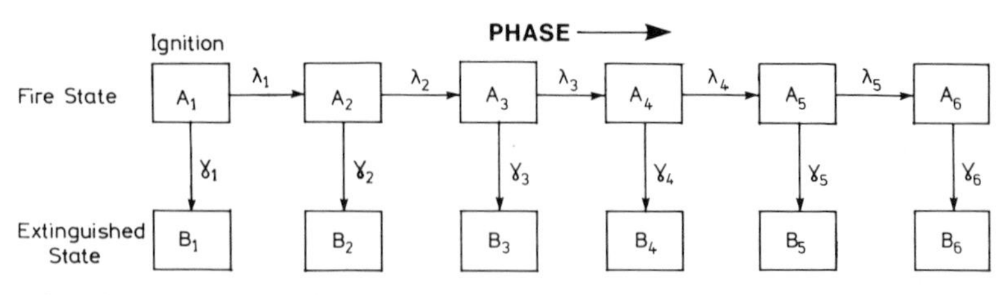

Fig. 13.10 State transition process

where ρ = occurrence rate
and Δt = small time interval.

Applying this to a fire engineering process of state transition, for fire spread shown in schematic terms in Fig. 13.10,

where λ_i = instantaneou rate of transition of a fire moving from state or phase to the next one $A_i \rightarrow A_{i+1}$
and γ_i = instantaneous extinguishment rate for a fire in phase $A_i \rightarrow B_i$

Defining $p_i(t)$ = probability of a fire existing in phase or state A_i at time t where $i = 1 \rightarrow 6$.

It is now possible to express the probability of the various states or phases after a short time interval Δt has elapsed from t.
If Δt is kept small then:

$$p_i(t + \Delta t) = p_i(t)[1 - \gamma_i \cdot \Delta t - \lambda_i \cdot \Delta t] + p_{i-1}(t) \cdot \lambda_{i-1} \cdot \Delta t$$

for $i = 1 \rightarrow 5$

Also
$$p_6 = (t + \Delta t) = p_6(t)[1 - \gamma_6 \cdot \Delta t] + p_5(t)\lambda_5 \cdot \Delta t$$

or in general terms

$$p_i(t + \Delta t) = p_i(t)[1 - a_i \cdot \Delta t] + p_{i-1}(t) \cdot \lambda_{i-1} \cdot \Delta t$$

where $a_i = \gamma_i + \lambda_i$

Rearranging this equation and letting $\Delta t \rightarrow 0$, it can be shown that:

$$p_i(t) = -a_i p_i(t) + \lambda_{i-1} p_{i-1}(t)$$

On solving this differential equation[9] it can be shown that the average arrival time $\bar{t}_n$, that is the time taken for a fire in state or phase i at zero time to reach state or phase n is

$$\bar{t}_n = \sum_{i=1}^{n-1} \frac{1}{a_i} \qquad \ldots [1]$$

Another important time interval is the average extinguishment time T_n which is the time taken for the fire to be extinguished in a particular phase or state n

$$T_n = \sum_{i=1}^{n} \frac{1}{a_i} \qquad \ldots [2]$$

Comparing eqn [1] with eqn [2] it follows that:

$$T_n = \bar{t}_n + 1$$

In the work carried out by Aoki,[9] data were collected from over 300 fires sampled in a random manner from all fire reports available. Estimations were made for both the transition rate λ_i and the extinguishment rates γ_i. This in turn allowed an estimation to be made of both T_n and t_n, Fig. 13.11(a), via eqns [2] and [1], respectively.

These estimations were then compared with evaluation of T_n and $\bar{t}_n$, respectively, from the real data, Fig. 13.11(b). Figures 13.11(a) and 13.11(b) show a good correlation between the estimated arrival times and the calculated arrival times for the house types shown.

It is clear from these graphs that the wooden house construction attains flashover faster than the fire-proofed house. This suggests a higher probability of loss of life after $\bar{t}_4$ from the wooden type construction.

Estimations for λ_i and γ_i are given in Table 13.1 for the two house types, for each phase of the fire.

In conclusion, this type of modelling using information collected from real fires in various building types can yield realistic and valid answers for fire spread problems.

13.4 THE RATIONALISED SYSTEMATIC APPROACH

This approach to fire safety evaluation of the effects of an existing or proposed technological change or innovation, is based upon the three primary steps of systems engineering. These steps involve the synthesis, analysis and interpretation of the impacts of alternative actions upon the levels of fire safety finally achieved. Each of these three steps, synthesis, analysis and interpretation are important in obtaining an understanding of the effects of potentially hazardous events. Each involve different components and different systematic activities.

In fire safety evaluation of buildings, for example, it is necessary to identify the types of risk or hazard likely to be encountered and their extent. A number of systematic approaches such as the nominal group technique, brainwriting or ideawriting, Chavette or Delphi are especially useful. Synthesis includes the identification of risk elements

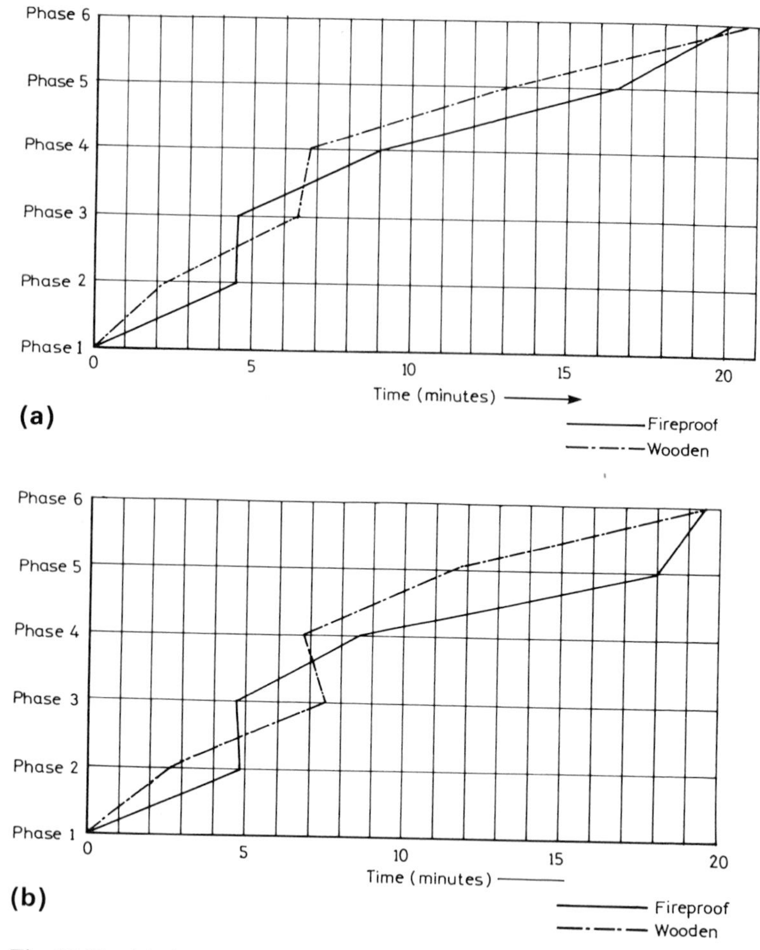

Fig. 13.11 (a), Average arrival time for each type of building (estimation); (b), average arrival time for each type of building (calculated from data directly)

as well as the identification of elements representing variables associated with risk mitigation, with and without technological innovation.

Using a Delphi technique (20), which is essentially a group of experts in the field, it is possible to identify risk factors, constraints, objectives and alternatives. It is also possible to assign values to these in order to obtain comparative levels of fire safety. Marchant[3] reviewed five points schemes.

Table 13.1 Estimation of λ and γ for both house types

PHASE	WOODEN		FIRE PROOF	
	λ_i	γ_i	λ_i	γ_i
1. Ignition	0·407	0·025	0·196	0·023
2. Firespread. No Ceiling Damage	0·184	0·063	49·448	116·203
3. <80% of Floor Area Damage	1·239	0·933	0·138	0·092
4. Flashover	0·156	0·004	0·121	0·011
5. Spread outside Room	0·084	0·005	0·000	0·282
6. Complete Burning of Section		0·246		0·312

13.4.1 Scheme 1: Means of escape from fire[10]

The critical components of escape routes can be quantified using simple algebraic expressions

$$A = \frac{G + H}{100 + B + C + D + E + F}$$

$$N = \frac{A + B + C + D + E + F}{H}$$

$$E = \frac{G}{B + D + D + F + 50}$$

Where:

N = number of persons permitted on each floor
A = number of units of stair width (22-inch units)
B = building construction,
ordinary $B = 4$
fire-resisting $B = 5$

C = protection of vertical openings

open stair	$C = 2$
stairs enclosed but other vertical openings not protected	$C = 4$
stair and other vertical openings protected	$C = 5$

D = automatic sprinklers

provided	$D = 2$
not provided	$D = 1$

E = horizontal exits

none	$E = 2$
one	$E = 3$
two or more	$E = 4$

F = occupancy

low hazard	$F = 3$
medium hazard	$F = 2$
high hazard	$F = 1\frac{1}{2}$

G = gross area (sq. ft.)
H = height of building (number of storeys)

13.4.2 Scheme 2: A fire safety evaluation for health care facilities[11]

In this scheme, risk factors are identified and values assigned, Table 13.2.

Table 13.2 Risk factors and assigned values

Risk factor	Range of values
1. Patient mobility	1.0–4.5
2. Patient density	1.0–2.0
3. Fire zone location	1.1–1.6
4. Ratio of patients to attendants	1.0–4.0
5. Patients average age	1.0–1.2

An increase in risk is reflected in an increase in the assigned factor value.

Similarly, safety factors are identified and values assigned, Table 13.3.

The concept of equivalency allows the risk and safety factors to be balanced using the NFPA Code[12] as the norm.

Table 13.3 Safety variance and range of parameters

Safety parameter	Range of variables
1. Construction type	(−13 to +4)
2. Interior finish (corridors and exits)	(−5 to +3)
3. Interior finish (rooms)	(−3 to +3)
4. Corridor partitions/walls	(−10 to +2)
5. Doors to corridors	(−10 to +2)
6. Zone dimensions	(−6 to +1)
7. Vertical openings	(−4 to +3)
8. Hazardous areas	(−11 to +0)
9. Smoke control	(−5 to +3)
10. Emergency movement routes	(−8 to +5)
11. Manual fire alarms	(−4 to +2)
12. Smoke detection and alarm	(0 to +5)
13. Automatic sprinklers	(0 to +10)

13.4.3 Scheme 3: Fire safety evaluation (points) scheme for patient areas within hospitals[13]

This scheme was produced as a tool to assist hospital administrators with the difficult task of allocating scarce resources in order to maintain an acceptable level of fire safety in patient areas in hospitals. The scheme requires the identification of fire safety objectives and components and the assignment of values such that the percentage contribution of each identified component can be determined, Table 13.4.

Using this scheme a maximum score (points) is possible which represents perfect fire safety. It is then possible to set levels of fire safety which must be attained and compare favourably with an acceptable norm, statutory or otherwise.

13.4.4 Scheme 4: Evaluation of fire hazard and determining protective measures[14]

Again in this scheme, fire risk is determined using a simple mathematical expression:

$$R = \frac{P \times A}{M} = \frac{P \times A}{N \times S \times F}$$

$$P = (q \times c \times f \times k)(i \times e \times g)$$

where:

A represents the activation hazard
N represents the standard fire measures

Table 13.4 Fire safety components and computed contribution to attainment of fire safety

Perfect fire safety component	Contribution (%)
Staff	9
Patients and visitors	6
Factors affecting smoke movement	7
Protected areas	6
Ducts, shafts and cavities	4
Hazard protection	7
Interior finish	5
Furnishings	6
Access to protected areas	4
Direct external egress	4
Travel distance	5
Staircase	5
Corridors	5
Lifts	3
Communication systems	5
Signs and fire notices	4
Manual fire fighting equipment	3
Escape lighting	5
Auto-suppression	3
Fire Brigade	4

S represents special fire measures
F represents constructional fire measures
q represents mobile fire load
c represents the ignitability and burning rate of available fuel
f is a smoke hazard
k represents the combination of corrosive and toxic products of combustion
i is a component representing the immobile fire load
e is a range of points between 1 and 3 representing the loss increase possible, related to building height and location of spaces above and below ground level
g represents the contribution to expected fire loss of the spatial geometry of the building

The acceptable residual risk is considered to be $R_{\text{acceptable}} = 1.3 \cdot$ A correction factor can be applied depending on the size of and population characteristics.

13.4.5 Scheme 5: Evaluation du risque incendre par le calcui[15]

This scheme is based upon the Gretner model,[14] but whereas in the latter the analysis of risks combines the assessment of life safety and property protection, this scheme distinguishes between life safety and property protection objectives and provides a method of assessing the residual risk related to either objective.

Risk factors and safety factors are identified, numerical values assigned and combined, for life safety and property protection separately, as ratios to give a measure of residual risk.

Risk to people is assessed from the relationship:

$$P_1 = E \times f \times i \times r \times c$$

where:

P_1 = assessible risk to people
E = co-efficient for total evacuation time
f = an assessment of smoke density in the occupied space
i = co-efficient for the toxicity of fire gases
r = the probability of the realisation of the risk, i.e. the occurrence of the unwanted event
c = co-efficient of risk associated with the combustibles in the building.

The time for evacuation T_{ev} is computed from the equation:

$$T_{ev} = \frac{P}{L_e \cdot C_c} + \frac{L_h}{V} \quad \text{(seconds)}$$

where:

P = the number of persons to be evacuated
L_e = number of escape routes and combined length of stairs and corridors (m)
C_c = co-efficient for circulation (persons per metre per second)
L_h = total distance of travel to a place of safety (m)
V = average velocity of persons ($m\ s^{-1}$)

The value of the co-efficients are based on a simple scoring system on the scale, e.g. 0–5, 0 representing no threat or danger and 5 representing immediate danger or untenable conditions. The various tables for determining the co-efficients are given below (Tables 13.5, 13.6).

(A) is an indication of the danger inherent in the activity pursued by the occupants and the probability of ignition.
(P) is an indication of physical and mental capability of the occupants.

Table 13.5
Value of co-efficient (f)

Factor (F)	0	1	2	3
Co-efficient (f)	1	1.1	1.4	1.6

Value of co-efficient (i)

	Factor (I)	Coefficient (i)
No danger	0	1
Danger	1	1.2

Value of co-efficient (C)

Factor (C) (classification of combustibles)	1	2	3	4	5	6
Co-efficient of combustibles (C)	1.6	1.4	1.2	1.0	1.0	1.0

Six classifications are taken from European Insurance Organisation classifications.

Table 13.6 Value of co-efficient (r)

A→ ***P***	**1**	**2**	**3**	**4**	**5**	**6**
0	0.85	1	1.20	1.45	1.85	2.60
1	0.95	1.15	1.40	1.75	2.35	3.70
2	1.0	1.25	1.55	2.0	2.90	5.20
3	1.1	1.4	1.75	2.35	3.70	8.70

Co-efficient r

The combination of the relative values of (A) and (P) give a relative value of the associated risk. The assessible risk to property is computed from the relationship:

$$P_2 = (q) \cdot (e) \cdot (g) \cdot (f) \cdot (k) \cdot (a) \cdot (c)$$

where:

(q) = co-efficient related to the mobile fire load ($\mathrm{kJ\,m^{-2}}$)
(e) = co-efficient which relates the height of the building, location of spaces above and below ground and expected fire loss
(f) = co-efficient for hazards associated with smoke
(k) = co-efficient which combines the toxic and corrosive products of combustion
(a) = co-efficient related to the occupancy of the building

(c) = co-efficient which represents the capability and burning rates of the fuel
(g) = co-efficient which relates spatial geometry to expected fire loss.

Tables, etc., are used to obtain the various co-efficients in a similar manner to that outlined for computing the P_1 factor.

Five principal safety factors are combined to give an assessment of their contribution to the prevention and protection from fire (M).

$$M_{1,2} = S_{1,2} \cdot T_{1,2} \cdot E_{1,2} \cdot \mathrm{DF}_{1,2} \cdot \mathrm{F}_{1,2}$$

Subscript 1 refers to the people safety values and subscript 2 to the values for property,

where:

S = co-efficient which represents the availability of water for fire fighting, the quality of installation and pump capacity
T = co-efficient which combines the elapsed times for detection and communication
E = co-efficient which represents various methods of extinction provided
DF = co-efficient related to smoke control
F = co-efficient which combines the fire resistance of components and compartmentation.

The relationship between the various factors is shown in Fig. 13.12.

A measure of the residual risk $R_{1,2}$ is computed on the basis of the ratio of risk factors ($P_{1,2}$) to safety factors ($M_{1,2}$)

$$R_{1,2} = \frac{P_{1,2}}{M_{1,2}}$$

It remains with the user of this scheme or building owner to predetermine permissible or acceptable levels of risk, the scheme in itself does not do so. It is simply an empirical technique which relates good and bad factors which contribute to the residual risk associated with fire.

13.4.6 Fire safety evaluation (points) scheme for dwellings

Theoretically it should be possible to develop a points scheme to assist with evaluation of the level of fire safety provision within and between identifiable building types.

A method of evaluating fire safety in dwellings has been developed[16] based on work originated by Marchant. This method seeks to assess which dwellings fall below an acceptable standard of

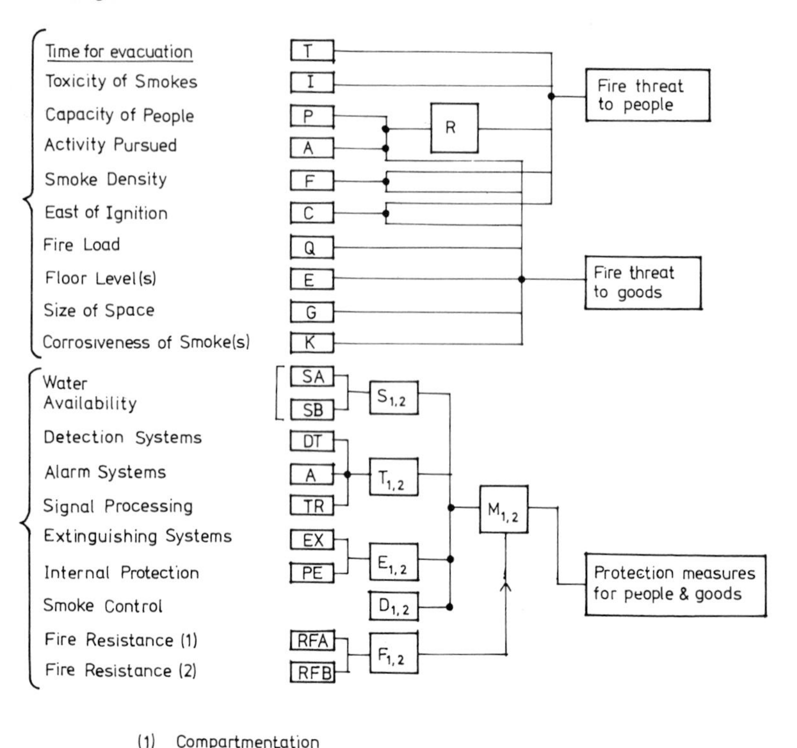

Fig. 13.12 Relationship of the factors (components) used in Scheme 5

fire safety and to indicate how cost-effective improvements might be made. Seventeen independent components of fire safety have been identified, Table 13.7.

Table 13.7 is accompanied by three sample worksheets (A, B and C), Tables 13.8, 13.9 and 13.10, designed for use by a housing manager, maintenance officer or a fire safety co-ordinator, and a detailed survey should take about thirty minutes per dwelling. Sub-components of the seventeen independent components have also been identified and are listed on the worksheets, together with an indication of the desired standard in order that the surveyor can grade each component on a scale between 0 and 5.

To determine a value for the overall fire safety of the survey volume, the seventeen survey grades are multiplied by the percentage contributions of the components. The percentage contributions of the seventeen components to the achievement of fire safety were determined by considering their relative contributions to the attainment of a set of fire safety tactics and objectives which constitute the overall fire safety policy. The method assumes linearity

Table 13.7 Components of fire safety

Components	Contribution (%)
1. Occupants and visitors	10
2. Internal planning	9
3. Interior finishes	6
4. Furnishings	8
5. Survey volume separation	7
6. Factors affecting smoke movement	6
7. Fire-resisting doors (self-closing)	7
8. Chimneys, flues, hearths, recesses	5
9. Travel distance	5
10. Communication systems	6
11. Communications vertical (stairways)	6
12. Hazard protection	6
13. Manual firefighting equipment	3
14. Escape lighting	4
15. External walls	5
16. Roofs	4
17. Fire Brigade	3
	100

throughout the matrices that have been used to solve the linear equations.

Using this method it is possible to grade dwellings on a 0–500 scale. A score of 0 representing no fire safety, whereas a score of 500 represents perfect fire safety. It is essential that an acceptable level of fire safety be determined. An acceptable level of fire safety for various dwelling types was established by means of a limited number of field trials, Table 13.11.

The evaluation scheme will identify those dwellings which fall below an acceptable level of fire safety and will assist management in the difficult decisions relating to the allocation of scarce resources and in particular the diversion of capital sums.

The method of evaluating fire safety in dwellings provides a systematic technique for determining fire safety which is simple to use, requires a relatively short time for the surveys, and is not design orientated. It provides for uniformity in the assessment of fire safety in dwellings.

13.4.7 The analytic hierarchical process

The authors are currently investigating the utilization of the analytic hierarchy process (A.H.P.)[18] as an aid to the determination of fire safety priorities. The main difference between it and the Delphi

Table 13.8

Worksheet A. Occupants and visitors

Definition: Those people who reside in the survey volume.
Visitors are those people other than the occupants of the survey volume, and their ability to assist or impeded evacuation must be assimilated in this assessment.

CONSIDER Risk Parameters			Risk Factor Values			
Occupant Mobility	(M)	Mobility Status	Not movable	Not mobile	Limited mobility	Mobile
		Risk Factor	0.7	1.0	1.2	1.38
Occupant Density	(D)	Occupant	as per bed spaces plus three	as per bed spaces plus two	as per bed spaces plus one	as per bed spaces
		Risk Factor	1.0	1.25	1.35	1.38
Survey Volume Location	(L)	Occupant	7th and above	3rd to 6th floors	1st to 2nd floors	Ground Floor
		Risk Factor	1.12	1.23	1.31	1.38
Ratio of adults to children	(R)	Children	4	3	2	1
		Adults	1	1	1	1
		Risk Factor	1.12	1.25	1.35	1.38
Occupants Average age	(A)	Age in years	60 and above	1–19	20–59	
		Risk Factor	1.0	1.27	1.38	

N.B. Children includes:

(a) Subnormal adults/children
1) Physically handicapped.
2) Mentally handicapped.

(b) Children
1) Able to move unaided when directed.
2) Unable to move unaided.

ASSESSMENT

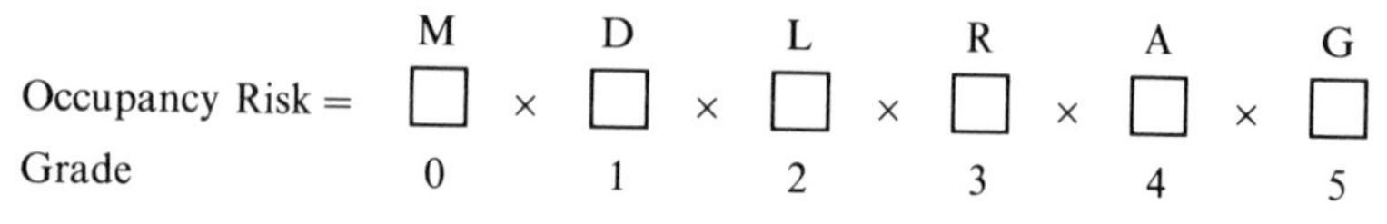

Table 13.9

Worksheet B. Furnishings

Definition: All furnishings associated with the various species within the survey volume.

CONSIDER
All furnishings including bedding, upholstery, curtains, blinds, and furniture.

Distance of fuel package from a potential ignition source	(M)	0–0.49 m	0.5–0.74	0.75–0.89	0.9–1.5	1 m
	(I) Risk Factor	0.12	0.54	1.25	2.1	2.5
Distance of fuel package from an adjacent fuel source	(M)	0–49 m	0.5–0.74	0.75–0.89	0.9–1.5	1 m
	(I) Risk Factor	0.12	0.54	1.25	2.1	2.5

Initial grade per fuel package = $\underset{I}{\square} + \underset{F}{\square} = \underset{R}{\square}$

Initial grade per room $= \dfrac{\sum R}{N}$ N = number of fuel packages

$= B$

Modify the initial grading per room B by consideration of any other relevant factors.

(a) *Traditional furniture*
Less easily ignited by flame
Worse for smouldering.
Slower burning (low temperatures)
Completely reduced to ash
(b) *Modern furniture*
More easily ignited by flame
Safe from smouldering
Rapid burning (high local temperature)
More smoke produced
(c) Position of the potential ignition source
(d) Portable ignition sources
(e) Nature of the potential ignition source
(f) Fuel height, e.g. curtains from floor to ceiling can contribute significantly to fire development within a room

Grading for the survey volume: $\dfrac{\sum B \text{ modified}}{N}$ N = No. of rooms and circulation spaces

GRADE 0 1 2 3 4 5

methodology for arriving at, and making decisions, is that it embodies within its framework an order which was not apparent in the Delphi technique.
The A.H.P. system is based upon pair-wise comparisons judgements that are incorporated in forced reciprocal matrices in order to consider factors that cannot be immediately and effectively quantified.

It is not the intention at this point to dwell on the details and mathematical background of this method which can be obtained from suitable texts[19] that describe the complete analysis of the A.H.P., but rather show the reader how it works in a practical sense.

Thus by way of an example, consider the fire safety provision for a public assembly building. The policy level 1 is Fire Safety (F/S); level 2 is tactics to attainment of pilicy, i.e. Life Safety (L/S) Property Protection (P/P); and level 3 gives the components to tactics namely ignition prevention (IGP) Fire Control (FC) and Egress from Building (EG).

Level 1	[F.S.]	- - - - pairwise comparisons
Level 2	[L.S.] [P.P.]	——— inter-level links
Level 3	[I G P] [F C] [E G]	

To enable the pair-wise comparison to be made the choice of assigned values for entry into the various matrices is limited to:

(1 2 3 4 5 6 7 8 9)

It is usual to use odd numbers since the level of fine tuning given by the use of all these numbers is seldom required.

The following meanings have been assigned to the odd numbers

1 X is as important to Y with respect to Z
3 X is slightly more important than Y with respect to Z
5 X is more important than Y with respect to Z
7 X is much more important than Y with respect to Z
9 X completely dominates Y with respect to Z

where X and Y are components or factors at the next lowest level compared to Z.

The reciprocals 1/3 etc. just mean Y more important than X with respect to Z.

The judgements of several experts is now aggregated by mutual agreement and given in a forced matrix form below.

Level 1 *Fire Safety* — L.S. P.P.

Level 2: Life Safety (L/S), Property Protection (P/P)

$$\begin{bmatrix} 1 & 5 \\ 1/5 & 1 \end{bmatrix}$$

Life Safety (L/S)

Level 3	IGP	FC	EG
IGP	1	1/3	1/7
FC	3	1	1/5
EG	7	5	1

Property Protection (P/P)

	IGP	FC	EG
IGP	1	1	7
FC	1	1	9
EG	1/7	1/9	1

The process now requires that the eigen-vectors corresponding to the matrices be evaluated.

These are computed and are given as follows with the corresponding eigen-vectors

Level (2) $\lambda_{max} = 2$ (F.S.) corresponding eigen-vectors $\begin{bmatrix} 0.84 \\ 0.16 \end{bmatrix}$ LS, PP

CR = 0

Level (3) $\lambda_{max} = 3.1$ (L.S.) $\lambda_{max} \simeq 3.0$ (P.P.)

Corresponding eigen-vectors $\begin{bmatrix} 0.19 \\ 0.08 \\ 0.73 \end{bmatrix}$ IGP, F.C., E.G. $\begin{bmatrix} 0.45 \\ 0.49 \\ 0.06 \end{bmatrix}$ IGP, F.C., E.G. Corresponding eigen-vectors

CR = 0.08 CR $\simeq$ 0

A check of the overall consistency of each matrix was determined using the consistency ratio (CR) which if less than 0.1 indicates that the matrix has attained an acceptable level of consistency. The CR values shown above indicate that the matrices are acceptable and can be used to determine the overall priority vector.

This is achieved by the matrix operation of these eigen-vectors which is given as follows.

$$\begin{bmatrix} 0.19 & 0.45 \\ 0.08 & 0.49 \\ 0.73 & 0.06 \end{bmatrix} \times \begin{bmatrix} 0.84 \\ 0.16 \end{bmatrix} > \begin{bmatrix} 0.23 \\ 0.15 \\ 0.61 \end{bmatrix} \begin{matrix} \text{IGP} \\ \text{FC} \\ \text{EG} \end{matrix}$$

(3 × 2) Resultant eigen-vector

Thus it follows from the resulting eigen-vector that Egress or Means of Escape is the most important component to the overall attainment of Fire Safety in the example given. It should be clear from this simple example that the AHP method gives a consistency which may not be perfect but is acceptable in the process of decision making.

This tool can be employed in situations where large numbers of components have to be prioritized as will be the case when determining a consistent and well ordered ranking of the factors contributing to the attainment of Fire Safety in Buildings, e.g. dwellings.

Table 13.10

Worksheet C. Chimneys, flues, hearths and fireplace recesses

Definition: Any construction designed to accommodate and convey the products of combustion to a safe area.

CONSIDER
(a) the gradient of the chimney or flue
(b) the integrity of the chimney or flue
(c) the presence of cracking in the chimney or flue
(d) the discoloration of applied finishes indicating leakage of smoke and toxic gases
(e) the use of materials other than non-combustible materials
(f) the dimensions of the flues
(g) the age of the property
(h) the presence of flue linings.

Make a preliminary grading by determining the approximate gradient of the chimney or flue by reference to the graph provided.

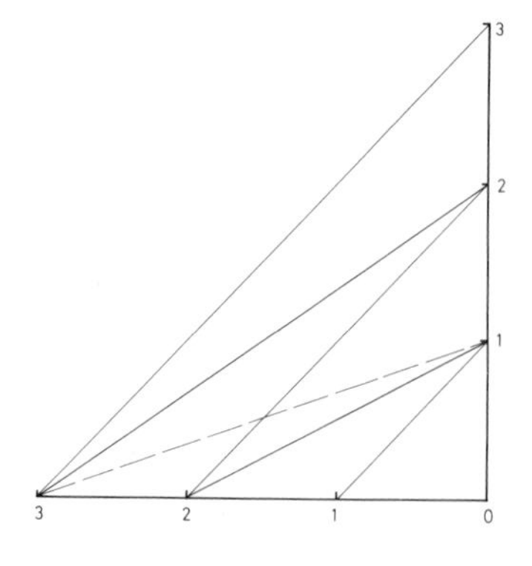

Vertical distance from the ceiling joint to exit through roof of chimney.

X	Y	SCORE
3	3	5
3	2	3-5
3	1	1-3
2	3	5
2	2	5
2	1	3-5
1	1	5
1	2	5
1	3	5

Horizontal distance from the centre line of the flue at ceiling level to the centre line of the flue at roof level.

Scores for gradient obtained by joining co-ordinates for all measurements taken to the nearest metre.

Make an assessment by raising or lowering the initial grade depending upon the relevance of the other considerations ((b)–(h)).

GRADE 0 1 2 3 4 5

Table 13.11 Acceptable level of fire safety related to dwelling types

No.	Survey volume type	No. of storeys	Norm
1.	Detached dwelling	Not exceeding two storeys	284
2.	Detached dwelling	Exceeding two storeys	345
3.	Semidetached/terraced	Not exceeding two storeys	324
4.	Semidetached/terraced	Exceeding two storeys	400
5.	Flats or maisonettes including roof	Any number of storeys	417
6.	Flats or maisonettes excluding roof	Any number of storeys	402

13.5 FUTURE TRENDS

Since current building legislation is by nature non-retrospective, future changes in the technical requirements, e.g. of building regulations, are unlikely to have much impact on existing buildings. The report on the Stardust fire[17] recognises this fact and recommends the introduction of management regulations for assembly-type buildings to ensure that acceptable levels of fire safety are maintained. Thus hazard assessment in existing buildings will fall within the compass of continuing control and three factors to be considered in hazard assessment are given:

1. Use category
2. Occupancy level
3. Vertical configuration of the building.

The report gives an example of a model which might be used when considering approving applications for relaxation of fire safety standards in public-assembly buildings. The model recommended is, in fact, a points scheme based on a matrix of nine variables in three categories, each of which may be assigned a value of 1, 5 or 10. These factors are listed in Table 13.12.

Table 13.12 Allocation of points based on building use, number of persons and vertical configurations (risk assessment)

Value	Occupancy	Occupancy level	Vertical configuration
1	Group occupancy A	Less than 50	Single-storey
5	Group occupancy B	Between 50 and 100	Multi-storey
10	Group occupancy C	More than 100	Basement

Table 13.13 Fire protection measures and assigned values

Fire protection measure	Value
Linings	3 to 7
Furniture (not controlled)	0 to 4
Furniture abutting on linings/separate from linings	0 to 5
Detection	0 to 5
Warning system	0 to 2
Extinguishers	0 to 3
Hose reels	0 to 9
Sprinklers	0 to 10

Thus it can be seen that the relative risk varies between a low values of 3 to the highest value of 30.

Eight fire protection measures have been identified and values assigned. These measures are listed in Table 13.13.

The total value or score of the protection is obtained by adding the values allocated to the various measures, which must not be less than the risk assessment made by reference to the building matrix, Table 13.12.

13.6 CONCLUSIONS

The development of fire safety evaluation models is not an easy task. The transient behaviour of fire introduces difficulties in assessing the continuing contribution of components over time, provided originally to enhance fire safety. The advent of the computer has led to some major developments in modelling optimum provisions for emergency evacuation from buildings (Chapter 12).

However, it does seem that at present it would be helpful if general agreement could be reached as to the precise meaning of some of the terminology currently being used.[20,21]

REFERENCES

1. Hamilton S G, '*A short history of structural fire protection of buildings*', National Building Studies, Special Report No. 27, HMSO, 1958.
2. Rasbash D J, 'The definition and evaluation of fire safety', *Fire Prevention Science and Technology*, **16**, pp. 17–22, 1977.
3. Marchant E M, 'A cost-effective approach to fire safety', Paper I – Points Scheme, *International Fire Security and Safety Conference*, 1984.

4. *Fire safety analysis for residential occupancies*, Department of Housing and Urban Development, Washington DC, 1977.
5. Drysdale D D and David G J, 'Hazard analysis for a storage sphere of pressurised liquified flammable gas', *Fire Safety Journal*, pp. 91–103, 1979–80.
6. Connelly E M and Swarty J A, 'Systems Concepts for Building Fire Safety', *Fire Protection Handbook*, National Fire Protection Association, Fifteenth Edition 1981.
7. Sphilbert D C, '*Statistical decomposition analysis and claim distribution for industrial fire losses*', 12 ASTIN Colloquium, 1975.
8. Cox D R and Miller H D, '*The Theory of Stochastic Processes*', Chapman and Hall, London, 1977.
9. Aoki Y, '*Study on probabalistic spread of fire*', B.R.I. Research Paper No. 80, Building Research Institute, Ministry of Construction, Japan, Nov., 1978.
10. BINC (Building Industries National Council), *Report on Means of Escape from Fire*, BINC, London, 1935.
11. Benjamin B R, 'A fire safety evaluation system for health care facilities', *Fire Journal*, pp. 52–55, 95, 96, Vol. 000, March 1979.
12. *Life Safety Code*, NFPA, Boston MA, USA, 1973.
13. *Fire Safety Evaluation (Points) Scheme For Patient Areas within Hospitals*, Department of Fire Safety Engineering, University of Edinburgh, 1982.
14. Gretner M, '*Evaluation of Fire Hazard and Determining Protective Measures*', edited by Association of Cantonal Institutions for Fire Insurance and the Fire Prevention Service for Industry and Trade, Switzerland, 1973.
15. Stuard P and Cluzel D, '*Evaluation du risque incende par le cacul*', Direction de la Recherche de l'UTI, France 1978–79.
16. Shields T J, 'Fire Safety Evaluation of Dwellings', Paper presented at Seminar, *Fire Safety in Dwellings*, Ulster Polytechnic, March 1984.
17. '*Report of the Tribunal of Inquiry on the Fire at the Stardust Artane, Dublin*, 1981', Government Publications, Stationery Office, Dublin, 1982.
18. Shields T J and Silcock G W, *An Application of the Analytic Hierarchical Process to Fire Engineering*, Fire Safety Journal, 1986.
19. Saaty T L, *The Analytic Hierarchy Process*, McGraw-Hill, New York, 1980.
20. Shields T J and Silcock G W, *Comments on the Confusion existing in the use of the Terms 'Model' and 'Scheme' in Fire Safety Modelling*, Fire Safety Journal (10), pp. 239–240, 1986.
21. Shields T J, Silcock G W and Bell Y, *An Introduction to Some of the Methodological Problems Associated with the Use of the Delphi Technique in Fire Engineering*, Fire Technology, 1987.

Index